RYE

Genetics, Breeding, and Cultivation

RYE

Genetics, Breeding, and Cultivation

Rolf H. J. Schlegel

CRC Press
Taylor & Francis Group
Boca Raton London New York

CRC Press is an imprint of the
Taylor & Francis Group, an **informa** business

CRC Press
Taylor & Francis Group
6000 Broken Sound Parkway NW, Suite 300
Boca Raton, FL 33487-2742

First issued in paperback 2016

© 2014 by Taylor & Francis Group, LLC
CRC Press is an imprint of Taylor & Francis Group, an Informa business

No claim to original U.S. Government works

Version Date: 20130621

ISBN 13: 978-1-138-03369-6 (pbk)
ISBN 13: 978-1-4665-6143-4 (hbk)

Library of Congress Cataloging-in-Publication Data

Schlegel, Rolf H. J.
    Rye : genetics, breeding, and cultivation / Rolf H.J. Schlegel.
        p. cm.
    Includes bibliographical references and index.
    ISBN 978-1-4665-6143-4 (hardcover : alk. paper) 1. Rye--Genetics. 2. Rye--Breeding.
    I. Title.

SB191.R9S34 2013
633.1'4--dc23   2013022915

Visit the Taylor & Francis Web site at
http://www.taylorandfrancis.com

and the CRC Press Web site at
http://www.crcpress.com

Contents

List of Figures

List of Tables

Preface

I grew up on a farm in the small village of Stadtlengsfeld in the volcanic Rhön Mountains region at the heart of Germany. Of all agricultural crops, rye has performed best in barren soil. Yet at that time I was aware of the robustness of this plant. Years passed before I became a student. More or less by accident, I came into contact with rye again. In 1969, Professor Dieter Mettin offered two other students and me the chance to produce trisomic plants from rye in order to improve genetic analysis. At that time almost nothing was known about it genetically.

Since then, I never dropped rye as a subject of cytology, genetics, and breeding. Over 50 years, my love was to it!

I was even able to inspire some of my students to study the same plant species. They continued this research using more sophisticated methods and molecular techniques.

Because rye still plays an important role as a bread food in many countries around the world, research and breeding are accomplished continuously. Over the years, much was studied and described about the exciting fields of rye research— sometimes in detail and sometimes collectively.

A monograph on rye has not yet come out even after 150 years. This was my incentive to take up this challenge. Approximately 10 years ago, a decision to compile and complete a manuscript on the most important facets of rye research matured. Now, it has been completed. I hope it brings readers benefit, encouragement, and inspiration.

Rolf Schlegel

Professor Wolf-Dieter Blüthner (right), Dr. Lutz Wölbing (center), and Professor Rolf Schlegel (left) as students in 1968, analyzing trisomic plants of rye at the Institute of Plant Breeding, Martin Luther University, Halle-Wittenberg, Germany. (Courtesy of Dr. Lutz Wölbing, Walbeck, Germany.)

Comin thro' the rye! (1782)

O, Jenny's a' weet (wet), poor body,
Jenny's seldom dry:
She draigl't (draggled) a' her petticoatie,
Comin thro' the rye!

Comin thro' the rye, poor body,
Comin thro' the rye,
She draigl't a' her petticoatie,
Comin thro' the rye!

Gin (if, should) a body meet a body
Comin thro' the rye,
Gin a body kiss a body,
Need a body cry? (call out for help)

Gin a body meet a body
Comin thro' the glen,
Gin a body kiss a body,
Need the warl' (world) ken (know)?

Gin a body meet a body
Comin thro' the grain;
Gin a body kiss a body,
The thing's a body's ain (own).

Ilka lassie has her laddie,
Nane, they say, ha'e I
Yet all the lads they smile on me,
When comin' thro' the rye.

Robert Burns (1759–1796)

"Comin thro' the rye!" is well known as a traditional children's song, with the words put to the melody of the Scottish minstrel *Common' Frae the Town*—a variant of the tune to which *Auld Lang Syne* is usually sung. The name of the poem also became the title of a British silent drama movie in 1923, directed by Cecil Hepworth which starred Alma Taylor and Ralph Forbes. A woman is prevented from marrying the man she loves by interference from another woman. Derived from the poem's name, a bestseller was titled *The Catcher in the Rye* (1951) by the American author Jerome David Salinger (1919–2010). Holden Caulfield, the protagonist, imagines children playing in a field of rye near the edge of a cliff and catching them when they start to fall off.

Acknowledgments

The author would like to thank Katrin Kumke, Dr. Andreas Houben, Dr. Helmut Knüpfer, and Michael Grau, Leibniz-Institut für Pflanzengenetik und Kulturpflanzenforschung, Gatersleben, Germany; Dr. Viktor Korzun, Lochow-Petkus GmbH, Einbeck, Germany; Dr. Ute Kastirr and Grit Lautenbach, Julius Kühn-Institut, Quedlinburg, Germany; Dirk Rentel, Bundessortenamt, Hannover, Germany; Dr. Ekatarina Badaeva, Engelhardt Institute of Molecular Biology, Moscow, Russia; Dr. Elena Khleskina, Institute of Cytology and Genetics, Novosibirsk, Russia; Professor Jaroslaw Dolezel, Institute of Experimental Botany, Olomouc, Czech Republic; Dr. Zofia Banaszak, DANKO Hodowla Roślin, Choryn, Poland; Dr. Klaus-Dieter Schmidt, Naumburg, Germany; Heidi and Dr. Lutz Wölbing, Walbeck, Germany; Professor Dr. Irmgard Müller, Ruhr Universität, Bochum, Germany; and Dr. Reiner Nürnberg, Institut für Angewandte Analysis und Stochastik, Berlin, Germany, for their substantial contributions to the manuscript and for proofreading as well as providing several references and photographs.

Author

Rolf H. J. Schlegel, PhD, DSc, is a professor of cytogenetics and plant breeding with over 40 years of experience in research and teaching advanced genetics and plant breeding in Germany and Bulgaria. Professor Schlegel has authored more than 200 research papers and other scientific contributions, co-coordinated international research projects, and been a scientific consultant at the Bulgarian Academy of Agricultural Sciences for several years. He earned his master's in agriculture and plant breeding and his PhD and DSc in genetics and cytogenetics from the Martin Luther University, Halle/S., Germany. Later he became the head of the Laboratory of Chromosome Manipulation and the Department of Applied Genetics and Genetic Resources at the Institute of Plant Genetics and

Crop Plant Research, Gatersleben, Germany, and the head of the Genebank at the Institute of Wheat and Sunflower Research, General Toshevo/Varna, as well as the Institute of Plant Biotechnology and Genetic Engineering, Sofia, Bulgaria. At the end of his career, he worked as the director of research and development at a private company in Germany.

His latest book contributions include *Dictionary of Plant Breeding* (2009), 2nd ed. (Taylor & Francis, Inc., New York), *Concise Encyclopedia of Crop Improvement: Institutions, Persons, Theories, Methods, and Histories* (2007) (Haworth Press, New York), and Rye *(Secale cereale* L.)—A Younger Crop Plant with Bright Future (2005), and with R. J. Singh and P. P. Jauhar, eds., *Genetic Resources, Chromosome Engineering, and Crop Improvement* (CRC Press, Boca Raton, FL), which received significant international attention. Moreover, he has written three bestsellers in German: *Vincent van Gogh—der Genetiker?* (2013), *Ikarus oder geflügelter Affe* (2011), and *Spermien mögen Maiglöckchen—Kurioses aus Botanik, Züchtung und Genetik I* (2010), BoD Verl., Norderstedt, Germany, http://www.bod.de.

Abbreviations

AFLP	amplified fragment length polymorphism
Al	aluminum
asl	about sea level
BAC	bacterial artificial chromosome
BAP	6-benzylaminopurine
bp	base pair(s)
BTT	balanced tertiary trisomic
°C	degree Celsius
cDNA	complementary DNA
cf.	confer, compare
Cl	chloride
cm	centimeter
cm³	cubic centimeter
cM	centimorgan
CMS	cytoplasmic male sterility
Cs	cesium
ctDNA	chloroplast DNA
Cu	copper
2,4-D	2,4-dichlorophenoxyacetic acid
Da	dalton
DAPI	4′,6-diamidino-2-phenylindole
DArT	Diversity Arrays Technology
DH	doubled haploid
DIBOA	2,4-dihydroxy-1,4(H)-2-benzoxazin-3-one
dwt	dry weight
ELISA	enzyme-linked immunosorbent assay
EMS	ethyl methanesulfonate
EST	expressed sequence tag
EU	European Union
f. sp.	forma specialis
f. w.	formular weight
g	gram
Gbp	giga base pairs
GCA	general combining ability
GM	genetically modified
GUS	β-glucoronidase
h	hour
HMW	high molecular weight (glutenin)
HPLC	high-performance liquid chromatography
Hz	hertz
IRAP	inter-retrotransposon amplified polymorphism

ISSR	intersimple sequence repeat
k	kilo
K	potassium
kb	kilobase
kbp	kilo base pair
kDa	kilodalton
kg	kilogram
km	kilometer
l	liter
LD_{50}	lethal dose 50
LMW	low molecular weight
μ	micro
m	meter
M	Morgan
MAS	marker-assisted selection
Mbp	mega base pairs
MI	metaphase I of meiosis
min	minute
ml	milliliter
μm	micrometer
mm	millimeter
Mn	manganese
mol	mole
mRNA	messenger RNA
ms	millisecond
m.w.	molecular weight
mya	million years ago
n	nano
n	haploid chromosome set, 2*n* diploid chromosome set, etc.
N	normal
NILs	near-isogenic lines
nm	nanometer
osm	osmolal
PAGE	polyacrylamide gel electrophoresis
PCR	polymerase chain reaction
PEG	polyethylene glycol
pg	picogram
pH	pH value
pl	plural
PMCs	pollen mother cells
PMSR	peptide methionine sulfoxide reductase
ppm	parts per million
QTL	quantitative trait locus
R	roentgen
RAPD	random amplified polymorphic DNA
rDNA	ribosomal DNA

REMAP	retrotransposon-microsatellite-amplified polymorphism
RFLP	restriction fragment length polymorphism
RILs	recombinant inbred lines
RNase	ribonuclease
rpm	rounds per minute
SBMV	soil-borne mosaic virus
SC	synaptonemal complex
SCAR	sequence-characterized amplified region
SD	standard deviation
SDS	sodium dodecyl sulfate
SDS-PAGE	sodium dodecyl sulfate polyacrylamide gel electrophoresis
sec	second
SNP	single-nucleotide polymorphism
sp.	species
SSC	standard saline citrate buffer
SSD	single-seed descent
SSLP	simple sequence length polymorphism
ssp.	subspecies
SSR	single-sequence repeat
STMS	sequence-tagged microsatellite site
STS	sequence-tagged site
syn.	synonymous
TBS	tris-buffered saline
TE	tris-ethylenediaminetetraacetic acid buffer
TET	transiently expressed transposase
TGW	thousand-grain weight
TKW	thousand-kernel weight
T_m	melting temperature of DNA
TRIM	terminal-repeat retrotransposon in miniature
UK	United Kingdom
URL	uniform resource locator
US/USA	United States/United States of America
UV	ultraviolet
V	volt
VIGS	virus-induced gene silencing
v/v	volume to volume
w/v	weight per volume
Xta	chiasmata
Zn	zinc
~	about, adequate
>>>	see

1 Introduction

Rye is a cereal that played a major role in the feeding of European populations throughout the Middle Ages, owing to its considerable winter hardiness. Recently, rye is cultivated on ~5.4 million ha (see Table 1.1). The world production is ~13 million tons (see Table 9.6). Cultivated rye is the result of crossbreeding *Secale vavilovii* and the perennial species, *S. anatolicum* and *S. montanum* (see Figure 1.1). *Secale montanum* plays a central role in the evolution of rye, but the origin of cultivated rye by hybridization of *S. silvestre* and *S. montanum* (Stutz 1957) was denied by Nürnberg-Krüger (1960). Turkey is the primer genus center of perennial rye, and it is agreed that the geographic origin of rye is the area around Mount Ararat and Lake Van.

Secale silvestre diverged early from the original rye type as a species with a particular characteristics. Later, *S. africanum* branched away from the main tree by continuously moving south along the African mountain chains and/or large geological rift valleys, finally arriving in the Cape region. Two hundred years ago, *S. africanum* was still common there at the so-called Roggeveldbergen, that is, frost regions up to 3000 m above sea level (m.a.s.l.). The early Dutch farmers used it as perennial forage for sheep. Later, the endemic places of *S. africanum* disappeared because of overgrazing and tillage. Until 1976, no endemic plants could be found.

Basically there are two centers of rye origin: (1) a primary region around Tabriz (Iran) and toward the Black Sea and (2) a second region east of Iran and toward Afghanistan. They coincide with the recent distribution of weedy ryes. From here, cereal rye was distributed to the western and middle Europe via Anatolia and along the Danube River, to Scandinavia via Russia, and to China, Korea, and Japan.

In Scandinavia, including Finland, rye domestication occurred in prehistoric time, north of the 60th parallel of latitude where weedy rye survived after some harsh winters when the winter wheat perished. The northern environment suppresses the brittleness of the rachis in weedy rye as shown by experiments conducted at the stations of the Vavilov Institute (Russia) in the 1920s. In 1924, a new subspecies of rye with a brittle rachis causing complete articulation at maturity was found by Vavilov (1927) in Afghanistan, where it occurred as a weed in the fields of wheat and barley. Between 1926 and 1927, Vavilov found additional forms of wild rye with a brittle rachis in the valley and on adjacent slopes of the Meander River (Lydia). This wild polymorphous rye possesses strong straw and is gigantic in size. Many forms show thick pubescence on the external flowering glumes, a rare occurrence in rye. It differs from cultivated forms in its brittle rachis, small-flattened grain with white apical brush hairs, broad empty glumes, and the pubescent leaf sheaths of young plants. It is distinguished from the Afghan form by its large size, pubescent leaf sheath, adhesion of flowering glumes to the grain, smaller size and flatness of the kernels, hairy brush, and winter habit. According to Vavilov, there are three subspecies of rye: (1) those

TABLE 1.1
Largest Rye Producing Countries in the World in 2010

Country	ca. 2010 (ha)
Albania	1,100
Argentina	24,100
Armenia	156
Australia	57,400
Austria	45,699
Belarus	342,542
Belgium	459
Bolivia	122
Bosnia	7,424
Brazil	2,947
Bulgaria	10,900
Canada	89,100
Chile	1,456
China	180,000
Croatia	1,035
Czech Republic	30,200
Denmark	52,100
Egypt	23,500
Estonia	12,600
Finland	25,200
France	30,000
Germany	627,100
Greece	15,600
Hungary	35,900
Ireland	180
Italy	4,513
Kazakhstan	43,500
Kyrgyzstan	30
Latvia	34,600
Lithuania	51,300
Luxembourg	896
Macedonia	3,590
Morocco	3,500
The Netherlands	2,252
North Korea	49,800
Norway	6,760
Peru	125
Poland	1,395,600
Portugal	20,400
Moldova	1,024
Romania	14,439

TABLE 1.1
(Continued) Largest Rye Producing Countries in the World in 2010

Country	ca. 2010 (ha)
Russia	1,367,500
Serbia	4,865
Slovakia	17,000
Slovenia	796
South Africa	3,700
Spain	133,300
Sweden	25,400
Switzerland	2,249
Tajikistan	387
Turkey	140,905
UK	6,000
Ukraine	279,100
USA	107,240
Uzbekistan	1,900
Total	**5,334,844**

Note: Bold text represents main rye-growing countries.
Source: Courtesy of FAO statistics; http://www.fao.org/corp/statistics/en/

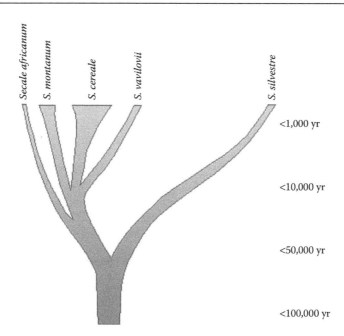

FIGURE 1.1 The phylogenetic relationships among rye species (*Secale* ssp.) and their evolutionary development. (After Jones, J. D., and Flavell, R. B., *Chromosoma*, 86, 613–641, 1982.)

with a brittle rachis, (2) those with a tough rachis, and (3) those with intermediate degrees of rachis brittleness.

Tough rachis was further selected unintentionally by the drying, hitting, threshing, and seed-cleaning techniques used by ancient Scandinavians. For instance, partial brittleness still occurs in Finnish landraces. The earliest known paleobotanical proof of rye in Finland was dated back to 4120 BC. The sites of the proofs of age 1200 BC or older are associated with ice margin eskers up to 600 km long, or some long glaciofluvial eskers, all formed during the Ice Age. The eskers provided suitable sandy soil and natural paths for movement and grain transport.

The Finns even had a god of rye, Rukiin Jumala or Ronkateus. The so-called *riihi* drying technique and the Finnish word *riihi* were widely borrowed by others, appearing among Russians, Latvians, Lithuanians, Belarusians, northern Tatars, Swedes, and Norwegians. Ancient Finns grew rye in the so-called *kaski*, that is, slash-and-burn land that fortified harvested grain with the good contents of macro- and microelements and, together with the smoky *riihi* heating before threshing, controlled pests (ergot) and ensured seed longevity.

The earliest transports of Finnish grain were made apparently by Russian troops returning from plundering raids, because the raids usually occurred after food shortages in Novgorod as described in chronicles since the 1000s. The Lithuanians evidently obtained rye from the Karelians on raids in ca. 1065–1085 and as "taxes" since 1333. The main destinations of Finnish grain for seed were Estonia, mainland Sweden, Russia, and the Baltic ports, from where the Hanseatic League traded seed to Holland, England, Portugal, and Norway. Gotland was probably the main transit port of the earliest trade.

Since the sixteenth century, the general sowing time of winter rye was in June, but it later shifted to autumn, so that in 1950 only 4.8% of the winter rye area was sown in June. In the past, winter rye was often sown together with spring barley, turnips (*Brassica rapa*), spring oats, buckwheat (*Fagopyrum esculentum*), or spring rye.

During the Russo-Swedish War (1570–1595), the Tatars and, in the 1702–1721 and 1741–1743 occupations, the Kalmyk or Zyungory soldiers in the Russian army plundered grain from Karelia. These Tatars and Kalmyks were nomadic peoples who gradually adopted field grain growing after these wars. The *zyungorka* rye, renamed *murav'yóvka* in Russia in 1839, was a typical Finnish *juureinen* rye, probably descended from grain seized by the Zyungory soldiers (Ahokas 2012).

Starting in 1638, the Finns brought their rye from Finland and from the areas settled by the so-called Forest Finns (also called Rye Finns) in Sweden and Norway to North America. The Finnish provincial rye strains Uusimaa and Vaasa were much sought-after export varieties for seed in the eighteenth and nineteenth centuries. The Finns forcibly transported by Russians to Siberia introduced winter rye to new regions, including some pioneer colonies in Siberia, where spring rye was earlier grown regionally, for example, by the aboriginal Finno-Ugric Udmurts.

Rye is part of quite young cultivated plants. Apparently, this cereal has not been comparatively long in cultivation. According to Derr (1892), it has not been found in Egyptian monuments. It has no name in any of the Semitic languages or in Sanskrit or in the modern languages directly derived from them. It is not mentioned in old Chinese or Japanese literature. Nevertheless, it is now much cultivated in Siberia. Unlike the cases of other cereals, no trace of rye has been found in the remains of the Lake Dwellings (Pfahlbauten) in Germany. The oldest rye excavation belongs to the stilts of Olomouc (Bronze Age), and the crop is more common in the medieval Slavic settlements. In any case, the Slavs brought rye from Eastern Europe (Russia) to the West. The German name Roggen for rye is of Slavic origin.

However, rye is mentioned in the English version of the Old Testament (Exodus 9:32), but in the opinion of Sprengel (1807), wheat is meant by it. Sprengel also states that Theophrastus (1916) is the earliest author who notes that *S. cereale*, but the word τιφη used by Theophrastus (ca. 372–287 BC), refers to *Triticum monococcum* rather than to rye. Galenius (born AD 131) (1965) mentions rye under the name βριζα, the term by which it was known then, as well as σικαλι. He saw it cultivated in Thrace and Macedonia under the name briza. Rye is still known in modern Greece. Pliny (AD 77) speaks of *Secale*, or rye, cultivated by the Taurisci. The Taurisci (taur = Celtic for mountain) were from the third to first century BC, a Celtic tribal group on the eastern Alps border, whose settlements were mostly restricted to Carinthia and Slovenia.

The Russian plant geneticist N. I. Vavilov called rye a secondary crop, meaning that it first appeared in farmers' fields as a weed among other crops, such as emmer wheat and barley of the Near East, and then subsequently underwent domestication and became a useful grain.

In the high Caucasian mountain regions of Georgia (1800–2200 m.a.s.l.), fields still exist where wild rye, *S. segetale* (called *svila*), is widespread in wheat and barley fields. It is harvested together with them. The bread of wheat with *svila* is considered to be very nutritious and has a good taste.

This domestication process was likely a side effect of the cultivation of the other cereals, as each harvest would direct the selection toward erect straws and nonshattering larger grains. Archeological remnants of rye are found among wheat and barley in the Anatolian regions of Turkey from at least 6000 years ago. The earliest cultivation began 4000 years ago in Persia, central Anatolia, and north of the Black Sea region. Domestication probably happened at several locations but, presumably, within the general area defined below. A tougher rachis, three-floret spikelets, looser glumes, and bigger seeds were the characteristics selected by the primitive populations. Rye as an original crop has been grown since the fourth millennium BC in Turkey.

Rye's primary center of origin appears to be southwestern Asia. From there, it was probably distributed to Russia and Scandinavia and from there into Poland and Germany. Then it gradually spread throughout most of Europe. Thus it became a European crop. Later, it was brought to North America and western South America. American rye was brought from Europe to the settlements established in the sixteenth and seventeenth centuries.

Rye was introduced into China from the Middle East. The earliest Chinese ryes originated from southwest Asia. Afterward, the species was introduced into Japan. Both American and Chinese ryes were planted mainly for forage.

However, rye could also have migrated from its center of origin to northern Europe via Turkey and across the Balkan Peninsula (Bushuk 1976).

Rye grains found in Neolithic sites in Austria and Poland are considered to be of "wild" origin (Vavilov 1917). The earliest seed finds of cultivated rye in central

Europe came from the Hallstatt period (1000–500 BC). From 2500 to 2000 BC, the cultivation of rye moved northwest toward Poland and Sweden. The first landraces were derived from those selections. Recent excavations in Berlin show that rye was widely grown by the end of the eleventh century in what is now central Germany. One of the earliest depictions of rye is found in the book of plants by Jacobus Theodorus (1590) (see Figure 1.2), which shows compact spikes with many awns.

During the sixteenth century, rye cultivation subsequently increased, possibly favored by cold winters during the so-called Little Ice Age. The Little Ice Age was a period of cooling that occurred after the Medieval Warm Period. While it was not a true ice age, the term was introduced into scientific literature by François E. Matthes (1874–1948) in 1939. It has been conventionally defined as a period extending from the sixteenth to the nineteenth century, or from ca. 1350 to ca. 1850, although climatologists and historians working with local records no longer expect to either agree on the start or end dates of this period, which varied according to local conditions.

At the beginning of the twentieth century, rye surpassed wheat in acreage. Professor C. H. E. Koch (1809–1879) described a first-mutant rye with forked

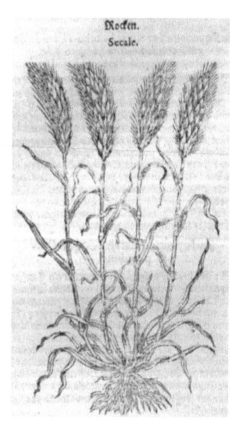

FIGURE 1.2 Presentation of rye in a medieval book. (Modified from Theodorus, J., *Eicones plantarvm, sev stirpivm, arborvm nempe, frvcticvm, herbarvm, frvctvvm, lignorvm,* Frankfurt, Germany, 1590.)

spikes in 1757 (Martiny 1870; see Figure 7.16). Another mutation found in 1832 led to the landrace Wollny's schlaffährigen Roggen (lax-eared rye). First targeted selections were made north of Germany by the Probsteier Seed Cooperative around 1850 (see Section 7.4). The Schlanstedter Roggen is the first modern rye variety with higher lodging resistance. It was bred by W. Rimpau (1842–1903), who started winter rye breeding at Schlanstedt (Germany) in 1875. Most successful were the varieties of F. von Lochow at Petkus (Germany), derived from the Probsteier gene pool (see Section 7.4). The village of Probstei is close to the Baltic Sea in the north of Germany. Since 1880, directed crosses led to the variety Bestehorn's Riesenroggen (G. Bestehorn, 1836–1889) at Biberitz (near Könnern), Germany. N. Rudnitsky began rye breeding at Vyatka station (central Russia near Kirov) in 1894. The Vyatka variety is still used to some extent in the agricultural areas to the northwest of Russia. Rye is cultivated also in the northern and western districts of the Ukraine, where it is the main grain used for bread making. Rye takes up 48% of the farmland in Chernihiv, Minsk, and Grodno; 38% in Volhynia; 3% in Poltava; 29% in Kharkiv; 28% in Kiev; 22% in the Don region; 19% in Katerinoslav and Podolia; 18% in Tauria; 17% in Kherson and Galicia; and 7% in Bessarabia.

Original rye was a meter-high grass. The long straw was used for roofing. Meanwhile, rye became a modern crop plant with technological and agronomic advantages. Subsequent increases in rye grain yield were caused by improved agronomy and new (hybrid) varieties.

Although rye cultivation acreage decreased by more than a half during the past four decades, the cool temperate zones of Europe remain as the major growing areas (see Figure 1.3). Approximately 94% of the world production is harvested in Poland (27% of the total acreage), 23% in Germany, 13% in Russia, and 16% in Belarus (see Tables 1.1 and 9.6). Highest average yields are achieved in Germany with ~8 tons/ha. Top yields with hybrid varieties under favorable agronomic conditions reach almost 11 tons.

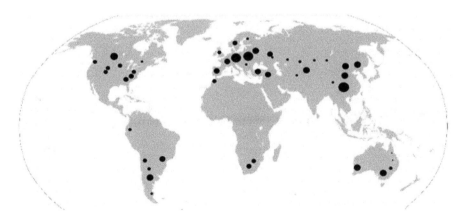

FIGURE 1.3 Main areas of rye cropping worldwide. (Courtesy of FAO statistics; http://www.fao.org/corp/statistics/en/.)

Acreage in the United States has been decreasing, down to ~107,000 ha. All US rye is winter rye, mostly used for grain fodder production. The average rye production in the United States in 2010 was ~189,000 tons. The leading states in rye production are South Dakota, Georgia, Nebraska, North Dakota, and Minnesota. In 1930, Minnesota grew 2.8 million ha of rye for grain, but in 2010 it harvested 12,000 ha and Wisconsin harvested 32,000 ha. The average yield in 1920 was 1.5 tons/ha, but in 2010 it was 1.9 tons/ha in the Upper Midwest.

Approximately 50%–75% of the yearly harvest is used for bread making, resulting in rich dark bread that holds its freshness for about a week. The rest is used for feeding and alcohol production for industrial and consumption purposes.

The rye market is increasingly shifting toward sustainable production. Rye produced for bread making has decreased or stagnated, but other market segments are getting bigger. Particularly, rye for feeding, ethanol processing, and biogas is most promoted in Europe. Depending on national prize regulations, it is becoming a cash crop. In 2007, during the European Rye Congress at Berlin (Germany), rye was seen to be an essential crop fit for food and bread and sustainable products (Anonymous 2007).

2 Botany

2.1 ORIGIN

Like wheat, barley, and oats, rye is a member of the grass family, the tribe "Gramineae." It is a typical allogamous plant species that shows a high degree of self-incompatibility. The divergence of the wheat and rye lineages from the "Pooideae" happened ~7 million years ago (mya) (Kingdom: Plantae, Division: Magnoliophyta, Class: Liliopsida, Order: Poales, Family: Poaceae, Subfamily: Pooideae, Tribe: Triticeae). The angiosperms evolved ~145 mya, during the late Jurassic period, and were eaten by dinosaurs. They became the dominant land plants ~100 mya (edging out conifers, a type of gymnosperm). Angiosperms are then divided into monocots (grasses) and dicots (see Table 2.1).

In the United States, the common name for rye is feral rye or cereal rye. The use of the phrase "cereal rye" reduces confusion with ryegrasses (*Lolium* spp.) in English-speaking countries. The English name is derived from the Old English word ryge (approximately twelfth century AD) and from the proto-German word *ruig* (related to Old Saxon *roggo* [later *rocken*], Old Norse *rugr*, or Old Frisian *rogga*). The name was introduced by Slavic tribes.

Its well-accepted scientific name is *Secale cereale* L., which is the name given by Carolus Linnaeus in 1753, from plants cultivated in northern Europe. Additional Latin names also were given to rye, such as *S. aestivum, S. ancestrale, S. ancestrale aidinum, S. ancestrale arenosum, S. ancestrale karaburun, S. ancestrale spontaneum, S. arundinaceum, S. cereale ancestrale, S. cereale brevispicatum, S. cereale indo-europaeum, S. cereale montaniforme, S. cereale rigidum, S. cereale spontaneum, S. cereale tetraploidum, S. cereale tsitsinii, S. cereale vernum, S. compositum, S. creticum, S. hybernum, S. spontaneum, S. triflorum, S. turkestanicum, S. vernum, Triticum cereale, T. cereale brevispicatum, T. cereale montaniforme, T. ramosum,* or *T. secale.*

Internationally, rye is called centeno (Spanish), centeio (Portuguese), rosh (Russian), segale comune (Italian), zyto (Polish), seigle (French), råg (Swedish), rug (Danish), ruis (Finnish), rogge (Dutch), rúgur (Icelandic), hei mai (Chinese), rai-mugi (Japanese), homil (Korean), or roggen (German). In former German times, rye was simply das Korn (*neutrum*). The sense of the word was "grain with the seed still in" (e.g., barleycorn) rather than a particular plant, and it was locally understood to denote the leading crop of a district. It is restricted to corn on the cob in the United States (during the sixteenth century, originally Indian corn, but the adjective was dropped) and usually wheat in England or oats in Scotland and Ireland.

Rye occurred in small quantities at a number of Neolithic sites in Turkey, such as Can Hasan III of the prepottery Neolithic B period, in that people living during this period began to introduce domesticated animals to supplement their earlier agrarian diet. But rye is otherwise virtually absent from the archeological record

---

TABLE 2.1
About the Evolution of Triticeae

Divergence	Million Years Ago (Approximately)
Earliest land plant fossils	420
Origin of angiosperms	200–340
Monocot–dicot divergence	160–240
Origin of grass family	65–100
Oldest known grass fossils	50–70
Divergence of the subfamilies	50–80
Pooideae (wheat, barley, oat)[a]	
Bambusoideae (rice)[a]	
Panicoideae (maize, sorghum)[a]	
Earliest fossil of the rice lineage	40
Divergence of maize and sorghum lineages	15–20
Divergence of wheat and barley lineages	10–14
Divergence of wheat and related lineages	7
Divergence of *Secale* lineages	1.7

[a] Panicoideae diverged from the Pooideae–Bambusoideae first, followed shortly by divergence of the latter two subfamilies. The authors suggest that either the Pooideae or the Bambusoideae branched off first. In general, it seems that all three subfamilies diverged about the same time. Some phylogenies divide Bambusoideae into two subfamilies, Oryzoideae and Bambusoideae.

---

until the Bronze Age of Central Europe (1800–1500 BC). Claims of much earlier cultivation of rye at the Epipaleolithic site of Tell Abu Hureyra in the Euphrates valley of northern Syria remain controversial (Hillman 1978). Wild rye was reported to have been cultivated in the Epipaleolithic (12,000 BC) period, that is, 700 years after the establishment of Abu Hureyra. Interestingly, this dating is 1000 years earlier than that previously accepted for the beginning of intentional plant cultivation. Hillman (2001) believes that wild rye was cultivated before its domestication, based particularly on the finding at Abu Hureyra I of layered remains of weed flora, which is indicative of cultivated fields. Nevertheless, rye is now considered as one of the founding agricultural crops in the Near East.

It is possible that rye traveled west from Turkey as a minor admixture in wheat and was only later cultivated in its own right. Although archeological evidence of this grain has been found in Roman contexts along the Rhine and Danube rivers and in the British Isles, Plinius Secundus Maior (1669: 77) is dismissive of rye, writing that it "is a very poor food and only serves to avert starvation" and wheat is mixed into it "to mitigate its bitter taste, and even then is most unpleasant to the stomach."

The most likely place of origin of the weedy rye is the central and eastern Turkey, northwest Iran, and Armenia. It is the area of maximum genetic diversity and coincides with the high degree of variability of the perennial *S. montanum*. The cold and harsh climate of this area possibly favored rye instead of the two main cereals, barley and wheat, which also spread into this region. It is without doubt that rye under such conditions could have become established as a cornfield weed. Weedy annual races with brittle and semibrittle rachises were and still are colonizers.

2.2 TAXONOMY AND CYTOTAXONOMY

Apart from artificial polyploids, all rye species are diploid ($2n = 2x = 14$). However, numerous reports about accessory or B-chromosomes are available in both wild and cultivated *S. montanum* and *S. cereale*. Roshevitz (1947) recognized 14 species, but it is questionable whether all these have to be given specific rank.

Two groups of species can readily be separated as being important in the evolution of cultivated rye. First, there is a group of annual weeds common in the Near East, such as *S. ancestrale*, *S. dighoricum*, *S. segetale*, and *S. afghanicum*, which cytologically resemble each other and cultivated rye (Schlegel and Weryszko 1979). They could be included as a subspecies of *S. cereale*. This group is confined to agricultural areas, the weedy types being widespread in cereal crops in Iran, Afghanistan, and Transcaspia (Zohary 1971).

Second, there is an aggregate of wild perennial forms of the *montanum* complex, widely distributed from Morocco through the Mediterranean area, Anatolia to Iraq and Iran. The group includes *S. ciliatoglumes*, *S dalmaticum*, and *S. kuprijanovii*. It has been separated into distinct species but is most probably described as a variant of a single species, *S. montanum* (syn. *S. strictum*). An isolated population also appeared in South Africa. Members of this group are cytologically similar and interfertile. However, they differ from the *S. cereale* complex by two major reciprocal translocations involving three pairs of chromosomes. In addition, two annual species, *S. vavilovii* and *S. silvestre*, have affected the rye evolution.

A compilation of most rye races and species is given by Mansfeld's World Database of Agricultural and Horticultural Crops, an online database since 1998 developed at the Institute of Crop Plant Research, Gatersleben, Germany. It reflects the contents of *Mansfeld's Encyclopedia of Agricultural and Horticultural Crops* (Hanelt 2001) and contains information on 6100 crop plant species, including rye, but excludes forestry and ornamental plants. Each species entry provides nomenclature and synonymy, common names in different languages, spontaneous distribution and regions of cultivation, uses, images, references, ancestral species, and notes on phylogeny, variation, and history (see Table 2.2).

Stutz (1972) and Evans (1976) proposed a stepwise evolution of *S. cereale* from *S. montanum*. The annual species are supposed to be derived from the introgression of *S. montanum* into *S. vavilovii*. In 1975, Kobyljanski (1975) described *S. iranicum*. Cytogenetic studies suggest it appears to be essentially identical to *S. vavilovii*. This annual, self-fertile species was in turn derived from the annual *S. silvestre* because of chromosome translocation (see Figures 2.1 and 2.2). *Secale silvestre*, a self-fertile, annual outbreeder, is distinguished from the above species by the presence of an additional awn on the outer glume and three seeds per spikelet (Figure 2.3). It is widely distributed from Hungary to the sandy steppes of southern Russia and from the Caucasus to northern Afghanistan. The karyotypes of *S. montanum*, *S. silvestre*, *S. africanum*, *S. vavilovii*, and *S. cereale* are as different as their chromosomal

TABLE 2.2
A List of Ryes So Far Described Worldwide

Serial Number	Naming of Rye
1	*Secale aestivum* Uspensky (1834)
2	*S. afghanicum* (Vavilov) Roshev. (1947)
3	*S. africanum* Stapf (1899)
4	*S. anatolicum* Boiss. (1844)
5	*S. ancestrale* (Zhuk.) Zhuk. (1933)
6	*S. arundinaceum* Trautv (1854)
7	*S. cereale* L. (1753),[a] *sensu lato*
8	*S. cereale* L. (1753), *sensu stricto*
9	*S. cereale* race *vavilovii*[a]
10	*S. cereale* ssp. *afghanicum* (Vavilov) Hammer (1987)
11	*S. cereale* ssp. *ancestrale* Zhuk. (1928)[a]
12	*S. cereale* ssp. *cereale*[a]
13	*S. cereale* ssp. *derzhavinii* (Tzvelev) Kobyl. (1975)
14	*S. cereale* ssp. *dighoricum* Vavilov (1939)
15	*S. cereale* ssp. *indo-europaeum* Antropov & Antropova ex Roshev. (1947)
16	*S. cereale* ssp. *primitivum* Kranz (1973)
17	*S. cereale* ssp. *rigidum* Antropov & Antropova ex Roshev. (1947)
18	*S. cereale* ssp. *sativum* Kranz (1973)
19	*S. cereale* ssp. *segetale* Zhuk. (1928)
20	*S. cereale* ssp. *tetraploidum* Kobyl. (1975)
21	*S. cereale* ssp. *tsitsinii* Kobyl. (1975)
22	*S. cereale* ssp. *vavilovii* (Grossh.) Kobyl. (1975)
23	*S. cereale* var. *montanum* (Guss.) Fiori
24	*S. cereale* var. *multicaule* Metzg. ex Alef. (1866)[a]
25	*S. cereale* var. *perennans* Dekapr[a]
26	*S. cereale* var. *vulgare*[a]
27	*S. chaldicum* Fed. (1938)
28	*S. ciliatoglume* (Boiss.) Grossh. (1924)[a]
29	*S. compositum* Poir. (1806)
30	*S. cornutum* Bald (1771)
31	*S. creticum* Siebold ex Kunth (1833)
32	*S. dalmaticum* Vis. (1842)
33	*S. daragalesi*[a]
34	*S. daralagesi* Tumanian (1938)
35	*S. dighoricum* (Vavilov) Roshev. (1947)
36	*S. hybernum* Poir. (1806)
37	*S. iranicum* Kobyl. (1975)
38	*S. kasakorum* Roshev. (1947)
39	*S. kuprijanovii* Grossh. (1928)
40	*Secale* L. (1753)[a]
41	*S. montanum* Guss.

TABLE 2.2

(Continued) A List of Ryes So Far Described Worldwide

Serial Number	Naming of Rye
42	*S. montanum* ssp. *anatolicum* (Boiss.) Tzvelev (1973)
43	*S. montanum* ssp. *kuprijanovii* (Grossh.) Tzvelev (1973)
44	*S. rhodopaeum* (Delip.) Kožuharov (1962) syn. *S. montanum* ssp. *rhodopaeum*
45	*S. segetale* (Zhuk.) Roshev. (1947)
46	*S. strictum* (C. Presl) C. Presl (1826),[a] *sensu lato*
47	*S. strictum* ssp. *africanum* (Stapf) Hammer (1987)[a]
48	*S. strictum* ssp. *anatolicum* (Boiss.) Hammer (1987)
49	*S. strictum* ssp. *ciliatoglume* (Boiss.) Hammer (1987)[a]
50	*Secale strictum* ssp. *kuprijanovii* (Grossh.) Hammer (1987)
51	*S. strictum* ssp. *strictum*[a]
52	*S. strictum* var. *ciliatoglume* (Boiss.) Fred and Peters (1998)[a]
53	*S. transcaucasicum* Grossh. (1949)
54	*S. triflorum* P. Beauv. (1812)
55	*S. triflorum* P. Beauv. (1812)
56	*S. turkestanicum* Bensin (1933)
57	*S. vavilovii* Grossh. (1924)
58	*S. vernum* Poir. (1806)
59	*Secale* × *derzhavinii* Tzvelev (1973)[a]

Source: Hanelt, P., *Mansfeld's Encyclopedia of Agricultural and Horticultural Crops*, Springer, Berlin, 2001. With permission.

[a] Accepted names according to the data sources.

behavior (see Figures 2.4 and 4.4; Table 4.6). The most pronounced terminal C bands are seen in *S. montanum* and *S. cereale*, followed by *S. vavilovii*, *S. silvestre*, and *S. africanum*. It seems during evolution that telomeric DNA (heterochromatin) was particularly amplified in cultivated rye.

Frederiksen and Peterson (1998) gave the latest revision of rye taxonomy. They recognized only three species of *Secale*, that is, *S. silvestre*, *S. strictum*, and *S. cereale*. *Secale strictum* has priority over *S. montanum*. It includes two subspecies, that is, ssp. *strictum* and ssp. *africanum*. *Secale cereale* also includes two subspecies, which is, the cultivated taxa, marked by tough rachises of ssp. *cereale* and wild or weedy taxa showing fragile rachis of ssp. *ancestrale*.

By using microsatellite markers, the average polymorphism is higher between species than within *S. cereale*. The mean genetic similarity (GS) index in the genus *Secale* was lower than that in the cultivated rye. The highest within-species GS index was observed for *S. silvestre* and the lowest for *S. strictum*, whereas the highest between-species GS index was found between *S. cereale* and *S. vavilovii* and the lowest between *S. silvestre* and *S. cereale*. There is no obvious difference

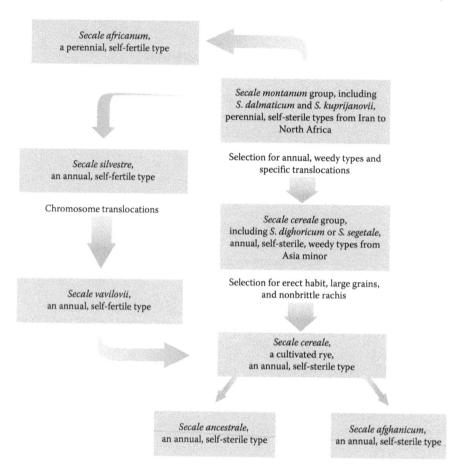

FIGURE 2.1 Possible evolutionary origin of cultivated rye (*S. cereale* L.), derived from *S. montanum*.

in GS levels in the cultivated rye accessions from Asia, Europe, North America, or South America. In molecular clusters, American cultivars were more closely related to Chinese cultivars than to European cultivars as a result of adaptation to their local conditions. Further random amplified polymorphic DNA–polymerase chain reaction (RAPD–PCR) analysis among the Chinese cultivars shows comparatively high genetic distances. Perhaps this was because only Chinese farmers in a few barren, geographically isolated areas planted rye and genetic material could not be exchanged. Genetic drift then resulted in independent subgroups of Chinese cultivars.

Cluster analysis indicated that all the *Secale* accessions could be distinguished by the 24 microsatellite loci (see Figure 2.5). The *S. silvestre* accessions are obviously divergent from the accessions of other species, and the *S. vavilovii* accessions are closely related to the *S. cereale* accessions. *Secale strictum* was heterogeneous and showed great within-species differences. The microsatellite- and amplified fragment length polymorphism (AFLP)-derived dendrogram faithfully reflected the

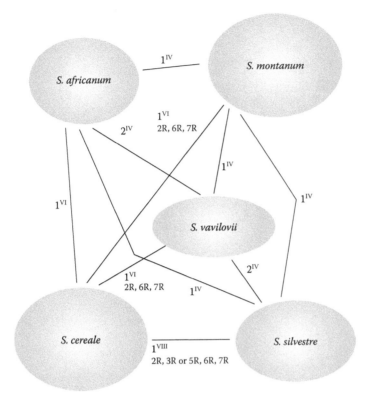

FIGURE 2.2 Schematic drawing of cytological differences between rye species in the presence of interchanges. IV, quadrivalent; VI, hexavalent; VIII, octovalent.

FIGURE 2.3 Spike morphology of *S. cereale*, *S. vavilovii*, *S. silvestre*, *S. montanum*, and *S. africanum* (from left to right).

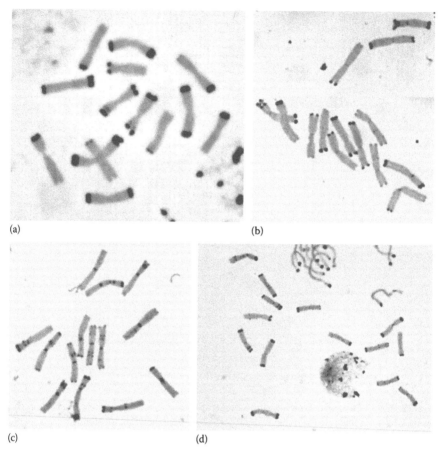

(a)   (b)

(c)   (d)

FIGURE 2.4 Comparison of karyotypes of (a) *S. montanum* (syn. *S. strictum* ssp. *strictum*), (b) *S. vavilovii*, (c) *S. silvestre*, and (d) *S. cereale* revealed by Giemsa C-banding. (Courtesy of Badaeva, E. Moscow.)

phylogenetic relationships between *Secale* species but did not indicate a possible domestication process of the cultivated rye based on the geographical sources of the accessions (Chikmawati et al. 2005, Shang et al. 2006).

Comparison of chloroplast DNA (ctDNA) variation confirmed the particular distinctness of *S. silvestre* from the remaining taxa. This basic differentiation between *S. silvestre* and *S. montanum* and/or *S. strictum* took place during the Pliocene Epoch or later. Skuza et al. (2007) applied restriction fragment length polymorphism (RFLP) markers of mitochondrial DNA in order to determine the species relationships. Stutz (1972) suggested that *S. silvestre* evolved from *S. strictum*, and these two species share a common ancestor. Molecular results showed that *S. silvestre* differs considerably in Austrian Agricultural Cluster (AAC) cluster distribution from *S. strictum* and from other species. Significant distinctions in patterns of AAC distribution between *S. silvestre* and *S. strictum* suggest an early separation of these two species (Achrem et al. 2013, Nürnberg-Krüger 1960). Cluster analysis of the GS

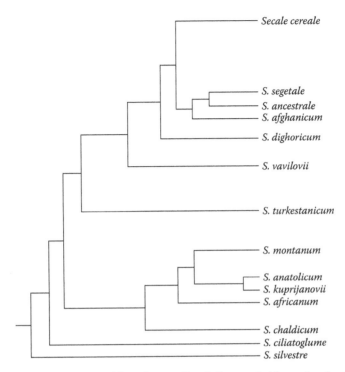

FIGURE 2.5 Genetic relationships of genus *Secale* L. revealed by molecular AFLP and RFLP markers. (After Chikmawati, T. et al., *Genome*, 48, 792–801, 2005; modified from Shang, H.-Y. et al., *Genet. Mol. Biol.*, 29, 685–691, 2006.)

coefficients clearly distinguished two groups of species. The first includes *S. cereale* ssp. *segetale* and *S. silvestre*, and the second group comprises the rest of the analyzed species. The second group has a subgroup of closely related species: *S. montanum* and *S. kuprijanovii*. The next highest GS was between those species and *S. vavilovii*, *S. cereale*, and *S. africanum*. The calculated GS between the species of rye was very high between *S. silvestre* and *S. segetale* as well as between *S. montanum* and *S. kuprijanovii*. After *S. kuprijanovii*, *S. cereale* is the next most genetically species similar to *S. montanum*. This corresponds with the suggestion that *S. silvestre* was the first species to separate from the others during *Secale* evolution and partially agrees with the statement that *S. vavilovii*, *S. africanum*, *S. strictum*, and *S. kuprijanovii* are evolutionarily the youngest. The distinctness of *S. silvestre* is also confirmed by molecular studies of ctDNA (Murai et al. 1989, Petersen and Doebley 1993), repetitive sequences (Cuadrado and Jouve 2002, Shang et al. 2006, Zhou et al. 2010), 18S–5.8S–26S ribosomal DNA (rDNA) (Bustos and Jouve 2002), as well as by AFLP-based analysis of genetic diversity (Chikmawati et al. 2005). The chloroplast genome size was estimated to be 136 kb. The ctDNA of all rye species except *S. silvestre* produced identical patterns with three restriction enzymes applied.

New molecular results match well with the morphological classification proposed by Hammer (1990) as well as Hammer et al. (1987) (see Table 2.3). According to

TABLE 2.3
Compilation of Rye Taxa

Species	Notes
Secale cereale L.	
ssp. *afghanicum* (Vavilov) K. Hammer syn. *S. afghanicum*	Used in breeding of cultivated rye, i.e., ssp. *cereale*
ssp. *ancestrale* Zhuk. syn. *S. ancestrale*	Progenitor of the cultivated ssp. *cereale*, a potential genetic donor for breeding
ssp. *cereale* syn. *S. cereale* var. *viride* syn. *S. cereale* var. *vulgare*	Major cereal with winter and spring cultivars; it is the youngest species of the genus.
ssp. *dighoricum* Vavilov syn. *S. segetale* ssp. *dighoricum* syn. *S. dighoricum* (Vavilov) Roshev.	A potential genetic donor for breeding
ssp. *rigidum* Antropov et Antropova ex Roshev. syn. *S. turkestanicum* Bensin syn. *Triticum cereale* (L.) Salisb.	Discovered in 1912 by B. M. Bensin (1933); local name in Turkestan *karabidaj* (black wheat); autogamous
ssp. *segetale* Zhuk. syn. *S. segetale* (Zhuk.) Roshev.	A potential genetic donor for breeding
ssp. *tetraploidum* Kobyl. syn. *S. cereale* syn. *S. cereale* ssp. *tetraploidum nudipaleatum* Kobyl.	Rye with four genomes instead of two; mostly artificial hybrids
ssp. *tsitsinii* Kobyl. var. *multicaule* Metzg. ex Alef. syn. *S. cereale* ssp. *cereale* syn. *S. cereale* var. *multicaule*	A perennial rye, formerly grown in Eastern and Central Europe in forest clearings or as a component of shifting cultivation for forage and grain (Waldstaudenroggen in German)
S. strictum (C. Presl) C. Presl syn. *S. dalmaticum* Vis. syn. *Triticum strictum* C. Presl syn. *S. montanum* Guss.	Wild perennial and open pollinating; possibly the progenitor of the annual *Secale* taxa
ssp. *africanum* (Stapf) K. Hammer	Most distant to *S. strictum* among the perennials; used as cereal food; a potential genetic donor for breeding
ssp. *anatolicum* (Boiss.) K., Hammer syn., *S. daralagesi* (Tumanian) syn., *S. anatolicum* Boiss. syn., *S. montanum* ssp. *anatolicum* (Boiss.) Tzvelev syn., *S. montanum* var. *anatolicum* (Boiss.) Boiss. syn., *S. rhodopaeum* Delip. syn., *S. cereale* var. *perennans*	Together with *S. kuprijanovii*, the most close to *S. strictum* among the perennials; caryopsis with adherent pericarp; donor for β-amylase; distributed across Asia-temperate, i.e., Caucasus and western Asia
ssp. *ciliatoglume* (Boiss.) K. Hammer syn., *S. strictum* var. *ciliatoglume* (Boiss.) Fred. & G. syn., *S. anatolicum* var. *ciliatoglume* (Boiss.) Ivanov & G. V. Yakovlev Petersen syn., *S. montanum* var. *ciliatoglume* syn., *S. montanum* var. *ciliatoglume*	An isolated weedy population with pubescent culms endemic to Turkey (around Mardin)
ssp. *kuprijanovii* (Grossh.) K. Hammer syn., *S. kuprijanovii* Grossh.	Together with *S. anatolicum*, the most close to *S. strictum* among the perennials

TABLE 2.3
(Continued) Compilation of Rye Taxa

Species	Notes
ssp. *strictum*[a] syn., *S. daralagesi* (Tumanian)	Wild stands provide a good pasture and hay
Secale × *derzhavinii* syn., *S. derzhavinii* Tzvelev	An artificial hybrid between *S. strictum* and *S. cereale*; introduced into cultivation as perennial rye
S. silvestre Host syn., *S. fragile* M. Bieb.	An annual and self-pollinating rye with characteristic long awns; native to the Aralo-Caspian basin; largely intersterile with *S. cereale* and *S. strictum*; a potential genetic donor for breeding, e.g., three low-molecular-weight glutenin-like genes (*Ssy1*, *Ssy2*, and *Ssy3*); molecular data show closer relationship to the perennials rather than to the annual taxa; it is the most ancient species of the genus
S. vavilovii Grossh. syn., *S. iranicum* Kobyl.	Armenian wild rye; a potential genetic donor for breeding; it is commercially grown in >30 countries, primarily for forage and feed, but also for unleavened bakery products.

Source: Hammer, K. et al., *Kulturpflanze*, 33, 135–177, 1987. With permission; Shang H.-Y. et al., *Plant Genet.*, 41, 1372–1380, 2005.

[a] Armenian populations have been interpreted as relics of former cultivation of perennial rye (as *S. daralagesi* Tumanian); introgressions to sympatric cultivated rye have been observed; in the United States, it has been tried as green manure, cover crop, and forage grass; in the western United States, it is recommended for erosion control, as early spring forage, and as a cover crop; in the past, it has been used for winter pasture in regions with winter rainfall.

Hammer (1990), *S. strictum* should be considered the most ancient species of *Secale*. The subspecies of *S. strictum*, that is, *S. strictum* ssp. *kuprijanovii* and *S. strictum* ssp. *africanum*, show high similarity to each other. The low degree of molecular similarity (42%–53%) between *S. strictum* subspecies and *S. strictum* species supports the hypothesis that *S. strictum* stopped evolving a long time ago, whereas its subspecies are still evolving (Achrem et al. 2012). Another evidence that *S. strictum*, *S. strictum* ssp. *kuprijanovii*, and *S. strictum* ssp. *africanum* evolved independently (Khush and Stebbins 1961) was provided by comparative sequence analysis of the internal transcribed spacer (ITS) of the rDNA performed by Bustos and Jouve (2002). This conclusion is supported by the differences in the pattern of AAC sequence distribution among *S. strictum* species and subspecies. *Secale strictum* and *S. strictum* ssp. *kuprijanovii* chromosomes showed the same pattern of AAC cluster distribution, while the pattern observed in *S. strictum* ssp. *africanum* chromosomes was found to deviate from all taxa studied. The distinctive distribution pattern of AAC repeat in *S. strictum* ssp. *africanum* chromosomes can be explained by the migration of this subspecies to South Africa during the Early Pleistocene epoch, where it evolved separately from the rest of the genus. This polymorphism could be due to deletion and/or amplification from the preexisting sequences.

On the other hand, *S. cereale* showed high similarity to its subspecies, that is, *S. cereale* ssp. *segetale* and *S. cereale* ssp. *afghanicum*, providing support for the classification proposed by Hammer (1990), who recognized these two taxa as subspecies. Very similar conclusions can be drawn from single-sequence repeat (SSR)-based studies (Shang et al. 2006), inter-SSR (ISSR)-based studies (Ren et al. 2011), and the analysis of ITS rDNA regions (Bustos and Jouve 2002). The physical localization of AAC clusters in *S. cereale* was identical to that in *S. cereale* ssp. *segetale*, while the differences in *S. cereale* ssp. *afghanicum* can originate from distinct deletions or rearrangements of sequences that occurred during taxa differentiation.

ISSR and interretrotransposon amplified polymorphism (IRAP) results individually and collectively revealed a high similarity between *S. cereale* and *S. vavilovii* (inbred lines). *S. vavilovii* has a semidwarf growth habit and is annual, cleistogamous, and self-fertile with brittle rachis. It was grown in eastern Anatolia and northern Iran, where it was collected by Grossheim in 1923 and by Kuckuck in 1953. The degree of similarity appeared to be higher between *S. cereale* and *S. vavilovii* than between *S. cereale* and its subspecies. The similarity of AAC repeat distribution between *S. cereale* and *S. vavilovii* indicates that these two taxa share a recent common ancestor. The pattern of AAC distribution is rather conservative in the majority of *Secale* taxa. Cuadrado and Jouve (2002) observed a more variable pattern of AAG repeat distribution. Their results point to a closer relationship between *S. cereale* and *S. vavilovii* and suggest the common origin and early separation from *S. strictum*. *Secale vavilovii* and *S. cereale* also share the same translocation that distinguishes these taxa from *S. strictum* (Singh and Röbbelen 1977) as well as the presence of a polymorphic locus 5S rDNA on the short arm of rye chromosome 3 (3RS) of *S. vavilovii*. These similarities led to the idea of classifying *S. vavilovii* as a subspecies of *S. cereale* (Frederiksen and Peterson 1998, Ren et al. 2011). In *S. cereale*, the 5S rDNA genes have been located on three pairs of chromosomes—1R(S), 3R(S), and 5R(S)—and 18S–26S rDNA genes have been found on one pair of the chromosome 1R(S) (see Section 5.3). The 5S rDNA genes were detected on the distal region of the secondary constrictions in nucleolus organizer regions (NORs) in the chromosome 1R(S) and the other was detected in the intercalary region on the short arm of the chromosome 5R(S).

The molecular results confirm basically three groups of taxa:

1. The *silvestre* genotypes close to *T. monococcum* (AA) and *Agropyron cristatum*
2. The annual genotypes (*S. afghanicum*, *S. segetale*, *S. dighoricum*, *S. cereale*, *S. vavilovii*, *S. ancestrale*, *S. turkestanicum*) close to *Aegilops tauschii* (DD)
3. Perennial genotypes (*S. montanum*, *S. kuprijanovii*, *S. anatolicum*, *S. chaldicum*, *S. africanum*, *S. ciliatoglume*)

One can also assume that most of the genetic diversity is found within genebank accessions rather than between those populations. Landrace from Germany, Sweden, Norway, and Finland shows particularly low variation as compared to other landraces worldwide. This type of information has to be considered in maintaining and evaluating genebank collections.

2.3 GROSS MORPHOLOGY

In comparison with cereal rye, the important wild relatives are characterized as follows (see Table 2.4).

Cereal rye is an annual organism and, like winter wheat, grows well when planted in the autumn. Only rarely is it sown in the spring. It can be described as an erect annual grass with flat blades and dense spikes. Types of opaline silica bodies (opal phytoliths) occur in the mature prophylls, radical and culm leaves, culms, and inflorescence bracts (Blackman and Parry 1968). Silica bodies are absent from the coleoptile and the adaxial epidermis of the prophylls, leaf sheaths, and inflorescence bracts.

The coleoptile of the germinating plant may be violet because of anthocyanins. The anthocyanins are most expressed by cool or cold environments. The water-soluble vacuolar pigments may appear red, purple, or blue to green, depending on the pH value. Anthocyanins have been shown to act as a sunscreen, protecting cells from high light damage by absorbing blue-green and ultraviolet light, thereby protecting the tissues from photoinhibition, or high light stress. This has been shown to occur in red juvenile leaves, autumn leaves, and broad-leaf evergreen leaves that turn red during the winter.

The culm is tall (1.5 m), erect, slender, and glabrous, except for pubescence near the spike, and has a hollow structure. The color is bluish-green with a coat of waxy bloom. Genetic analysis showed that the development of a waxy bloom on all parts of the plant is controlled by one pair of alleles, *Epr/epr*. Two further complementary dominant genes, *Es1* and *Es2*, are required in addition for wax formation on the ear. The genes are independent of each other and of those for dwarfing and anthocyanin pigmentation (Fedorov et al. 1967).

The number of nodes is 6 or 7 for spring rye and 10 to 12 for winter rye. One leaf is formed at each leaf base. The leaf sheaths are long and loose. The ligulae are short.

TABLE 2.4
Gross Growth Habit of Basic Species of Rye, *Secale* L.

Species	Growth Habit	Flowering	Rachis	Agriculture
Secale cereale	Annual	Open pollination	Nonbrittle	Cultivated
S. ancestrale	Annual	Open pollination	Semibrittle	Wild, weedy
S. vavilovii	Annual	Open pollination	Brittle	Wild, weedy
S. montaum	Perennial	Open pollination	Brittle	Wild, cultivated
S. africanum	Perennial	Self-pollination	Brittle	Wild, grazing
S. silvestre	Annual	Self-pollination	Brittle	Wild

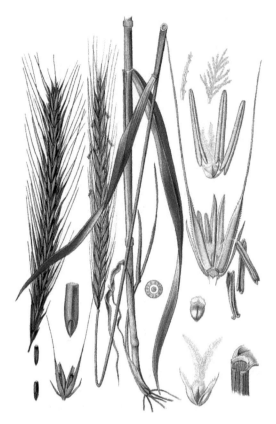

FIGURE 2.6 Morphological characteristics of common rye, *S. cereale* L. (After Thomé, O. W., *Flora von Deutschland, Österreich und der Schweiz*, Köhler Verl., Gera, Germany, 1885.)

Auricles are small and short (see Figure 2.6). The leaf blade is linear–lanceolate. The habit resembles that of wheat, but usually taller, even modern cultivars. The flowering stems (culm) are jointed.

2.3.1 PLANT HEIGHT

The cultivated rye plants are 80–180 cm high. On the whole, rye height may vary with its different varieties from 10 or 15 to 300 cm. It is known to be a complex trait under the control of more contributing genes and environmental effects. However, major genes exist that may decrease plant height. The latter is usually achieved by the introduction of height-suppressing genes or the removal of height-promoting ones. The plant hormone gibberellic acid (GA) is involved in the determination of plant height. It has been shown that elevated GA levels are associated with taller plants, while reduced GA levels are associated with dwarfism. Natural or induced dwarf mutants are classified into groups: (1) GA-sensitive and (2) GA-insensitive genotypes. Both types of mutants are described in rye.

Numerous GA-sensitive mutants were identified in cereals, but GA-insensitive ones were more often used in breeding. GA-sensitive mutants show defects in GA synthesis, that is, GA supply is blocked. They respond to exogenously applied GA by elongation of stem and/or restoration of normal growth habit. GA-insensitive mutants show almost no reaction to exogenously added GA (Börner and Melz 1988). They are characterized by reduced coleoptile length, leaf size, and internode length. Such mutants lack a transcriptional repressor, which limits the expression of GA biosynthetic enzymes. For that reason, they have higher rates of endogenous GA synthesis and do not react with GA added.

In rye, dwarfing genes are distributed throughout the genome. The EM1 mutant—a single dominant dwarfing gene—causes the most common dwarfness in rye. Also known as *Dw1* or *Ddw1*, the gene was localized on the long arm of chromosome 5R (Korzun et al. 1996). A second dominant dwarfing gene *Dw2* or *Ddw2* was found on chromosome 7R (Melz 1989). All the other dwarfing genes showed recessive inheritance (Schlegel and Korzun 2013) (see Table 2.5).

Studies on various progenies of the rye Carsten's Kurzstroh (short-strawed) grown in close proximity to Petkuser Normalstroh (normal-strawed) showed high correlations between plant height and grain yield (Rehse 1960). Values based on the longest tiller were found to correspond closely to those for the entire plant. The degree to which the values could be modified by environmental effects was investigated, and the results showed that the variance due to environment was greater than the residual variance only for the number of tillers per plant and spike density, whereas for the height of plant, length of spike, number of spikelets, yield and weight of grain, and number of grains per spikelet (fertility), the differences were primarily genetic. In the

TABLE 2.5
Short-Strawed Genes in Rye and Their Sensitivity to Exogenous GA

Designation	Chromosomal Location	Inheritance	GA Response	References
dw4	1R	Recessive		Melz (1989)
dw2, d2	2R	Recessive	Sensitive	de Vries and Sybenga (1984)
dw3, d3	3R	Recessive		de Vries and Sybenga (1984)
dw5	4R	Recessive		Melz (1989)
np	4RL	Recessive	Sensitive	Malyshev et al. (2001)
dw6	5R	Recessive	Insensitive	Melz (1989); Börner et al. (1992)
Dw1, Ddw1, Em1	5RL	Dominant		Korzun et al. (1996)
ct2	5RL	Recessive	Insensitive	Plaschke et al. (1993)
dw7	6R	Recessive		Melz (1989)
Dw2, Ddw2	7R	Dominant		Melz (1989)
ct1	7R	Recessive	Insensitive	Plaschke et al. (1993)
ct3	7R	Recessive		Malyshev et al. (2001)

Source: Schlegel, R., and Korzun, V. Genes, markers and linkage data of rye (*Secale cereale* L.), 9th updated inventory, V. 03.13, pp. 1–115. http://www.rye-gene-map.de. 2013. With permission.

group of progenies with medium–long straw, though there was a general positive regression between length and grain yield of the longest tiller, several exceptional individuals were found in which the correlation was broken. This occurred in 57 of 93 families examined. In most individuals, the linear regression between height and grain yield of the longest tiller was present, but the yield was greater and the height was less than expected. These data implied no difficulty in breeding short-strawed rye with high fertility and high 1000-grain weight.

The stem of vegetating rye is green with a bluish-gray shade due to a waxy film. There are rye varieties with no waxy film. The complete genomic and complementary DNA (cDNA) sequences of the *waxy* gene encoding granule-bound starch synthase I (GBSSI) is isolated from the rye genome and characterized. The full-length rye *waxy* genomic DNA and cDNA are 2767 and 1815 bp, respectively. The genomic sequence has 11 exons interrupted by 10 introns. The rye *waxy* gene is GC-rich, with a higher GC frequency in the coding region, especially in the third position of the codons. Exon regions of the rye *waxy* gene are more conserved than intron regions when compared with the homologous sequences of other cereals. The mature rye GBSSI proteins share more than 95% of their sequence identity with their homologs in wheat and barley (Xu et al. 2009). A phylogenetic tree based on sequence comparisons of the available plant GBSSI proteins shows the evolutionary relationship among *waxy* genes from rye and other plant genomes. The identification of the rye *waxy* gene enables the manipulation of starch metabolism in rye and triticale. The gene is most related to the *wx* locus on wheat chromosome 7A, that is, on homoeologous rye segments of 2R, 4R, 6R, or 7R, but most likely on 7RL since a recessive allele is located there.

This waxy film itself does not protect the stem tissues from penetration of mycosis pathogens. Mature stems are yellow or tinged with anthocyanin to different degrees. The stem surface is glabrous. It is slightly downy below the spike; however, varieties without down are quite common. Branches arise singly from nodes and are subtended by a leaf sheath and two-keeled prophyll.

Tillering is common depending on the genotype and growing conditions. The main primary tillering nodes are located at a depth of 0.5–2 cm, which depends on the length of the mesocotyl (the underground internode) that joins the caryopsis with the node.

Rye has the longest stems of all cultivated small grains (35–300 cm) and these provide most of the photosynthetic area. The average height of older ryes is 150 cm, for example, the landrace Vyatka (Russia) 190 cm, Wojcieszyckie 160 cm, Dańkowskie Złote (Poland) 145 cm, or Kustro (Germany) 35 cm. Plant height has a complex inheritance with many loci spread across the entire genome (Miedaner et al. 2011a). Modern, high-yielding varieties, in particular hybrid varieties, are much shorter. Their long compact ears, possessing seeds of equal size, distinguish the hybrids (see Figure 2.6).

During grain formation, stems with sheaths account for 60%–80% of the total plant area. Leaves are arranged alternately in two ranks, differentiated into sheath, blade, and an adaxial erect appendage at sheath/blade junction (ligule). Ligules are membranous or a line of hairs. They are horizontally truncated and located where the sheath turns into the leaf lamina. They tightly envelop the stem, protecting it from the penetration of moisture and insects. There are rye varieties with an oblique ligula or no ligula at all. Sometimes ligula can be absent.

For instance, a liguleless line was selected from a spring-type rye introduced from the Pamir Mountains. When crossed with a common spring-type rye, it resulted in a normal-to-liguleless ratio of 1:3 in the F_1. The same ratio occurred in all but one cross with normal winter-type ryes. For the character, the gene symbol *el* was assigned to describe *eligulatum*. As for the *el* locus, a monogenic segregation can be obtained for the spring:winter-type habit. The locus was assigned as *ae* by Fedorov et al. (1970b). However, there is no genetic linkage between *el* and *ae*, *vi* (for absence of anthocyanin pigment), or *epr* (for waxless surfaces), but there are indications of a weak linkage between *el* and *ct* (for a dwarf-growth habit) and a strong linkage between *el* and *S*, governing self-sterility.

Compared to wheat, the flag leaf is smaller and less important in photosynthesis (blades are 4 mm–12 mm wide). It has been determined that the size of the second leaf from the top allows prediction of the foliage of the plant as the size of this leaf is equal or close to that of the average leaf. Long and narrow leaves are typical of more drought-resistant varieties (see Section 3.4); short and wide leaves are typical of low yielders, which are relatively late maturing and prone to suffer from mildew (see Section 8.2.2). Leaves (stems and spikes) have a bluish color owing to their waxy surface. A leaf sheath is surrounding and/or supporting the culm internode. It is split at the base or infrequently tubular with partially or completely fused margins. The leaf blades are divergent, usually long, narrow, and flat, but varying from enrolled and filiform to ovate. Veins are parallel, sometimes with cross-connecting veinlets. The spike is longer and more slender (2–19 cm), somewhat nodding when mature.

2.3.2 Spike

Rye's reproduction is annual, except in some perennial wild relatives. Each productive stem usually forms one spike (see Figure 2.6). Rye inflorescence is a compound and slender spike of imperfect type (without the topmost spikelet). The terminal spike is ~10–15 cm long and more slender and nodding than wheat. From the average of many varieties, the length of rachis is 9.8 cm and the number of spikelets per spike is 32.4, considering more than 50,000 plants. Ryes with looser spikes are Rimpau's Schlanstedter Roggen, Chrestensen's Riesenroggen, Askanischer Riesenroggen, Aderstedter Roggen, or Heine's Hadmerslebener Klosterroggen; semi-looser spikes show Saaleroggen, Friedrichswerther, Pirnaer, or Hanna (see Table 7.6).

Cultivated rye varieties have white (straw-colored) spikes. Some old local populations have reddish-brown spikes. Wild-growing rye has white, reddish-brown, brown, and black spikes. Rye spikes are covered with a waxy film, which can be pronounced better or worse depending on the features and climatic conditions of specific varieties. A well-pronounced waxy film is typical of cultivars bred in areas with hot summers as this film protects the spikes from overheating by reflecting some sunlight.

2.3.3 Spikelet

Spikelets (see Figure 2.6) are solitary and sessile at each node and alternately arranged on a zigzag rachis, that is, on each nodus of the rachilla segment, there is one biflorous or, more rarely, triflorous spikelet. The spikelets lie flat against the rachis so the empty glumes and lemmas have an outer (abaxial) and inner (adaxial) margin with respect to the rachis. The rachis terminates in an apical spikelet, which is usually rudimentary and sterile. Each spikelet usually consists of only two fertile florets, with supernumerary florets vestigial or absent. As in other small-grained cereals, a modified leaf (collar) subtends the basal spikelet. A ridge representing a subtending leaf develops from the node beneath the second spikelet but rarely beneath the third.

The rye spike normally bears one spikelet per rachis node, and the appearance of supernumerary spikelets is rare. The gene responsible for the so-called monstrosum spike trait in rye was mapped by genotyping F_2 populations with microsatellite markers (Dobrovolskaya et al. 2009). The trait is under the control of a recessive allele at a single locus. The *Mrs1* locus located on the short arm of chromosome 2R, cosegregating with the microsatellite locus *Xwmc453*, ~10 cM from the centromere. A similar effect on the phenotype of *mo1* and *mrs1*, together with their presence in regions of conserved synteny, suggests that they may well be members of an orthologous set of Triticeae genes governing spike branching.

2.3.4 Peduncle

The peduncle of the common winter rye is covered with short fine hairs just beneath the collar. The pubescence is thickest at the top of the peduncle and decreases downward for 2–7.5 cm. The peduncle is glabrous (see Figure 2.6). Some rye varieties, however, have a glabrous peduncle throughout their length. A dominant gene controls the hairy neck character. It is localized on the distal part of the long arm of chromosome 5R (Schlegel et al. 1993).

2.3.5 Rachis

The rachis is the backbone of the ear (inflorescence), which bears the lateral spikelets. The internode of the rachis is slightly concave on the side next to the spikelet. Rye rachilla is not ramose, but there are some rye varieties with hereditarily ramose rachillae.

Wild Near Eastern cereals—such as barley, wheat, and rye—are characterized by a brittle inflorescence, which leads to a shattering of the dispersal units (i.e., spikelets) upon maturity. This seed dispersal mechanism is essential under natural conditions, in which the arrow-like morphology of the spikelet mediates their penetration through surface litter into the soil.

In the fragile types, a breaking tissue originates from a meristematic zone on the basis of spike segments. Fragility is greater at low atmospheric humidity. The potential fragility is genetically unifactorial and caused by the recessive gene *firm*. In the evolution of nonfragile races, cultures with irrigation and in maritime climates may have played a role.

The disarticulating (brittle) rachis (*Br*) character is of evolutionary significance because of its adaptive value as an effective seed dispersal mechanism. Under domestication, however, *Br* poses considerable difficulty for harvest operations as it leaves only a few spikelets on the rachis to collect. Thus, a nonbrittle rachis spike is thought to have been the first symptom of rye domestication. Kranz (1963) suggested one locus to be responsible for brittle and/or nonbrittle rachis. Later, spikelet shattering in rye has been reported to be a dominant character. Nonbrittleness, *br1*, was found to be controlled by a single gene/allele that mapped to the short arm of chromosome 5R (de Vries and Sybenga 1984).

2.3.6 NODE AND INTERNODE

Rye stems are hollow straws consisting of three to seven internodes separated by nodes (see Figure 2.6). Some varieties have been observed to possess a filled upper internode. The ability of rye and other cereal crops to resist lodging is related to the height of the plant, the strength of the stems, the size of their root system, and the weight of their ears. Four kinds of short rye stems have been distinguished. These are QTLs, *Sil2* with the flanking marker *Dw1* on chromosome arm 5RL coding for the length of the second internode, and *Sit3* with the flanking *Xpsr928* marker on chromosome arm 7RS coding for the thickness of the second internode (Milczarski 2008).

Long silky hairs extend from node to node along the lateral margins of all of the internodes except two or three basal internodes, which have short hairs like the peduncle. The hairs increase in length from the base of the internode upward and terminate in a tuft of hairs beneath the empty glumes at the node. There are no hairs beneath the node or on the surface of the internode between the margins, except that the first and second internodes are pubescent, as mentioned earlier.

The internode of the rachilla between the second and third florets is longer than any other internode. This internode has short, sharp hairs on the surface beneath the floret. The internodes beneath the rudimentary fourth and fifth florets are short, which can only be seen by a stereoscopic microscope. The internodes of the rachilla between the empty glumes and the first and second florets are so short that they are difficult to identify.

2.3.7 GLUME

There are two empty glumes at the base of the spikelet, one on each side. They are subulate (awl-shaped) structures about two-thirds as long as the lemmas of the fertile florets above them. The glumes are narrow, rigid, acuminate, or subulate-pointed (see Figure 2.6). Sometimes glumes can be pubescent. This pubescence was studied by Fedorov et al. (1970c) using F_2 segregation analysis. It is simply inherited. A dominant allele controls the character though 3 out of 18 combinations showed a deviation from a 3:1 ratio. The symbol *V* (*velutinum*) was assigned to the character. It proved to be independent of *Vi* for anthocyanin pigmentation, *Epr* for glaucous stems, *M* for branched ear, *Es* and *Es2* for glaucous ear, *ct* for dwarf stature, *R* for auricle pigmentation, and *el* for absence of ligule and, apparently, genes determining self-fertility.

Lemmas are broader than glumes, often sharply keeled, five-nerved, and ciliate on the keel and exposed margins, tapering into a long awn. The middle vascular bundle extends throughout the length of the keel into the awn. The lemma and palea that enclose the floret are free threshing, and the lemma often bears barbs on the keel and an awn of intermediate length. Two thin margins, which curve inward toward each other, extend laterally from the keel. The outer (abaxial) margin is wider and slightly thicker in cross section than the inner (adaxial) margin. There is a vascular bundle in each margin that terminates in the middle bundle. The lateral vascular bundle in the outer margin is larger than the one in the inner margin. A few small transverse (commissural) vascular bundles connect the lateral vascular bundles with the middle bundle. A band of cells containing chlorophyll (chlorenchyma) on each side of the middle vascular bundle extends laterally almost to the vascular bundles in the margin. There is no chlorenchyma along the lateral vascular bundles on the side next to the edge of the margins. A row of stomata is present in the outer and inner epidermis over the collenchyma. In cross section, the chlorenchyma is wider at the keel and tapers to the edge of the margins. In longitudinal section, the empty glume is thick at the base and very thin at the apex. A cross section shows that in the thin portion toward the apex, the glume consists of an outer epidermis of cells with thick, lignified walls and an inner epidermis of thin-walled cells.

Near the edge of the margins is a band, several cells wide, that extends from the base to the tip of the lemma, consisting only of thick-walled cells of the outer epidermis and collapsed cells of the inner epidermis. On the keel are several rows of sharp, curved hairs that extend along the keel from the base of the glume to the tip and throughout the length of the awn. Sharp, curved hairs are present along the edge of the margins and on the outer and inner surfaces. They are more numerous near the tip of the glume and on the outer than on the inner surfaces (Bonnett 1940).

Opaline silica bodies (opal phytoliths) occur in mature prophylls, radical and culm leaves, culms, and inflorescence bracts of rye. They are absent from the coleoptile and the adaxial epidermis of the prophylls, leaf sheaths, and inflorescence bracts (Blackman and Parry 1968). Anthers are ~7 mm long with light yellow to purple color. Rye anthers are markedly longer than wheat anthers. When the seven chromosomal addition lines of Imperial rye are compared to Chinese Spring wheat, the 4R addition has significantly longer anthers than the remaining rye chromosomes, although anthers are significantly shorter in rye. It was suggested that this intermediate anther length could be used to improve production of hybrid wheat (Plaha and Sethi 2000).

2.3.8 LEMMA

The lemma is a lanceolate-shaped structure that terminates in an awn about one and a half times as long as the lemma. The lemma has a keel that extends from the base to the tip. Rows of sharp, curved hairs extend the length of the keel and into the awn (see Figure 2.6). Lateral margins extend from the keel, curving inward and almost enclosing the palea, stamens, and pistil. The lateral margins are wider and thinner than those of the empty glumes. The outer (abaxial) margin is wider and thicker than the inner (adaxial) margin. There is no distinct tip to the lemma because the lateral margins extend upward, gradually merging with the awn. The lemma has five vascular bundles, one in the keel and two in each margin. Those nearest the keel are the largest and extend into the awn. The two vascular bundles in the margins nearest the edge terminate near the tip of the lemma. Small, transverse vascular bundles connect the two lateral vascular bundles and also connect the adjacent lateral vascular bundle with the middle vascular bundle in almost all of the lemmas examined. Occasionally, transverse bundles are few or absent. Bands of chlorenchyma extend along each side of the middle vascular bundle across the length of the lemma and into the awn. Chlorenchyma is also present on either side of the lateral vascular bundles adjacent to the middle bundle, but it may not be present at the termination of the outermost vascular bundles or along the outermost vascular bundle of the inner margin. Stomata are present over the chlorenchyma in both the outer and inner surfaces of the lemma. Sharp, curved hairs extend halfway down the edge of the outer margin, but they are absent from the edge of the inner margin. Some sharp, curved hairs are present on the outer and inner surfaces of the lemma, but these are fewer on the empty glumes.

2.3.9 Awn

An awn is present at the tip of the lemma and at each of the empty glumes. Rye shows medium–long awns in comparison with wheat (short) and barley (long). The awns of the lemma are much longer than those of the empty glumes. In each case, the awn is an extension of the basal structure. The keel and margins continue into the awn, forming a triangular-shaped structure with rounded corners. The middle vascular bundle and two lateral vascular bundles continue into the awn (see Figure 2.6). A row of stomata is on each side of the main vascular bundle. Hairs from the margins and the keel continue along the corner of the triangular-shaped awn to the tip. The awn terminates with a group of four sharp, pointed hairs, one of which projects above the other three. The awn is a stiff, rigid structure covered by lignified, thick-walled epidermal cells. Beneath the epidermis are several layers of thick-walled fibers that enclose the vascular bundles and the chlorenchyma. The only openings through the mechanical tissue are at the stomata where the substomatal chamber extends into the chlorenchyma.

2.3.10 Palea

In contrast to the empty glumes and lemma, the palea (inner bract) has two keels. These lie on either side of the floret, touching the lateral margins of the lemma at the position of the lateral vascular bundles nearest the edge of the margin. The tip of the palea is divided into two obtuse-shaped apexes in which the keels and their vascular bundles terminate (see Figure 2.6). Two margins, one from each keel, extend laterally, curve inward, and partly enclose the lateral stamens. At the base of the palea, the edges of the two margins have several thick and rounded cells where they fit into the lodicules. A thin membranous tissue connects the two keels. The tissue between the keels is folded inward so that it separates and partially encloses the two lateral stamens. Two vascular bundles, one in each keel, are the only ones in the palea. A band of chlorenchyma and a row of stomata are on either side and beneath each vascular bundle. A few stomata are found in the inner epidermis. Pointed hairs, varying in length, are found throughout the length of the keels, along the edges of the margins at the apexes, and on the outer and inner surfaces of the palea. The palea does not have as much mechanical tissue as the lemma and the empty glumes. No hairs are found on the palea.

2.3.11 FLOWER

There are three stamens in each floret. One is opposite the keel of the lemma. The other two are on either side of the pistil in the folds of the palea. A stamen consists of an anther and a filament. Each anther has four locules that contain pollen and are joined to a connective in the center. The filament is joined to the base of the connective and to the base of the floret just above the insertion of the lemma and palea (see Figure 2.6). The filament of the stamen opposite the lemma is inserted between the two lodicules, and the filaments of the two lateral stamens are inserted just behind the lodicules. The two lodicules are placed one on either side, anterior to the pistil. They are thick and fleshy at the base but thin and membranous at the apex and margins. The thick, fleshy base is in contact with the base of the lemma. There is a notch along the posterior edge of the thick base in which the thick, rounded edge of the palea fits. The thin, membranous margin extends from the notch and along the inner surface of the palea. The lodicules swell at anthesis and force the lemma and palea apart so that the stamens can be exerted to shed pollen. The pistil, as in other cereals and grasses, is composed of an ovary and two styles at the top. The ovary is oval in shape and terminates the floral axis. Long hairs cover the upper portion of the sides and the top of the ovary around the base of the styles. The styles extend from the top of the ovary. They are larger at the base and taper to a blunt point. Stigmatic branches cover the exterior surface of the styles from the base to the tip, forming a plume-like group.

2.3.12 FLORET

A rye spikelet usually has two florets and a rudimentary third floret (see Figure 2.6). Examination of the rudimentary portion of the spikelet above the two fertile florets, however, shows the remnants of more than one floret. In fact, early stages of spikelet development show six or more florets per spikelet. The florets are perfect, each containing three anthers and a pistil consisting of two feathery stigmas and an ovary (see Figure 2.6). The stigma is ~5 mm long. The two lower flowers are sessile. The third flower has a pedicle that usually consists of a lemma with a short awn, a palea, and remnants of the stamens. The florets above the third flower consist of shapeless remnants of lemmas and paleas enclosed by the lemma of the third floret.

2.3.13 SPIKE DIFFERENTIATION

Spike differentiation starts with elongation of the shoot apex (Bremer-Reinders 1958). Double ridges appear on the elongated shoot, which is a definite indication of spike initiation. The upper ridge is the spikelet primordium. Spikelet initiation occurs first in the middle of the spike and proceeds toward the tip and the base. The first indication of spikelet development is the appearance of the empty glumes as ridges on both sides of the spikelet primordium, parallel to the rachis. Soon after, the lemma of the two basal florets and the apex of the floral shoot are clearly delimited. The florets are initiated above the two basal florets. They develop acropetally. In the early stage of spikelet development, there is no indication that the florets above the two basal florets will abort, but by the time the awns have begun to elongate, the aborting florets can be easily identified. The awns elongate more rapidly than the basal part of the glumes or lemma. Initiation of the stamens and the palea occurs at about the same time. The stamens appear as three papillae above the primordium of the lemma. The anthers grow rapidly and protrude above the enclosed lemma and the palea. The pistil is the last part of the floret to develop.

2.4   ROOT SYSTEM

Rye has a fibrous root system with no defined taproot and consists of three to four embryonic (primary) roots, which are formed at the moment of seed germination, and nodouse (secondary) roots shooting off the underground stem nodes in the area of the tillering node (see Table 8.4). Continuous perception of ethylene is an indispensable step, permanently required for the regulation of gravitropic growth in germinating primary shoots of rye, within the process of graviperception and/or the transduction of the gravity signal (Kramer et al. 2003). Even gravi-incompetent coleoptile-less seedlings exhibit gravicompetence after exogenous application of ethylene. Aminocyclopropane-1-carboxylic acid, as a precursor of ethylene, has similar gravicompetence-inducing effects and also appropriate conditions of light, which strongly enhances ethylene synthesis. Both effects can be inhibited by the ethylene-perception blocking agent methylcyclopropene or inhibitors of ethylene synthesis such as aminovinylglycine, indicating that light exerts its gravicompetence-generating effect via induced/enhanced ethylene synthesis.

Upon germination, it rather regularly produces a whorl of four roots, which constitutes the primary root system, thus differing from the other cereals, which usually have only three. The mature root system is very similar to that of oats and spring wheat. Rye has the best-developed root system among the annual cereal crops, compared with other grasses. It is the cereal plant with the world's longest root system.

A quantitative study on the number, length, and total exposed surface area of the roots and root hairs of one rye plant (*S. cereale* L.) was made by Dittmer (1937). Counts were made by categories (main, secondary, tertiary, and quaternary) to determine the total number of these roots by ranks. The 13,815,672 roots had a surface area of 2554.09 square feet. Diameters of roots were characteristic of the categories to which they belonged. Living root hairs on this plant numbered 14,335,568,288 and had a total surface area of 4321.31 square feet. They covered all roots of each category, but they occurred in greatest number on the roots of the quaternary division. The root hair surface combined with that of the roots gave a total of 6875.4 square feet, which was the area of possible soil contact for this plant. The total external surface of the 80 shoots with their 480 leaves was 51.38 square feet. The surface area of subterranean parts was therefore 130 times that exposed at the top.

After an investigation by Russel (1977), the root length of rye in a ground cylinder of 7.6 cm diameter and 12.2 cm depth amounted to 64 m. In 1954, in the examined ground cylinder, Russel was able to find 12.5 million root hairs. In the literature, statements are found about an overall length of the roots of a rye plant of up to 80 km, whereas in wheat crumb it is up to 50 km. The larger part of the roots (75%–90%) is found in the upper 30–35 cm layer of soil.

Roots are so massive that their total dry weight may exceed that of the entire plant body. Another quantitative investigation revealed that a single rye plant that was 4 months old had a total root length of 623 km(!) or an average root growth of ~4.83 km per day(!). It consisted of some 14 million separate branch roots with more than 14 billion root hairs. All the roots and root hairs convert to an equivalent total absorptive surface area in contact with the soil of almost 640 m^2, all contained within a limited volume of ~0.057 m^3 of soil (Went 1957).

Root growth is greatest from the seedling to flag leaf stages for all three cereals. Rye generally roots to a depth of 90–230 cm; thus, it roots deeper than other small grains. The development of both shoots and roots of rye is greatly affected by differences in soil. Plants in silty clay loam of upland soil had a height growth of 130 cm, a working depth of 140 cm, and a maximum penetration of 200 cm. However, a crop from the same lot of Rosen rye seed (see Section 7.4 and Table 7.6) planted at the same time on alluvial soil of mellow silt loam was very different. The height of the tops was 115 cm, which was also the working depth of the roots. Practically, all of the roots ended abruptly at a level below which nearly pure sand occurred. In both cases, the roots were exceedingly well branched to the working depth. In fact, branching is usually better developed in rye than in wheat or oats when growing in the same kind of soil and under the same conditions of moisture. This is one reason why rye is better adapted to drier climates than wheat and will thrive on poorer and sandier soils than any of the other cereals. It sometimes produces a fair crop under adverse conditions in which other small grains would fail completely. In this connection, the work of Nobbe (1869) is interesting. He compared, measured, and counted the roots of winter wheat and rye plants grown in soil when 55 days old. He found that the roots of the first to the fourth order numbered 16,000 in rye and 10,700 in wheat. The combined lengths of these roots measured 118 and 82 m, respectively. The resistance of different rye varieties to lodging to a great degree depends on the development of the root system and its cohesion with the soil.

The extensive root system including its specific rhizosphere enables it to be the most drought-tolerant cereal crop (see Section 3.4) and makes it among the best green manures for improving soil structure. Field studies revealed that decomposition of incorporated ^{14}C-labeled rye residue was accelerated through having rye plants growing in the soil. This was believed to be due to the rhizosphere microbial complex. Comparison of shoot and root dry weight and soil moisture during progressive growth stages for wheat, triticale, and rye in greenhouse, pot, and field studies demonstrated significantly higher root dry weight in rye when grown in pots, but wheat showed greater root mass from depths of 20 to 50 cm in field studies.

Cereal roots have two main classes: seminal roots and nodal roots. Seminal roots originate from the germinating embryonic hypocotyls, and nodal roots emerge from the coleoptile nodes at the base of the apical culm. The genetic control of root characteristics is poorly understood as the growth pattern changes greatly depending on the environment and it is obscured from direct observation. Root traits are believed to be complex and controlled by many genes, each with a small effect. With the advent of molecular markers, it has become possible to estimate the genome location and size of quantitative trait locus (QTL), including those for root characters. Sharma et al. (2011) used a high-resolution chromosome arm-specific mapping population to locate and detect the gene for different root traits on 1RS in bread wheat. A total of 15 QTL effects, 6 additive and 9 epistatic, were detected for different traits of root length and root weight in 1RS-translocated wheat. Epistatic interactions were further partitioned into intergenomic (wheat and rye alleles) and intragenomic (rye–rye or wheat–wheat alleles) interactions affecting various root traits. Four common regions were identified involving all the QTLs for root traits. Two regions carried QTLs for almost all the root traits and were responsible for all the epistatic interactions.

2.5 SEEDS

The seed is relatively large. The 1000-kernel weight is ~40–50 g, thus somewhat lower than in wheat (45–55 g). Seeds in rye are popularly called grains. The term grain is often used in a more general sense as synonymous with cereal, for example, as in "cereal grains," which include some non-Gramineae. It is the fruit typical of the family Poaceae or Gramineae, such as wheat, rice, and maize. Considering that the fruit wall and the seed are intimately fused into a single unit, and the caryopsis or grain is a dry fruit, little concern is given to technically separating the terms "fruit" and "seed" in these plant structures. In many grains, the "hulls" that need to be separated before processing are actually flower bracts. The rye caryopsis is a type of simple dry fruit—one that is monocarpelate (formed from a single carpel) and indehiscent (not opening at maturity), and resembles an achene, except that in a caryopsis the pericarp is fused with the thin seed coat (see Figure 2.7 and Table 2.6).

The rye grain is longer and more slender than that of wheat. It is an oblong or oval caryopsis flattened at the sides (Figure 2.8). The caryopsis may be white, yellow, light-blue, violet, and brown, or has different shades of green. The 1000-seed weight is 30–45 g. Bigger seeds contribute to higher yields (see Table 2.7).

In 1910, the German, K. von Rümker, Director of the Agricultural Institute, Breslau (now Wrocław, Poland), described the grain color of rye as a hereditarily transmissible characteristic, capable of fixation by selection, and therefore of direct practical value. The bluish-green color of the grain transmits more intensively and regularly than the grass-green or yellowish-green shades. A characteristic of the grains of the latter two colors is that the resulting plants are coarse, the grain is heavier but softer, and the plants are inclined to lodge. The green-grained varieties were observed to stool more freely than the yellow-grained varieties. The brown-grained varieties lacked hardiness and were low yielders of both grain and straw. The darker the shade of brown, the lower the yields. This also applies to varieties with brown-tipped germ ends. There is objection also to short-grained ryes, because shortness of grain is intensified by heredity and is correlated in direct ratio of the yield of straw to the percentage of grain. No relation could be established between the color of the

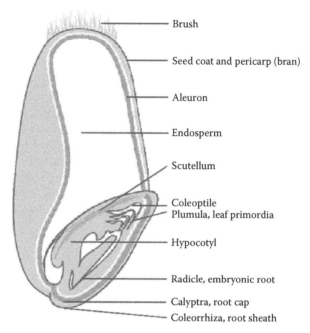

- Brush
- Seed coat and pericarp (bran)
- Aleuron
- Endosperm
- Scutellum
- Coleoptile
  Plumula, leaf primordia
- Hypocotyl
- Radicle, embryonic root
- Calyptra, root cap
- Coleorrhiza, root sheath

FIGURE 2.7 Longitudinal section of a rye caryopsis: Scutellum, coleoptile, plumula, hypocotyl, and radicle form the embryo.

TABLE 2.6
Mean Morphological Components of Rye Caryopsis as Compared to Bread Wheat

Cereal	Pericarp	Aleuron	Endosperm	Embryo	Scutellum
Rye	11–13	11–12	71–75	1.4–1.8	1.7–2.1
Wheat	8	6.5–7.0	81–84	1.2	1.5

grain and the shape of the spike, and the length of the growing and ripening periods. The percentage of grain to straw decreased almost regularly as stooling propensity increased. Where a variety shows inclination to stool abundantly, it is important, in making selections, to give preference to the heaviest yielding stalks. Protein content has less relation to the color of the grain than to plumpness, which increases as it increases. The yellow-grained varieties were more difficult to develop than the green-grained, but once developed transmitted that characteristic with greater regularity and certainty and much sooner reached the point of gametic purity than the green-grained varieties.

Later it became clear that the pale grain is genetically coded by recessive gene loci. The yellow-grained varieties of winter rye seem convertible into spring rye more easily and quickly than the green-grained (Rümker 1909). Purple seeds were found in mutant plants (Ruebenbauer et al. 1983). By different pigmentation of aleuron, testa, and pericarp, grain color is determined (see Table 2.8).

FIGURE 2.8   Seeds of diploid spring rye, *S. cereale* L., var. Petka.

TABLE 2.7
Relationship between Seed Size and Germination Ability as Well as Yield Characters in Diploid Rye

Character	Seed Fraction (mm)			
	2.2	**2.5**	**2.8**	**Unsieved**
1000-Grain weight (g)	24.30	33.40	40.40	32.70
Germability (laboratory) (%)	92.50	95.75	95.75	94.75
Germ length (mm)	33.00	46.50	62.70	46.70
Rootlet length (mm)	71.10	106.10	144.10	112.20
Germability, field condition (plants per m²)	498	519	527	520
Grain yield (dt/ha)	62.00	61.00	58.91	59.16

Source: Martincic, J. et al., *Rostl. Vyroba*, 43, 95–100, 1997. With permission.

Rye grain has in its composition water, proteins, lipids, carbohydrates, mineral substances [sodium, potassium, magnesium, calcium, manganese (Figure 3.6), iron, cobalt, copper, zinc, chromium, phosphorus, iodine, boron], vitamins (B1, B2, B3, B5, B6, B8, B9), and amino acids. These elements make rye indispensable for a balanced alimentation; it prolongs the digestion process and has an anticarcinogen effect. Dark rye flour contains several components (Table 2.9).

The crude protein content is the lowest among the cereals (8%–10% of dry matter), while the starch content ranges from 50% to 70% of dry matter again between barley and wheat. Rye has seed storage proteins called secalins similar to prolamins in wheat, which were classified into high-molecular-weight secalins (≥100 kDa),

TABLE 2.8

Composition of Grain Components and Its Consequence for Secondary Grain Color in Rye

Aleuron	Testa	Pericarp	Grain Color
Colorless →	Colorless →	Colorless →	White
		Brownish →	Brownish
	Brown →	Colorless →	Yellow
		Brownish →	Brown
Blue →	Colorless →	Colorless →	Blue
		Brownish →	Green
	Brown →	Colorless →	Greenish
		Brownish →	Dark brown
Anthocyanins →	Phlobapens →		Secondary color

Source: Hoffmann, W. et al., *Lehrbuch der Züchtung landwirtschaftlicher Kulturpflanzen*, P. Parey Verlag, Berlin, Germany, 1970. With permission.

TABLE 2.9
Mean Components of Rye Flour

Component	Value (%)
Water	10–15
Protein	8–13
Fat	2–3
Starch	56–70
Fiber	3–5
Minerals	~2

γ-secalins (40–75 kDa), and ω-secalins (48–55 kDa) (Kreis et al. 1985). The ω-secalin gene region (*Sec-1* locus) is located at 1RS (see Chapter 5). The DNA sequence of the ω-secalin gene is 5′-AACATGAAGACCTTCCTCATC-3′; the gray-marked ATG is an open codon (Yamamoto and Mukai 2005). Clarke et al. (1996) and Clarke and Apples (1999) analyzed the structure of the *Sec-1* locus by pulsed field gel electrophoresis and sequence analyses. The tandem repeat organization of the *Sec-1* locus has 15 copies of the gene. The fiber fluorescence *in situ* hybridization (FISH) analysis shows that ω-secalin genes are arranged in a head-to-tail fashion separated by 8 kb of spacer sequences with a total length of 145 kb.

The fat content is ~2% in seed. Pentosans play a particular role in rye utilization. Their content is between 9% and 10% of the dry matter, depending on the variety. Approximately 25% of the pentosans are water soluble. Pentosans show soaking effects and decrease digestion when fed to animals. The lysine content is ~3.7%, the methionine content 3.6%, and the threonine content ~3.0% of the mean crude protein content, respectively. Thus, rye shows the highest lysine content among the cereals mentioned above (wheat ~2.5%, barley ~3.5%, triticale ~3.0%).

There are also free amino acids and sugars in rye flour. Free asparagine is the main determinant of acrylamide formation in heated rye flour, as it is in wheat. However, in contrast to wheat, sugar, particularly sucrose, concentration correlated with both asparagine concentration and acrylamide formed. Free asparagine concentration is under genetic, environmental, and integrated control. The same is true for glucose, whereas maltose and fructose are affected mainly by environmental factors and sucrose is largely under genetic control. The ratio of variation due to genotype to the total variation for free grain asparagine concentration is ~23%. Free asparagine concentration is closely associated with bran yield, whereas sugar concentration is associated with low-Hagberg falling number (see Section 3.6). Rye seeds are found to contain much higher concentrations of free proline than wheat grain, and less acrylamide formed per unit of asparagine in rye than in wheat flour.

2.5.1 SEED FORMATION, GERMINATION, AND XENIA

2.5.1.1 Seed Formation

The pattern of early seed development is essentially the same in wheat, rye, triticale, and barley, although some interspecific variation in the rate between genotypes is noted. Fertilization occurs in some florets within 40–60 min of pollination. Mitosis in the primary endosperm nucleus is completed within ~6–7 h of pollination. During the next 24–48 h, the number of endosperm nuclei increase geometrically, doubling about every 4–5 h. The endosperm is coenocytic at first, but usually at ~72 h after pollination, it becomes cellular. The rate of nuclear development in the endosperm declined on each successive day, the greatest fall occurring at the time of cell wall formation. Mitosis in the zygote happens ~18–30 h after pollination, which is later than mitosis in the primary endosperm nucleus. The cell cycle time in the embryo varies between species from ~12 to 18 h, and is similar to its duration in cells of other meristematic tissues in the same species. The cell cycle time in the embryo remained fairly constant during the first 5 days of seed development unlike the rate of nuclear development in the endosperm. Thus, at first the rate of embryo cell development is very slow compared with that of the endosperm nuclei; however, by the end of day 5, the cell cycle time in the endosperm had increased to become equal to or longer than that of the cell cycle in embryo cells. The rates of embryo and endosperm development are much faster in wheat species than in rye. By comparison, the rates in hexaploid triticale genotypes are usually much slower than in wheat, and sometimes even slower than in rye. Results for wheat–rye chromosome addition lines, disomic for each rye chromosome, show that most rye chromosomes apparently have a pronounced effect on slowing both embryo and endosperm development. Indeed, rye chromosomes 6R and 7R apparently have an effect equal to that of the presence of a whole rye genome (Bennett et al. 1975). Additionally, in crosses between hexaploid wheat and inbred lines of rye, sometimes a small number of rye genotypes produce seeds carrying undifferentiated, nonviable embryos. By analysis of F_2 hybrids a single gene could be determined, *Eml-R1b*, causing this character (Tikhenko et al. 2011). *Eml-R1b* maps to chromosome arm 6RL, along with two cosegregating microsatellite loci, *Xgwm1103* and *Xgwm732*.

2.5.1.2 Germination

In most northern countries, rye can be established when seeded as late as October 1. Minimal temperatures for germination have been variously given as 1°C to 2°C (see Table 8.4). In the autumn, rye grows more rapidly than wheat, oat, or various other annual grasses. Although rye is usually regarded as a winter crop, several spring-sown varieties are available and also sown, for example, in Siberia. Germination depends on seed size (Martincic et al. 1997). Seed size and 1000-grain weight are positively correlated. When seeds were fractioned by size, germability is lowest in the smallest seed fraction (~2 mm) and vice versa. Similar results were achieved with germ and rootlet length. Seed fraction of ~3 mm showed the longest germs (~63 mm) and longest rootlets (~144 mm). The early growth influences the field performance later. The largest number of emerged plants per square meter is attained with the largest seed fraction (~3 mm) and the final grain yield as well (Table 2.7). It is concluded that 20%–30% higher grain yields can be attained by higher seed size, that is, 1000-grain weight.

Germination can be stimulated by either chemicals, for example, GA, or temperature or even ultrasound. Ultrasonic waves activate the enzyme and accelerate starch metabolism (α-amylase activity) and the overall biochemical processes that occur during seed germination. This technology has been increasingly used to obtain higher percentage of germination, faster growth, development of healthy seedlings, and potentially a good yield. For practical use it is questionable. Nevertheless, the development of longer coleoptile protects the seedling inside and roll-curved first leaves, and allows larger amount of absorbed water and essential materials for seedling growth and development (Kratovalieva et al. 2012).

2.5.1.3 Sprouting

Rye has the ability to germinate in the ear and the α-amylase activity related to it is a typical feature of many cultivars (Figure 2.9). This preharvest sprouting, due to low harvest dormancy, reduces grain quality at harvest and therefore serious economic losses happen. Rye seeds may germinate in storage or even while still in the ear. Seeds will germinate at temperatures as low as 3°C to 5°C, but optimal range is 25°C to 31°C. Thus, a defined level of seed dormancy is an essential component of seed quality. QTL analysis of rye dormancy is one way of identifying genes that underlie these physiological problems with preharvest sprouting. In wheat, red-grained cultivars are generally more dormant and sprout resistant than white wheats. The major genetic components used to decrease preharvest sprouting are the red grain color loci located on the homoeologous 3A, 3B, and 3D chromosomes. In rye, QTLs for preharvest sprouting loci associated with α-amylase genes were localized on the chromosome arms 2RL, 3RL, 5RL, and 6RL (see Section 3.6 and Chapter 5). The complex genetic control of the trait is demonstrated by numerous loci on almost all chromosomes (Masojć and Milczarski 2009, Mysków et al. 2010) (Table 2.10).

2.5.1.4 Xenia

A situation in which the genotype of the pollen influences the developing embryo of the maternal tissue of the fruit (endosperm) is called "xenia." Focke (1881) first coined

FIGURE 2.9 Preharvest sprouting in spikes of diploid rye, *S. cereale* L.: Comparison of two resistant varieties, Amilo and Dańkowskie Diament, with a susceptible strain. (Courtesy of Banaszak, Z. Choryn, Poland.)

TABLE 2.10
QTLs Controlling Genes Affecting Preharvest Sprouting in Rye

Character	Chromosomal Location	Phenotypic Variance (%)	Reference
Seed dormancy	1RS, 2RS, 5RL	11.4–53.0	Masojć et al. (1999)
α-Amylase activity	1RS, 1RL, 2RS, 2RL, 3RS, 4RL, 5RS, 5RL, 6RS, 6RL, 7RS, 7RL	3.0–59.8	Masojć and Milczarski (2009)
Sprouting damage	1RS, 1RL, 2RS, 2RL, 3RS, 3RL, 4RL, 5RS, 5RL, 6RS, 6RL, 7RS, 7RL	4.0–63.8	Masojć and Milczarski (2009)

Note: See Section 5.5.

the term. It produces an observable effect on the seed. By crossing pale-grained rye (*an1*) with a green-seeded male parent (*An1*), green-grained xenia can be selected already on the ripened spike. Green seed color dominates recessive pale seed color. Therefore, hybrid seeds can easily be distinguished from selfings.

During fertilization, the first sperm nucleus fuses with the egg cell. The second sperm nucleus associates with the secondary embryo sack nucleus that is just formed, and/or was formed by the fusion of two haploid nuclei, leading to a triploid nucleus. In this way, the paternal genome is included in the endosperm, that is, grain color and size are also influenced by the male genome. In rye, the grain color is determined by the gray-yellow to transparent pericarp, the yellow-brownish seed coat (testa), and the pigments of aleuron layer (see Figure 2.7). If the latter appears paler than the seed coat, the seed color is determined by the color of seed coat, that is, it is poorly maternal. Paternal genes do not influence the seed coat. However, if the color of aleuron is darker than that of the seed coat, a complementary interaction between colors of both seed coat and aleuron layer happens. A blue aleuron plus a yellowish seed coat result in green grains.

The missing blue aleuron color, and therefore yellow, grains are coded recessive (anthocyaninless = *an1*). One locus is mapped on chromosome 7RL, close to a biochemical marker, *Aat1* (9.1 cM), and the *compactum* gene, *ct1* (0 cM).

In rye, Rümker [cited after Römer and Rudorf (1950)] described first color xenia in 1911. During the Fourth International Congress of Genetics, Paris, France, he demonstrated a spike of rye showing two rows with yellow seeds and two rows with green seeds. It was produced by emasculation of a yellow-grained plant spike and pollination on one side of spike with yellow-grained pollen and the other side with green-grained pollen, respectively.

This method was applied for detection of spontaneous F_1 autotetraploids after valence crosses (see Sections 4.5.1 and 7.2), that is, tetraploid pale-grained into diploid green-grained rye (Schlegel 1973). The color of the rye grains depends on the combination of the color, thickness, and transparency of the seed and fruit coats, and of the color of the aleuron layer (Melz 1989; see Chapter 5). It may vary between bright yellow, dark yellow, bright red-brown, brown, dark brown, and violet. Ryes with brown and black grains or seeds with brown tips are not suitable for practical utilization. Steglich and Pieper (1922) were the only ones who described black xenia in crosses with Pirnaer Roggen, and even its 3:1 F_2 segregation.

Nicolaisen (1931) analyzed xenia for seed weight in Petkus rye. He studied more than 1 million seeds. Green seeds showed ~10% higher seed weight than yellow ones, although the seed weight of green-grained parent Petkus was not completely reached. This phenomenon has to be considered in yield testings. If varieties with identical flowering time but notably different seed size are grown together, the pollen of the heavier seeded variety can wrongly increase the yield of the smaller seeded variety. It would not happen when the two varieties are tested under isolated cropping. Tschermak (1932) discussed the so-called false xenia, that is, greater seed size of crossed seeds than selfed seeds, as reported by Nicolaisen (1931) in rye and pea.

Nevertheless, the results inspired Frimmel and Baranek (1935) to produce the so-called Bastard-Roggen (hybrid rye). He proposed crosses between small-seeded and heavy-seeded varieties. After harvest, the bigger F_1 seeds could be sieved out from the smaller maternal parent—a first proposal for hybrid seed production in rye (see Section 7.1.3). The authors also observed xenia for seed shape. Therefore, crosses of short- and plump-seeded and long- and narrow-seeded varieties could be used for hybrid seed separation as proposed above, provided that a Trieur device can separate small grains from plump grains. Much later those experiments were repeated by Grochowski et al. (1995). They again claimed the use of xenia for breeding of hybrid varieties and testing of combining ability.

2.5.2 Long-Term Storage of Seed

There are big differences in storage response, for both species and varieties. In general, hybrid varieties are less sensitive to storage than common varieties. The latter are better than inbred lines (Stahl and Steiner 1998).

Rye breeders and genebank operators require seed storage systems, which preserve highly vigorous seeds for several years to several decades. A prolongation of the storage time decreases the frequency of regeneration, minimizes the costs of maintenance, and reduces the risk of possible genetic changes due to frequent seed sample regeneration. Despite optimal storage conditions, a natural seed aging process accompanies long-term storage. There is strong support that composition of the gaseous environment (oxygen) during hermetic storage has a significant influence on seed viability.

For long-term storage of rye, grain moisture should be 12%. During storage, the grain needs to be aerated to control the temperature of the stored grain so as to avoid moisture buildup in bins during changing outdoor temperatures. This is for farmers that store grain and sell at peak markets. When rye seeds are kept as genetic resource in a genebank, the conditions have to be improved. The interactions between temperature and moisture regulate seed aging.

Seed deterioration is always slower at lower storage temperatures. Significantly faster deterioration is observed in seeds stored at <5% relative humidity (RH). Decrease in germination percentage and radicle length of germinated seedlings with storage time can be modeled using Avrami kinetics and longevity. It is expressed as the time for a 50% reduction (P50). Maximum P50s can be observed between 5% and 6%, 8% and 13%, and 8% and 20% RH at 45°C, 35°C, and 25°C, respectively. It is reflected by the increased optimum water content for storage with decreasing storage temperature. The aging rate followed Arrhenius behavior, but Arrhenius parameters varied by storage RH. Intraspecific variation of seed longevity corresponds to the pre-exponential factor of the Arrhenius equation, and this provides a fundamental insight into the mechanisms of seed aging and the differences in seeds' responses to changing water contents (Walters et al. 2008).

The Arrhenius equation is a simple, but remarkably accurate, formula for the temperature dependence of reaction rates. The equation was first proposed by Svante Arrhenius in 1884. Years later in 1889, Dutch chemist J. H. van 't Hoff provided a physical justification and interpretation for it. Currently, it is best seen as an empirical relationship. It can be used to model the temperature variance of diffusion coefficients, creep rates, and many other thermally induced processes/reactions.

Genotypes may also influence storage ability. Seeds of population varieties, hybrid varieties, cytogenetic male sterility (CMS) single cross lines, and CMS inbred lines of rye (*S. cereale*) were stored to determine the rates of germination loss. The seeds were kept under controlled conditions at seed moisture content of 14% and a temperature of 30°C for periods of up to 80 days. The storage potential decreased in the order: hybrid varieties > population varieties > CMS single cross lines > CMS inbred lines. However, the rates of germination loss showed no differences within the categories. This behavior seems to be related to seed vigor of the entries included (Stahl and Steiner 1998).

An extreme long-term seed storage experiment was initiated in 1978 at Gatersleben genebank (Germany). Seeds of 3.7% and 5.5% moisture content were kept at −15°C, 0°C, and 10°C in hermetically sealed glasses filled with air, CO_2, and N_2, and also evacuated (Specht and Börner 1998). The initial germination was measured as 72%. Subsequently, germination was studied after 1, 5, 15, and 17 years of storage. After

17 years at −15°C, the smallest losses in germination and the smallest interactions with the gas atmosphere were observed. At 0°C and 10°C, there are greater but variable losses in germination and inconsistent interactions with the gas atmosphere. Generally, air promoted and N_2 reduced germination loss (see Table 2.11).

After 26 years of storage, the analysis showed the highest germability and viability at −15°C followed by 10°C and 0°C, and also the highest under vacuum and N_2 as opposed to air and CO_2. For the seed vigor traits, the temperature of −15°C together with vacuum gave the best results. For the traits obtained from the investigation of 7-day-old seedlings, again seeds stored at −15°C gave the highest values for shoot dry weight, seedling dry weight, shoot length, root length, and seedling length. Considering the storage media, CO_2 gave significantly lower values compared to all other atmospheres (Barzali et al. 2005).

Highest germability and viability of seeds, however, do not guarantee genetic integrity. A study by Chebotar et al. (2003) demonstrated the changes in allele frequencies as a continuous process during storage and seed multiplication. Rye seeds available from a herbarium collection and the cold store were multiplied 2–14 times and fingerprinted using microsatellite markers. Four accessions had significantly different allele frequencies. These were multiplied 7–13 times. Nearly 50% of the alleles discovered in the original samples were not found in the material present in the cold store. Consequences of the occurrence of genetic changes for managing allogamous rye maintained in *ex situ* genebanks have to be considered.

TABLE 2.11

Results of Standard Germination (%) of Rye Samples Maintained at Different Temperatures and Atmospheres and Studied after 1, 5, 15, 17, and 26 Years

Medium			Years		
10°C	**1**	**5**	**15**	**17**	**26**
CO_2	75	64	25	29	7
N_2	74	76	21	58	40
Vacuum	76	69	69	63	57
Air	70	68	28	25	9
0°C					
CO_2	77	70	31	17	26
N_2	78	67	60	58	12
Vacuum	79	76	52	37	29
Air	79	70	28	29	17
−15°C					
CO_2	79	78	59	50	29
N_2	77	78	65	71	54
Vacuum	75	80	75	76	59
Air	72	78	54	55	37

Source: Barzali, M. et al., *Seed Sci. Technol.*, 33, 713–721, 2005. With permission.

When the genetic integrity of seeds of Polish rye cultivar Dańkowskie Złote stored for 25 years at 3.7% or 5.5% moisture content under vacuum or in air and regenerated in two successive seasons was assessed using 24 SSR markers, seeds also showed significantly different germination ability. Decrease in seed viability due to aging resulted in increased homozygosity. Principal coordinate analysis showed a clear segregation of the population with the lowest seed viability in relation to the other tested samples. Inbreeding coefficient values indicated a small variation of the seed samples related to natural aging (Boczkowska and Puchalski 2012).

2.6 FLOWERING, FERTILIZATION, AND APOMIXIS

2.6.1 FLOWERING

Rye is an allogamous (cross-pollinating) anemophilous plant. Under favorable conditions, it begins blossoming on days 7–10 after ear formation. As a long-day plant, flowering is induced by 14 h of daylight when temperatures from 5°C to 10°C accompany this. Shortened day length can cause rye plants to remain vegetative for up to seven and more years. This behavior is sometimes used for maintaining or multiplying specific genotypes by cloning under short-day condition.

There is a gene locus, *Vrn1* syn. *Sp1*, that controls flowering. Recessive alleles promote flowering after vernalization. A gene cluster consisting of four QTLs controlling flowering time and yield components was discovered in the centromere region of chromosome 2R (Börner et al. 2000). Further loci are distributed on chromosomes 1R (one), 4R (one), 6R (three), and 7R (one).

In rye cultivars, the anthers usually burst after 1–2 min upon their emergence from the flower. Rye pollen is carried away by the wind. As with all the other anemophilous plants, rye produces lots of pollen (each flower producing up to 60,000 pollen grains). Feil and Schmid (2002) estimated ~17 trillions of pollen grains per hectare. In rare cases, anthers burst prior to their emergence from the flower and self-pollination occurs. Each flower remains open for 12–30 min, but the pollen takes only 2–4 min to spill out from it. The spike begins to blossom in its middle part and blossoming gradually spreads upward and downward lasting for 4–5 days. The upper flowers cease to blossom earlier than the lowermost ones. Each plant blossoms for 7–8 days. The first to blossom is the spike of the main stem. In the field with optimal air temperature (12°C to 15°C), blossoming usually begins at 5 or 6 o'clock in the morning in southern and central areas of Europe, while in northeast and northwest areas blossoming begins at 7 or 10 o'clock in the morning. Blossoming has two or three peaks during the day, but it is exceptionally intensive early in the morning. When rye fields blossom in dry and warm weather, there is a cloud of pollen floating over them. In direct sunlight, rye pollen is viable for 15 min, but in the shade for 4–8 h. In an artificial environment with lower temperatures and higher moisture, it is viable for 1–3 days. In rainy and cloudy weather, rye pollen is poorly transported by the wind. It does not reach flowers greatly increasing incomplete setting of grains, which may spread to 30%–40% of flowers. Against the presumption as a wind pollinator, rye pollen is not transported as far. By using recessive traits/genes (pale grain) as indicator, Römer and Rudorf (1950) demonstrated that a seed row distance of ~3.50 m reduces cross-pollination to less than 5%. Of course, the wind direction has to be considered. Test plots should be sown parallel to the main wind direction. Isolation means between plots should be built in the same way. Recent studies show that pollen may drift away up to 600 m (Feil and Schmid 2002). There is still inter-crossing of 3% when the 2 ha field was that distance away.

2.6.2 FERTILIZATION

Incomplete setting of grains may also be accounted for by genetic factors. Self-fertility with rye is low, covering on average 0%–6%. Approximately 30 min elapse from the moment when rye pollen falls onto the pistil's stigma (see Figure 2.6) until the moment when the pollen tube penetrates into the cavities of the embryo sac while the whole fertilization process lasts for 6–8 h. Ovaries that have not been fertilized remain fertilizable for a relatively long time—up to 14 days. It has been determined that the critical factors for high winter rye yields are the number of its productive stems per square meter and the weight of grains in one ear. The total number of stems and the number of productive stems per unit area are the adaptability features that characterize biological resistibility of rye cultivars, which depends on their hardiness, drought resistance, resistance to disease, and pests (see Section 3.4).

Certation—the competition in growth rate between pollen tubes of different geno-types resulting in unequal chances of accomplishing fertilization—of rye pollen can be observed, particularly between euploid and aneuploid pollen grains. Aneuploid microspores have a delayed development, whereby genetic factors determining male transmission rate are primarily expressed during pollen germination and tube growth rather than before anthesis (Janse 1987).

2.6.3 INCOMPATIBILITY

Self-incompatibility is widespread in the grasses and is under the control of a series of alleles at each of the two unlinked loci, S and Z. Specification of the pollen grain is gametophytic and depends upon the combination of S and Z alleles in the pollen grain. Both of these must match in the pistil for the pollen grain to be incompatible (see Table 2.12). In the system of gametophytic self-incompatibility, pollen tube growth is inhibited and the pollination is completely restricted, when the pollen genotype and the pistil genotype match at all gametophytic self-incompatibility loci (Yang et al. 2008).

Several lines of research suggest that this two-locus self-incompatibility system is shared by all grass species of the Pooideae and possibly by the entire Graminae. The existence of self-incompatibility in the grasses has been known for over a century. For example, Lundqvist (1954, 1956, 1957, 1958) reported that Rimpau (1877) proved the predominance of self-sterile individuals in populations of rye (see Chapter 7).

Meanwhile, there are three loci determining self-fertility at the S, Z, and $S5$ (T) self-incompatibility loci on chromosomes 1R, 2R, and 5R of rye, respectively. The isozyme loci $Prx7$ and β-Glu are linked to the two loci S and Z. In addition, there

TABLE 2.12
Possible Gametophytic Segregation of Self-Incompatibility Considering Genes S and Z and Two Alleles

Possible Genetic Constitution of Diploid Pistil Tissue	Genetic Constitution of Parental Genotype $(2n = S_1S_2Z_3Z_4)$			
	Possible Genetic Constitution of Haploid Pollen Grains (n)			
	S_1Z_3	S_1Z_4	S_2Z_3	S_2Z_4
$S_1S_2Z_3Z_4$	–	–	–	–
$S_1S_1Z_3Z_4$	–	–	+	+
$S_1S_1Z_3Z_3$	–	+	+	+

Note: –, noncompatible; +, compatible.

is a close linkage to four RFLP markers (Voylokov et al. 1998). These markers are *Prx7*, *Xiag249*, and *Xpsr634* for the S locus (1R), *Xbcd266* for the Z locus (2R), and *Xpsr100* for the S5 locus (5R). Different alleles of S and Z strongly influence the pseudocompatibility. A ubiquitin-specific protease gene is expressed in pistils but not leaves. It is a good candidate for the Z gene. The effect of the latter can also be stimulated by high temperature (30°C–35°C) shortly before anthesis (Wricke 1978, Gertz and Wricke 1991). However, just a low seed set is induced. Most genotypes do not respond at all.

For the locus S at 30°C, a scale of varying efficiency of the S alleles is found. In two experiments, Trang et al. (1982) estimated the number of alleles at the two incompatibility loci of rye in the variety Halo. In one experiment, "I1" progenies from enforced selfing under controlled conditions were isolated. In the other experiment, a genotype, homozygous at both incompatibility loci, was used as pollinator for a sample of the Halo population, which was regarded as an equilibrium population. Genotypes, which are homozygous at both incompatibility loci, can be found after selfing. The estimate for the number of alleles was 6–7 at one locus and 12–13 at the other locus.

2.6.4 APOMIXIS

Apomixis and apogamy are synonymous terms for sexual reproduction and asexual reproduction without fertilization. Despite different definitions, in flowering plants the term "apomixis" is commonly used in a restricted sense, that is, asexual reproduction through seeds. Apomixis can be found in two systems. Apomictic seeds can arise from a plant's sexual cells, which fail to go through the cellular mechanism underlying sexual reproduction (meiosis). Alternatively, seeds can be generated from nonsexual (somatic) cells. Sometimes, both sexual and asexual seeds develop from the same flower. Apomictic plants produce cloned seed, enabling them to reproduce asexually. However, their pollen is often viable, so that apomixis can also be transmitted through the more common mechanism of sexual reproduction. Many flowering plants with pseudogamous apomixis require fertilization to produce the endosperm of the seeds. However, it was recently shown that pollination with compatible pollen could be required even in some species where endosperm development is autonomous. Apomicts are most frequent in Gramineae (the cereal family), Compositae (which includes sunflowers), Rosaceae (which includes many fruit trees), and Asteraceae (the dandelion family). In rye, this appearance is rather rare. The Russians Meister and Tyumyakov (1927) were the first to report matromorphic development of rye caryopses after pollination with wheat pollen. Apomictic grain formation was also found after hybridization of tetraploid rye with winter durum (Zolotova and Zolotov 1977).

In similar experiments, Kyzlasov (2005) confirmed matromorphic development after pollination with pollen from related species. The percentage of seeds formed in these experiments varied from 6.9% to 7.9% of the total number of pollinated flowers. A strain of winter rye R-1 was pollinated by winter wheat Nemchinovskaya 24, spring wheat A-1, and triticale Victor. By comparison, seed set after self-pollination was 19%. The maternal offspring varied in quantitative characteristics, such as stem length or spike productivity since the pollinated female plants were heterozygous. Although low inbreeding depression was observed in F_0, plants with sterile pollen and abortive anthers were found in the next generation. The absence of wheat protein fractions in the endosperm of matromorphic caryopses demonstrated true apomicts. Polyhaploid offspring with $2n = 4x = 28$ chromosomes sometimes arises in the reciprocal combination. Pseudogamy was not observed. Thus, diploid rye can be reproduced by fertilization with soft wheat pollen. Apomictic reproduction and sterile pollen were inherited by the rye offspring as linked characters. Pistils of apomictic plants show a normal germinating ability. F_1 plants established after pollination of paternally sterile flowers by pollen from other plants produce fertile pollen. It indicates that the embryo sac cells are reduced. Diploid plants produced without a paternal parent inherit the sterile pollen feature in a recessive manner (Kyzlasov 2008). Haploids were not revealed among plants with both sterile and fertile pollen, while polyembryony happened in the apomictic offspring. Up to five viable seedlings sometimes could be seen from one caryopsis (Figure 2.7). Therefore, apomictic plants with sterile pollen can be reproduced without pollination on isolated plots and can be used for production of hybrid seeds (see Section 7.1.3). The plant breeder's dream in using apomixis for fixation of heterotic effects is one of the benefits. Another benefit would be to expand the range of wild relatives that could be integrated into breeding programs. This is because asexual seeds can contain two sets of chromosomes of different sizes and still be viable, while equivalent sexual seeds would probably not develop.

Except sexual hybridization, transferring of alien genes, responsible for apomixis, could be a choice. Somatic hybridization enlarges the circle of the potential donors of apomixis, but difficulties may arise with the regeneration of the fused protoplasts. The molecular transfer of the genes, especially by means of cDNA, seems to be the more acceptable way for creation of the apomictic rye. Attention is paid also to ameiotic and asynaptic mutations. *Calamagrostis purpurea* and *Poa palustris*, autonomous diplosporous apomicts, are the most suitable donors of apomixis from graminaceous species for rye.

2.7   RYE GENEBANKS AND COLLECTIONS

Despite several difficulties of alien germplasm introgression into advanced breeding material, for a number of breeding approaches genetic information of landraces or wild species is useful for improvement of oligo- and polygenically inherited traits in highly selected breeding populations.

The exploitation of related gene pools can be achieved by advanced backcross methodology considering QTLs or by molecular introgression libraries. The latter consist of a series of rye strains, each carrying a single molecularly marker-defined region of a chromosome. When these regions are introgressed from useful gene pools into elite recipient ryes, they can be traced molecularly during the following crossing generations.

Until now, three introgression libraries are established with the Iranian primitive rye Altevogt 14160 as donor by marker-assisted selection using AFLP and SSR markers (see Section 7.1.6.1). The objectives were to detect candidate introgression lines with a better testcross performance than the recurrent parent and identify donor chromosome segments responsible for the improved performance. The marker libraries cover ~70% of the total donor genome. The phenotypic evaluation of the libraries already revealed considerable genetic variation for quantitatively inherited traits, such as baking quality and pollen fertility restoration (Falke et. al. 2008, 2009; Geiger and Miedaner 2009). Donor chromosome segments from Iranian rye were introduced into the genetic background of an elite inbred line "L2053-N" by marker-assisted backcrossing. The linked marker *Xiac76* is ~0.8 cM away from the target gene. The line L2053-N is used as recurrent parent.

Testcross performance for three agronomic and six quality traits was evaluated in replicated field trials across two testers at five locations over 2 years. The phenotypic effect of the donor chromosome segments was analyzed for all traits. The candidate introgression lines had on average a testcross performance comparable to that of the recurrent parent. Significant differences between the individual candidate introgression lines and the recurrent parent were detected for all traits except for heading date. For more than 60% of the significant differences, the candidate introgression lines were superior to the recurrent parent. For some candidate introgression lines, specific donor chromosome segments were identified containing presumably QTLs responsible for the superior hybrid performance. The study revealed that the development and employment of introgression libraries offers the opportunity for a targeted increase of genetic diversity of elite rye for hybrid and/or population performance.

In addition, the precise localization of favorable donor chromosome segments allowed a detailed analysis of pleiotropic effects and the study of the consistency of effects per se and testcross performance. In many cases, the linear model analysis allows the assignment of donor effects to individual donor chromosome segments even for introgression lines with long or multiple donor segments. This may considerably increase the efficiency of producing subintrogression lines, because only such segments need to be isolated which are known to have a significant effect on the phenotype (Mahone et al. 2013).

These results demonstrate that introgression experiments can contribute to broadening the genetic base of rye breeding. Several taxa of *Secale* are cultivated, and others are used as genetic resources in breeding. Rye displays a broad range of genetic diversity based on the great ecological differences among the various growing areas and the center of origin. The intravarietal variation is sometimes higher than the intervarietal one. Large collections exist in various countries comprising international cultivars; landraces from Europe, Asia, and South America; primitive populations from the Near East; and wild *Secale* species. Worldwide, there are more than 22,200 accessions documented from 94 genebanks.

The biggest rye collections are maintained in Radzikow and Warsaw (Poland), St. Petersburg (Russia), Gatersleben (Germany), Aberdeen (USA), Saskatoon (Canada), and Sadovo (Bulgaria) (Table 2.13). Five species and various wild, segetale, and cultivated taxa of the genus *Secale* are kept in genebanks as well (Kleijer et al. 2007). Thirty-three percent of accessions maintained can be assigned as duplicates (Podyma 2003). The European *Secale* Database (ESDB) is maintained at Instytut Hodowli i Aklimatyzacji Roślin (IHAR) in Blonie (Poland; http://www.ihar.edu.pl/national_centre_for_plant_genetic_resources.php), agreed during the first European Cooperative Programme for Plant Genetic Resources (ECPGR) *Secale* Working Group meeting, held in Jokioinen, Finland, in 1982.

2.7.1 Reproduction of Genebank Accessions

Identical multiplication of rye accessions is carried out on 1-m^2 field plots. Before flowering, the plants either are isolated by bags or tends, usually a 2-m-high metal frame covered with a pollen-proof tissue. A mechanical shaking of the frame improves the pollination. In this way ~200 accessions can be propagated every year. The method is space-saving and guarantees sufficient seed per accession for long-term storage. Depending on capacity from each accession of the collection, a 3-year microtrial with drill rows can be prepared in order to evaluate data of breeding and agronomic performance, such as yield characters, field resistance, and quality traits.

The introduction and maintenance of plant germplasm became a very important task of early US breeding research as well. In 1898, the Congress appropriated $20,000 for the introduction, from foreign countries, of rare and valuable seeds as well as plants to be tested in cooperation with the US Department of Agriculture (USDA). This was the beginning of the organized introduction of wheat and rye germplasm. In 1898, M. A. Carleton was sent to Russia to obtain cereals resistant to cold, drought, and fungal diseases. In 1897 and 1898, N. E. Hansen (USDA horticulturist) made a trip to Russia, Siberia, and Turkestan under the auspices of the USDA. He collected many samples of cereals. Later, in 1948 the Division of Cereal Crops and Diseases hired D. J. Ward to develop, maintain, and distribute seed from the USDA Small Grains Collection at Beltsville. The collection contained wheat and other small grains (rye, barley, and oats) that had been collected worldwide by scientists in the Cereal Division. Between the years 1948 and 1958, D. J. Ward

TABLE 2.13
Main Collections of *Secale* Accessions

Country	Location	Institute	Content	Web Address	E-Mail
Bulgaria	Plovdiv, Sadovo	Institute of Plant Introduction and Genetic Resources	337 entries	http://www.ipgrbg.com	office@ipgrbg.com
Canada	Saskatoon	Plant Gene Resources Center	1461 entries	http://pgrc3.agr.gc.ca/about-propos_e.html	info@agr.gc.ca
Czech Republic	Praha, Ruzyně, Kroměříž	Crop Research Institute and Agricultural Research Institute	659 entries	http://genbank.vurv.cz/genetic/resources/asp2/default_a.htm	cropscience@vurv.cz
Germany	Gatersleben	Institut für Pflanzengenetik und Kulturpflanzenforschung	~2400 entries	http://www.ipk-gatersleben.de/abt-genbank/	graner@ipk-gatersleben.de
Italy	Bari	C.N.R. Instituto del Germoplasma	382 entries	http://users.ba.cnr.it/igv/germgc21/	germgc21@area.ba.cnr.it
Poland	Warszawa, Powsin, Radzików	Polish Academy of Sciences, Botanical Garden, Plant Breeding and Acclimatization Institute (IHAR)	2954 entries	http://egiset.ihar.edu.pl	kubickah@gmail.com
Russia	St. Petersburg	N.I. Vavilov Research Institute of Plant Industry	2685 entries	http://vir.nw.ru/index.htm	postal service@vir.nw.ru
Spain	Madrid	Instituto Nacional de Investigación y Tecnología Agraria y Alimentaria (INIA)	428 entries	http://apps3.fao.org/wiews/institute.htm?i_1 = EN&query_INSTCODE = ESP003	rosa@inia.es
Sweden	Alnarp	Nordic Genetic Resource Center	63 entries	http://www.nordgen.org/index.php/en/content/view/full/562	morten.rasmussen@nordgen.org
Switzerland	Nyon, Changins	Station Fédérale de Recherches Agronomiques	63 entries	http://www.agroscope.admin.ch/ressourcesgenetiques-qualite-boulangere/index.html?lang = de	geert.kleijer@acw.admin.ch
Turkey	Izmir, Menemen	Aegean Agricultural Research Institute	512 entries	http://www.tagem.gov.tr/eng/priorty.htm	webadmin@tagem.gov.tr
Ukraine	Kharkov	Ukrainian Centre for Plant Genetic Resources	171 entries	http://www.aginukraine.com/UAAS/directory.htm	leader@kharkov.ukrtel.net
USA	Beltsville, MD	Henry A. Wallace Beltsville Agricultural Research Center	1892 entries	http://www.ars.usda.gov/main/site_main.htm?modecode = 12-75-15-00	dbmu@ars-grin.gov

Note: See ECPGR European *Secale* database.

mainly contributed to the USDA Small Grain Collection, while J. C. Craddock did it between the years 1958 and 1972, respectively.

The viability of rye seeds can be decreased by storage mainly as a result of natural aging. It also leads to significant changes in allele frequency and genotypes. Deviation from Hardy–Weinberg equilibrium can be observed. It reduces heterozygosity, while homozygosity is increased. Molecular markers, such as expressed sequence tag (EST)-derived SSRs, are not yet efficient in assessing the effects of natural aging of rye seeds, probably due to the induction of repair processes during the germination of grains.

3 Physiology

3.1 LIFE CYCLE

The developmental cycle of rye can be divided into 12 stages: During stage 1, the growing point is not differentiated. In stage 2, the primordia of stems, nodes, and internodes are formed in the growing point. Winter rye, usually planted in autumn and moderate climate, enters the winter period in stage 2. In stage 3, the growing point differentiates into further segments, which are primordia of spikelets. During this period, nitrogen supply has a positive effect on the formation of a large number of spikelets, which leads to the subsequent formation of longer spikes with a greater number of flowers and grains. A further differentiation of growing points takes place during stages 3 and 4, in which flower primordia are formed. This process takes place in early spring. During the formation of spikelet primordia in the upper part of the spike, organs of generative reproduction (flowers) are formed in the middle portion. The plants then enter stage 5 of organogenesis. Under conditions of long days and with a poor nitrogen supply, this process is relatively fast. Meiotic divisions of pollen mother cells and the formation of tetrads, the embryo sac, and the egg take place during stages 6 and 7 of organogenesis. Stage 7 is characterized by extensive elongation growth during which pronounced elongation of shoot internodes takes place. In stage 8, the plants ear and subsequently flower. Fertilization and maturation of caryopses and plants then follow in the remaining four stages of development.

In 1989, a joint proposal for a so-called BBCH code was developed by the European chemical and/or breeding companies BASF, Bayer AG, Ciba Geigy AG, and Hoechst AG. BBCH officially stands for "Biologische *B*undesanstalt, *B*undessortenamt und *CH*emische Industrie." The abbreviation of the scale is also said to unofficially represent the four companies mentioned earlier that initially sponsored its development. According to the BBCH key, the growth cycle in cereals is divided into nine macrophases, and each macrophase is subdivided into nine microphases (see Table 3.1). The BBCH code is a scale that is also used to identify the phenological development stages of a plant. The phenological development stages of plants are used in a number of scientific disciplines, such as crop physiology, phytopathology, entomology, and plant breeding as well as in agriculture and industry (e.g., timing of pesticide application, fertilization, or agricultural insurance). The BBCH scale uses a decimal system, which is divided into primary and secondary growth stages, and is based on the cereal code system (Zadoks scale) developed by Zadoks et al. (1974). Meanwhile, the BBCH code is well accepted by a wide range of users. It matches well with the development of rye.

3.1.1 GENERATIVE FEATURES

Rye is the most widely adapted of the cereals because of its extreme winter hardiness and ability to grow in marginal soils. Rye is quite drought resistant and taller than

TABLE 3.1
The European BBCH Code for Characterization of Growth Cycles in Cereals Dividing Nine Macro- and Nine Microphases Each

BBCH Code	Developmental Stage
0	*Germination/sprouting/budding*
00	Dry seed
01	Beginning of seed imbibition
02	
03	Seed imbibition completed
04	
05	Radicle emerged from seed
06	Elongation of radicle, formation of root
07	Coleoptile emerged from caryopsis
08	Hypocotyl growing toward soil surface
09	Emergence, that is, coleoptile breaks through soil surface (cracking stage)
1	*Leaf development (main shoot)*
10	First leaf emerges from coleoptile
11	First leaf unfolds
12	Two leaves unfolded
13	Three leaves unfolded
14	Four leaves unfolded
15	Five leaves unfolded
16	Six leaves unfolded
17	Seven leaves unfolded
18	Eight leaves unfolded
19	Nine and more leaves unfolded
2	*Formation of side shoots/tillering*
20	Beginning of side shoot formation
21	One side shoot visible
22	Two side shoots visible
23	Three side shoots visible
24	Four side shoots visible
25	Five side shoots visible
26	Six side shoots visible
27	Seven side shoots visible
28	Eight side shoots visible
29	Nine and more side shoots visible, maximum number of tillers detectable
3	*Stem elongation or rosette growth/shoot development (main shoot)*
30	Beginning of stem elongation
31	One visibly extended internode
32	Two visibly extended internodes
33	Three visibly extended internodes
34	Four visibly extended internodes
35	Five visibly extended internodes

TABLE 3.1
(Continued) The European BBCH Code for Characterization of Growth Cycles in Cereals Dividing Nine Macro- and Nine Microphases Each

BBCH Code	Developmental Stage
36	Six visibly extended internodes
37	Seven visibly extended internodes, flag leaf just visible, but still rolled
38	Eight visibly extended internodes
39	Flag leaf stage: flag leaf fully unrolled, ligule just visible
4	*Development of harvestable vegetative parts*
40	Reproductive organs begin to develop
41	Flag leaf sheath extending, early boot stage
42	
43	Flag leaf sheath just visibly swollen (mid-boot stage)
44	
45	Flag leaf sheath swollen (late-boot stage)
46	
47	Flag leaf sheath opening
48	
49	First awns visible
5	*Inflorescence emergence (main shoot)/heading*
50	
51	Beginning of heading, tip of inflorescence emerged from sheath, first spikelet just visible
52	20% of inflorescence emerged
53	30% of inflorescence emerged
54	40% of inflorescence emerged
55	Middle of heading: half of inflorescence emerged, first individual flowers visible
56	60% of inflorescence emerged
57	70% of inflorescence emerged
58	80% of inflorescence emerged
59	Inflorescence fully emerged, that is, end of heading
6	*Flowering (main shoot)*
60	First flowers open sporadically
61	Beginning of flowering, first anthers visible
62	
63	Flowering, ~30% flowers open
64	
65	Full flowering, ~50% flowers open
66	
67	Flowering finished
68	
69	End of flowering, all spikelets have completed flowering but some dehydrated anthers may remain

(Continued)

TABLE 3.1

(Continued) The European BBCH Code for Characterization of Growth Cycles in Cereals Dividing Nine Macro- and Nine Microphases Each

BBCH Code	Developmental Stage
7	*Development of seeds*
70	
71	Watery ripe: first grains have reached half their final size
72	
73	Early milk stage
74	
75	Medium milk stage: grain content milky, grains reached final size, still green
76	
77	Late milk stage
78	
79	Nearly all seeds reached final size
8	*Ripening or maturity of seed*
80	
81	Beginning of ripening, seed color appears
82	
83	Early dough formation
84	
85	Soft dough: grain content soft but dry; fingernail impression not held
86	
87	Hard dough: grain content solid; fingernail impression held
88	
89	Fully ripen, grain hard, difficult to divide with thumbnail
9	*Senescence, beginning of dormancy*
90	
91	
92	Overripe: grain very hard, cannot be dented by thumbnail
93	Grains loosening in daytime
94	
95	
96	
97	Plant dead and collapsing
98	
99	Seed dormancy, harvested product

Source: Lancashire, P. D., *Ann. Appl. Biol.*, 119, 561–601, 1991. With permission.

wheat and tillerless (see Section 3.4). After sowing, rye needs a temperature sum of 90°C for field emergence, provided that there is sufficient water. Earlier than 280 grade days after field emergence, the four-leaf phase is reached and tillering starts. Once a shoot has formed six leaves and it is vernalized enough, the formation of spikelets begins (double-ring stadium). This phase (BBCG 25) should be reached before

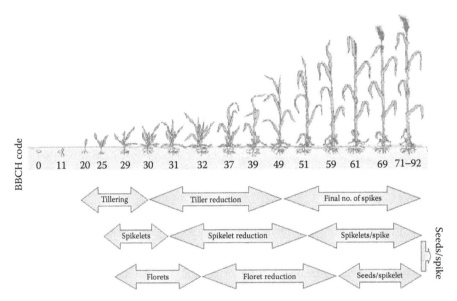

FIGURE 3.1 Different growth phases in rye and critical stages of tillering, spikelet, spike, and floret formation.

December/January, at least in the stronger tillers. Later-formed shoots, which begin spikelet formation only in the spring months, are usually unstable and are quickly reduced again (see Figure 3.1). With the beginning of shooting (BBCH 30–31), the reduction of tillers starts. This is reinforced by nutrients, especially nitrogen deficiency and/or drought. Overall, rye should not have to reduce more than 50% of the preformed tillers. Depending on the desired number of spikes (450 spikes per square meter in light soils, and up to 550 spikes per square meter in better soils, respectively), ~700–1200 culms per square meter should thus exceeding during early heading.

The flag leaf is smaller and less important in photosynthesis. Based on the data of Gawronska and Nalborczyk (1989), rye has the longest stems of all cultivated small grains, which provide most of the photosynthetic area. During grain formation, stems with sheaths account for 60%–80% of the total plant area. At grain set, leaf blades provide 15%–20% of the photosynthetic area, which is much lower than that for maize, wheat, and oat. Stems and sheaths have lower rates of photosynthesis and export of assimilates than leaves. The most important periods of yield formation are the flowering and grain filling. Successful pollination depends on sufficient spreading of the husks. Cold and rainy weather at flowering hamper the opening of the husks, and thus the pollen distribution.

Cereal rye is an annual grass. Winter, spring, and intermediate varieties are available. It is a long-day plant; that is, it requires increasing day length to induce flowering. Flowering is induced by 14 h of daylight accompanied by temperatures of 5°C to 10°C (see Table 8.4). It shows a shorter growth period than other winter and spring cereals that also differ considerably in the vernalization period (see Section 3.7). The winter types of rye require 40–60 days, whereas the spring types require only 10–12 days of cold temperature to shift into the reproductive stage.

TABLE 3.2

Duration of Developmental Stages (Days) in Diploid Winter Rye Compared to Some Other Cereals

Developmental Stage	Winter Wheat	Spring Wheat	Winter Barley	Winter Rye	Winter Triticale
Germination	30	19	9	11	20
Tillering	150	35	190	172	160
Stem development	48	29	31	30	38
Spike emergence	6	5	6	9	8
Flowering	11	14	12	21	12
Seed ripeness	50	40	34	47	48

Source: Compiled from different sources.

The maturation date of rye varies according to soil moisture, but vegetative growth stops once reproduction begins. In general, rye matures earlier than oat. The growth period, that is, the time from sowing to harvest, is ~300 days in winter wheat, ~175 days in spring wheat (alternating wheat ~280 days), ~285 days in winter barley, ~295 days in winter rye, and ~290 days in winter triticale (see Table 3.2).

At seed set, leaf blades provide 15%–20% of the photosynthetic area, which is much lower than those for maize, wheat, and oat. Stems and sheaths have lower rates of photosynthesis and export of assimilates than leaves. For winter rye, the photosynthetic area decreases rapidly after seed setting and does not achieve a plateau near the maximum, as seen with other grains. The photosynthetic area decreases rapidly after grain set and does not achieve a plateau near the maximum, as seen with other grains. Its grain formation occurs under unfavorable physiological conditions for yielding.

The vegetation period lasts for 120–150 days (their autumnal vegetation period lasts for 45–50 days, while their spring and summer vegetation periods last for 75–100 days). In northeast Europe, the rye crop can be established when seeded as late as October 1. In the autumn, it grows more rapidly than wheat, oat, or various other annual grasses, and in the autumn and early spring, it grows more rapidly than oat. Minimal temperatures for germinating cereal rye seed have been variously given as 1°C to 5°C. The structure of the rye plant enables it to capture and hold protective snow cover, which enhances its winter hardiness (Stoskopf 1985).

3.1.2 Vegetative Behavior

Rye tillers grow profusely. Individual plants can easily be split into several clones. Genotypes can differ in cloning ability. The production of tillers can be stimulated by cutting back the plants, which helps to retard the development of the spikes. Cool, moist, and short-day conditions increase the tillering capacity. Continuous light prevents jointing. Clones can be used in rye improvement if portions are preserved vegetatively long enough for evaluation and use in breeding (Bühring 1960).

Since 1935, Riebesel (1937) at Salzmünde (Germany) vegetatively propagated partially sterile F_1 wheat–rye and wheat–*Agropyron* hybrids for safe seed production.

For genetic studies, the Russian Kovarsky (1939) split the tillered rye plants into three parts in order to test different environments on plant development of a specific genotype. At Müncheberg (Germany), in 1937 H. P. Ossent began to multiply elite strains of rye by vegetative growing (Aust 1941). Breeding material with low seed sets was sown in September, kept in pots frost-free until December, and then moved to greenhouses for cloning at temperatures between 10°C and 12°C. By the end of April, clones were replanted in the field and harvested in August. On average, from 1 seedling 21 clones with 100 culms each and 604 seeds per seedling could be produced. Vegetative propagation is still an adequate method of recent breeding approaches (see Chapter 7).

3.2   COLD TOLERANCE

Rye is the most frost-tolerant cereal species. During winters with little snow, it can survive frosts of −30°C to 35°C (Stoskopf 1985). Winter hardiness is a complex feature, which includes cold resistance and resistance to damping, and is often related to resistance to snow mold, ice crust, and lifting. Therefore, frost tolerance may be increased by land treatment (melioration, high quality of tillage, and timely sowing).

Cold-resistant rye plants have some typical morphological and biological features. They have narrow and short rosette leaves with a microcellular structure, spreading bushes, a thicker outer epidermis wall, a short mesocotyl, and, therefore, a deeper tillering node. Frost-resistant plants grow more slowly in autumn and have a relatively higher concentration of dry matter in their cell sap. They expend this dry matter in their growth processes and respiration in a more economical way.

3.2.1   CYTOPLASM

Cytoplasm has little direct effect on cold hardiness, or on the nuclear expression of cold hardiness. Although many changes occur within the cytoplasm of plant cells during cold acclimation, the cause-and-effect relationship between cytoplasmic response to low temperature and the development of cold hardiness in cells is difficult to determine (Limin and Fowler 1984). The cold hardiness of octoploid triticale produced from hardy rye and nonhardy wheat was similar to that of the wheat parent, demonstrating a complete suppression of the cold hardiness genes of rye. Similar observations were made for wheat–rye amphiploids from reciprocal crosses, indicating that this suppression was not due to cytoplasmic effects. The cold hardiness of alloplasmic rye with *Triticum tauschii* cytoplasm was similar to that of the rye parent, indicating that the cold hardiness genes of rye have normal expression in the *T. tauschii* cytoplasm.

3.2.2   HARDENING

Much higher levels of malate are detected in cold-hardened leaves than in nonhardened leaves. When malate metabolism was studied in more detail by biochemical assays, the activities of several enzymes of malate metabolism, NADP-malate dehydrogenase, NAD-malate dehydrogenase, phosphoenolpyruvate carboxylase, and

NADP-malic enzyme were also increased. Short exposures to low temperature for 1–3 days did not induce increases in the malate content or the enzymatic activities of malate metabolism. The malate content and the enzymatic activities decline within 1–2 days after a transfer from their growing temperature of 4°C to 22°C.

Malate serves as an additional carbon sink and a CO_2 store. It can further function as a vacuolar osmolyte balancing increased concentrations of soluble sugars previously observed in the cytosol (Crecelius et al. 2003). Shang et al. (2003) observed an increased capacity for synthesis of the photolabile "D1" protein and catalase at low temperature in the leaves of cold-hardened winter rye.

3.2.3 INHERITANCE

Genes with mainly additive effects control cold hardiness, but other factors may also be involved. Multiple alleles, or modifiers, for this major gene are also present. Similar to cold hardiness, frost tolerance is a complex trait with polygenic inheritance. A large number of genes are up- and downregulated when plants are exposed to cold/frost stress. In wheat, transcriptome analyses have shown between 5% and 8% of transcripts represented on microarrays to be regulated under cold stress. The dehydrin gene family (*Dhn1-13*) is one group among the cold-responsive genes that has been characterized in barley. Six members of the dehydrin gene family, including *HvDhn1* and *HvDhn3*, were induced under mild frost stress in barley. Expression of cold-responsive genes under cold stress in *Arabidopsis* is regulated through the binding of the C-repeat binding factor (*Cbf*) gene family to the *cis*-regulatory element present in the promoter region of cold-responsive genes. Most members of the *Cbf* gene family are closely linked and map to loci on homoeologous group 5. Twelve members of the *Cbf* gene family have been assigned to the long arm of chromosome 5R in rye.

There are changes in gene expression during dehardening of cold-hardened winter rye leaves. Suppression subtractive hybridization and differential display polymerase chain reactions were used to identify nuclear genes that were differentially expressed in cold-hardened and dehardened leaves of winter rye (In et al. 2005). The transcripts of nine genes declined during dehardening at 22°C of cold-hardened 4°C-grown leaves, indicating some role in cold acclimation. Among the genes that were strongly expressed in cold-hardened leaves were five genes of photosynthetic metabolism, the gene of the antioxidative enzyme peptide methionine sulfoxide reductase (PMSR), and three genes of RNA and protein metabolism. Four genes were identified that were more strongly expressed during dehardening of cold-hardened leaves at 22°C. A full-length complementary DNA (cDNA) for a presumed cytosolic PMSR of rye leaves was identified. After heterologous expression in *Escherichia coli*, an antiserum against the PMSR was produced. The content of the PMSR protein, visualized by immunoblotting, was much higher in cold-hardened than in nonhardened leaves and declined during dehardening. In nonhardened leaves, the messenger RNA (mRNA) of PMSR is slowly increased during exposures to light at 4°C and is not affected by exposure to darkness at 4°C. However, the PMSR mRNA was also induced by prolonged exposure (48 h) to intense light at 22°C, or by treatment with 2 μM paraquat. Consequently, the induction of cytosolic PMSR is a late response to prolonged photooxidative stress conditions, as expected during growth at low temperature in light. In cold-hardened leaves, PMSR can protect proteins from photodamage, and thus prevent their degradation and the need for repair.

As an allogamous species, rye exhibits high levels of intraspecific diversity, which makes it suitable for molecular allele mining in genes that are involved in cold response. Among several populations of Europe, a total of 147 single-nucleotide polymorphisms (SNPs) and 9 insertion–deletion polymorphisms were described within 7639 bp of DNA sequence from 11 candidate genes, resulting in an average SNP frequency of 1 SNP/52 bp (Li et al. 2011b). According to an analysis of molecular variance, most of the genetic variation was found between individuals within populations. *ScCbf14*, *ScVrn1*, and *ScDhn1* were dominated by a single haplotype, while the other eight genes (*ScCbf2*, *ScCbf6*, *ScCbf9b*, *ScCbf11*, *ScCbf12*, *ScCbf15*, *ScIce2*, and *ScDhn3*) had a more balanced haplotype frequency distribution. Middle European populations do not differ substantially from Eastern European populations in terms of haplotype frequencies or in the level of nucleotide diversity. However, the nucleotide diversity is reduced by advanced breeding material. There is a selective sweep toward reduced polymorphisms in flanking regions of advantageous genes.

In 2011, Li et al. (2011a) associated two SNP markers, *XScCbf15* and *XScCbf12*, all leading to amino acid exchanges, with frost tolerance. Two-way epistasis was found between 14 pairs of candidate genes. Results support the findings of biparental linkage mapping and expression studies that the *Cbf* gene family plays an essential role in frost tolerance.

Broad-sense heritability estimates were generally high (48%–82%), indicating that selection for cold hardiness should be effective in breeding programs (Brule-Babel and Fowler 1989).

3.2.4 ANTIFREEZE ACTIVITY

Cold tolerance and antifreeze activity are induced by cold temperatures in winter rye. Antifreeze proteins are found in a wide range of overwintering plants where they inhibit the growth and recrystallization of ice that forms in intercellular spaces. Unlike antifreeze proteins found in fish and insects, plant antifreeze proteins have multiple, hydrophilic ice-binding domains. Surprisingly, antifreeze proteins from plants are homologous to pathogenesis-related proteins and provide protection against psychrophilic pathogens. In winter rye, antifreeze proteins accumulate in response to cold, short-day length, dehydration, and ethylene, but not pathogens. Transferring single genes encoding antifreeze proteins to freezing-sensitive plants lowered their freezing temperatures by approximately 1°C.

Calcium interacts with antifreeze proteins and chitinase from cold-acclimated winter rye. Antifreeze activity is unaffected by salts before freezing, but decreases after freezing and thawing in $CaCl_2$ and is recovered by adding a chelator. Ca^{2+} enhances chitinase activity three- to fivefold in unfrozen samples, although hydrolytic activity also decreases after freezing and thawing in $CaCl_2$ (Stressmann et al. 2004).

The activity arises from six antifreeze proteins that accumulate in the apoplast during cold acclimation. The individual antifreeze proteins are similar to pathogenesis-related proteins, including glucanases, chitinases, and thaumatin-like proteins. Immunoblotting revealed that six of the seven accumulated apoplastic proteins consisted of two glucanases, two chitinases, and two thaumatin-like proteins. Glucanases are expressed at cold temperatures and display antifreeze activity. Both basic β-1,3-glucanases and an acidic 1,4-glucanase can be expressed in *E. coli*, purified, and assayed for their hydrolytic and antifreeze activities *in vitro*. All were found to be cold active and to retain partial hydrolytic activity at subzero temperatures (e.g., 14%–35% at −4°C). The two types of glucanases have antifreeze activity as measured by their ability to modify the growth of ice crystals. Structural models for the winter rye β-1,3-glucanases were developed by Yaish et al. (2006) on which putative ice-binding surfaces (IBSs) were identified. Residues on the putative IBSs were charge conserved for each of the expressed glucanases, with the exception of one β-1,3-glucanase recovered from nonacclimated winter rye in which a charged amino acid was present on the putative IBS. This protein also had a reduced antifreeze activity relative to the other expressed glucanases. The results support the hypothesis that winter rye glucanases have evolved to inhibit the formation of large, potentially fatal ice crystals, in addition to having enzymatic activity with a potential role in resisting infection by psychrophilic pathogens.

The ethylene-releasing agent "ethephon" and the ethylene precursor 1-aminocyclopropane-1-carboxylate also induced high levels of antifreeze activity at 20°C, and this effect could be blocked by the ethylene inhibitor $AgNO_3$. When intact rye plants are exposed to 5°C, endogenous ethylene production and antifreeze activity were detected within 12 and 48 h of exposure to cold, respectively. Rye plants exposed to drought produced both ethylene and antifreeze activity within 24 h. Thus, ethylene is involved in regulating antifreeze activity in winter rye in response to cold and drought (Xiao-Ming et al. 2001).

3.3   DROUGHT TOLERANCE

Drought is one of the major environmental factors reducing grain production in rainfed and semiarid regions. Plants cope with drought stress by manipulating key physiological processes such as photosynthesis, respiration, water relations, antioxidant, and hormonal metabolism. There exist multiple and often redundant stress sensors, which transduce the stress signal through secondary signaling molecules to the nucleus, where the expression of stress-response genes is regulated. Transcription factors play an important role in regulating the expression of the stress-response genes. Another level of regulation of gene expression is at the epigenetic level and involves modifications either at the chromatin level or at the mRNA level. Cereal plants show various adaptive and acclimatization strategies to drought stress, which range from seemingly simple morphological or physiological traits that serve as important stress tolerance markers to major upheavals in gene expression in which a large number of transcription factors are induced.

As the most drought resistant of the cereals, rye has an extensive root system and adjusts maturity to moisture. It uses 20%–30% less water per unit of dry matter formed than wheat (Starzycki 1976). Tetraploid varieties are more sensitive to drought than diploids.

In Australia, with its extensive arid regions, rye withstands adverse conditions better than other cereals. Its drought resistance and ability to withstand sand blasts enable it to produce a soil-binding cover on land where other cereals will not grow. Under conditions where wheat, oat, or barley will grow only a few centimeters high, or may even be completely blown away, rye often grows vigorously and reaches a height of a meter or more. A further reason for using it in erosion-prone soils is the fact that the grain and the straw are the least preferred fodder by sheep. Sheep provided with more than one choice of stubble within a paddock will preferentially graze other stubbles before they will eat rye stubble. After the crop is harvested, the tough, resilient stubble is generally left as a protective cover to reduce blowing of the soil and to assist colonization by other species. The stubble of rye breaks up more slowly than that of other cereals, ensuring soil cover for a long period.

The morphological and physiological bases of drought resistance in wheat, rye, barley, and mosquito grass (*Dasypyrum villosum* syn. *Haynaldia villosum*), including their derived disomic addition lines, were studied by Shakir and Waines (1994) in the field and greenhouse under nonstressed and stressed conditions. The four parental genotypes were first compared with each other for their performance in the field. Results from water-stressed conditions showed that Imperial rye was superior for plant height, above-ground biomass, grain yield, and its components such as number of spikelets and florets per main spike. Chinese Spring wheat and Betzes barley were intermediate in performance for most of the plant characters studied. *Dasypyrum villosum* G870 was the poor entry for the characters studied.

To locate the genes controlling drought tolerance, disomic addition lines of Imperial rye (donor) in the genetic background of Chinese Spring (recipient) were tested in a completely randomized design in controlled condition and in a randomized complete block design in field condition each with three replications. The materials were tested under two different water regimes (irrigated and rainfed) in the Agricultural Research Station of College of Agriculture, Razi University, Kermanshah, Iran (Mohammadi et al. 2003). The potential yield, stress yield, stress tolerance index, coleoptile length, primary root length, and primary root number were recorded. The results of analysis of variance exhibited highly significant differences among the lines for all the traits, indicating the presence of genetic variation and possible chromosomal localization of the genes controlling the traits investigated. Mean comparison of different traits revealed that most of the genes controlling yield and drought tolerance-related traits are located on chromosomes 7R, 3R, and 5R. The contribution of addition line 7R to the multiple selection index was 47%. Further evaluation revealed that most of the quantitative trait loci (QTL) involved in these quantitative criteria of drought tolerance are located on 3R and 7R. Correlation analysis indicated strong association between stress tolerance index with germination stress index, coleoptile length, and primary root length. Therefore, these criteria may be screened for indirect selection of drought tolerance in the initial stages of the crop growth.

The glasshouse, gravimetric-pot studies of root dry matter, total biomass, amount of water used, and water-use efficiency also confirmed the field results. "Imperial" rye was assumed to be the most drought resistant because it produced larger root dry matter and total biomass, while it consumed a smaller amount of water.

Dasypyrum villosum was assumed drought sensitive because it showed lower means for water-use efficiency and root dry matter, and consumed a larger amount of water. Chinese Spring wheat and Betzes barley were intermediate. Among Chinese Spring wheat and the disomic addition lines, those that had Imperial rye chromosome 2R and Betzes barley chromosome 4H showed pronounced performance over the other genotypes for above-ground biomass, number of tillers, number of spikes, dry matter, amount of water consumed, and water-use efficiency. Both chromosomes seem to carry loci that are responsible for the expression of plant characters that confer drought resistance. However, grain yield data suggested that rye chromosome 2R also carries undesirable genes in wheat, which cause partial floret sterility.

The short arm of chromosome 1RS seems also to carry gene-influencing drought stress. Hoffmann (2008) studied the effect of 1BL.1RS wheat–rye translocation (see Section 4.8.3.4) on drought tolerance. Mv5791-1BL.1RS translocation and the sister line Mv5791-1B.1B nontranslocation were examined among other cultivars in greenhouse under well-watered and drought conditions. Data were obtained for anthesis and maturity date, plant height, root/shoot ratio, components of grain yield, harvest index, and water-use efficiency. The translocation line had higher root-and-shoot dry weight in both treatments and an increased root/shoot ratio, which was more than the sister line was when in dry treatment (69% and 38%, respectively). The larger root biomass of the 1RS translocation line contributed to increased harvest index and water-use efficiency under drought conditions, which resulted in less yield loss (23% and 32%, respectively). Ehdaie et al. (2003) confirmed the data. The 1RS translocations of hexaploid wheat, in general, contribute to higher tolerance to field environmental stresses. These results encourage the development and use of the 1RS.1AL and 1RS.1DL translocations at least in wheat breeding programs.

Gordon-Werner and Dörffling (1988) investigated a *Secale montanum* hybrid. The variety Permontra resulted from a cross of *S. cereale* with *S. montanum*. Root length, surface area, and lateral density were measured on plants at the three-leaf stage. Tolerance to drought, acute heat, and chronic heat stress was assessed in culture chamber tests on control and drought or heat-conditioned potted plants, and membrane stability assessed by measuring electrolytic leakage. In comparison with other rye varieties, Permontra had the most extensive root system and was the most drought-tolerant variety measured. It was also the most tolerant of the varieties to chronic, though not to acute, heat stress. However, there was no correlation between drought tolerance in potted plants and membrane stability following either heat or dehydration stress.

Progressing climatic change will increase dry periods during vegetation. There is already a call for varieties with enhanced tolerance to drought. Even rye varieties and genotypes react differently to drought stress. The breeder, KWS Lochow, Bergen, Germany, carried out complex drought experiments; for example, an area of approximately 5 ha, which had approximately 3000 scattered rye plots, was used for measurement on which several drought stress trials were undertaken. The complete trial is split into a watered and a nonwatered variant, with both variants adjoining in a replicate to be able to compare both variants adequately. For the irrigation, drip hoses of 12–15 m length per plot have been installed to secure an even irrigation of the target plot. The water supply is controlled via a central distribution unit. The improvement will be demonstrated in first hybrid rye varieties with distinct drought tolerance, which could be released in 2018.

First studies revealed that drought stress leads to an average grain yield reduction of ~25% for rainfed compared to irrigated regime in drought stress environments. In addition, a decomposition of the variance revealed significant genotypic and genotype–environment interaction variances but only a minor effect of drought stress on the ranking of the genotypes with regard to grain yield. Therefore, separate breeding programs for drought-tolerant genotypes seem not superior to the currently practiced selection under rainfed conditions without irrigation in hybrid rye breeding, at least in Central Europe (Hüber et al. 2013).

3.4 NUTRITION

3.4.1 MACRONUTRIENTS

Rye is known to be tolerant not only to acidic soils. Because of the huge root system of rye (see Section 2.4), it is able to take up nutrients most efficiently if they are available. Since rye is mostly grown in poorer soils, additional fertilization is required. As a means, the following nutrients are uptaken by rye (see Table 3.3).

Macronutritional deficiency, for example, low nitrogen and potassium, causes plant stress in rye as well. Coleoptiles are shorter; roots are longer and more numerous. When cytological analyses of the elongation zone of the seminal roots of seedlings are carried out, fluorescence of dichlorofluorescein in the epidermal zone is observed, which proves the presence of reactive oxygen species. The appearance of reactive oxygen results from the activation of mechanisms for sensing and responding to induced stress (Smolik et al. 2012).

TABLE 3.3
Mean Nutrient Uptake of Rye Depending on Different Yield Levels

	Yield (dt/ha)		
Nutrients	40 dt/ha	60 dt/ha	80 dt/ha
N (kg/ha)	110	150	200
P_2O_3 (kg/ha)	40	60	80
K_2O (kg/ha)	75	115	150
MgO (kg/ha)	15	22	30
CaO (kg/ha)	15	22	30
S (kg/ha)	10	15	20
B (g/ha)	40	45	50
Cu (g/ha)	30	40	50
Zn (g/ha)	120	150	200
Mn (g/ha)	350	400	600

Source: Slotta, M., *Roggen Anbau und Vermarktung*. Verlag Roggenforum, Bergen, Germany. 1992. With permission.

Recent studies also demonstrate genotypic differences in nutritional efficiency. By application of intersimple sequence repeat (ISSR) primers, a rye inbred line "Ot0-6" was identified showing specific tolerance to nutrient deprivation (Smolik 2012). One R-ISSR product was considered as a molecular marker putatively linked to one of the loci determining rye tolerance to nutrient deficiencies. Sequential analysis and sequence alignment performed using basic local alignment search tool (BLAST) have shown its high homology to various expressed sequence tags (ESTs) in response to abiotic stress caused by different treatments: aluminum, drought, salinity, and cold, where the criterion of morphological assessment of response to stress is, among others, root morphology.

3.4.2 MICRONUTRIENTS

Important micronutrients in rye are boron, copper, iron, manganese, zinc, chloride, and molybdenum. Rye response to micronutrients is generally low, except zinc that needs somewhat more. Nevertheless, genotypes differ in their requirements for certain micronutrients.

3.4.2.1 Iron

Iron is involved in the production of chlorophyll, and iron chlorosis is easily recognized on iron-sensitive crops grown in calcareous soils. Iron also is a component of many enzymes associated with energy transfer, nitrogen reduction and fixation, and lignin formation. Iron is associated with sulfur in plants to form compounds that catalyze other reactions. Rye takes Fe up as Fe^{2+} and Fe^{3+} ions. Iron deficiencies are mainly manifested by yellow leaves due to low levels of chlorophyll. Leaf yellowing first appears on the younger upper leaves in interveinal tissues. Severe iron deficiencies cause the leaves to turn completely yellow or almost white, and then brown as leaves die. Iron deficiencies are found mainly in high pH soils, although some acidic, sandy soils low in organic matter also may be iron deficient. Cool, wet weather enhances iron deficiencies, especially in soils with marginal levels of available iron. Poorly aerated or compacted soils also reduce iron uptake. Uptake of iron decreases with increased soil pH, and is adversely affected by high levels of available phosphorus, manganese, and zinc in soils.

Graminaceous plants secrete mugineic acids (MAs) into the rhizosphere from their roots, although the amount differs among species in the order: barley > wheat = rye > oat > maize > sorghum > rice. While the amount of MAs secreted increased dramatically on account of iron and copper deficiencies, it was not clear whether zinc deficiency increases the secretion of MAs. Zn deficiency has been reported to increase or have no effect on the secretion of MAs from wheat and barley roots into the rhizosphere. However, recent studies suggest that deoxymugineic acid (DMA) in Zn-deficient rice plants has an important role in the distribution of Zn within the plant rather than in the absorption of Zn from the soil. The discovery of MAs as phytosiderophores has shown that some graminaceous monocotyledonous plants, such as rye, have a different iron acquisition strategy (strategy II; see Figure 3.2) from dicotyledonous and nongraminaceous monocotyledonous plants (strategy I). The process of iron acquisition by strategy II plants can be divided into four main steps: biosynthesis, secretion, solubilization, and uptake, all of which are

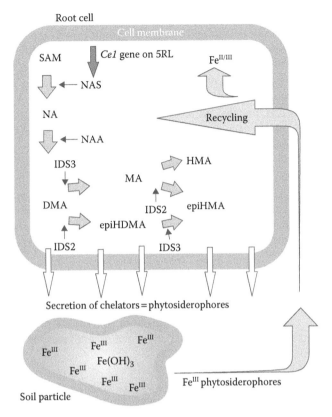

FIGURE 3.2 Simplified pathway of phytosiderophores in *Secale cereale* and the suggested action of *Cel* gene. DMA, 2'-deoxymugineic acid; epiHDMA, epi-hydroxydeoxymugineic acid; epiHMA, epi-hydroxymugineic acid; IDS, indoleamine 2,3-dioxygenase; NA, nicotianamine; NAAT, nicotianamine aminotransferase; SAM, *S*-adenosyl-methionine.

effectively regulated by different systems. The biosynthesis of MAs is controlled by an on–off system, which is operated under the control of iron demand in the plant. All MAs share the same biosynthetic pathway from L-methionine to 2'-DMA, but the subsequent steps differ among plant species and even cultivars. These results revealed that the biosynthetic pathways of both 3-epihydroxymugineic acid and 3-hydroxymugineic acid (HMA) are as follows: L-methionine > 2'-DMA > MA > 3-epihydroxymugineic acid in barley > HMA in rye. The biosynthesis of MAs is associated with the methionine-recycling pathway.

Nicotianamine aminotransferase (NAAT), the key enzyme involved in the biosynthesis of MAs, catalyzes the amino transfer of nicotianamine (NA). This amino transfer reaction is the first step in the unique biosynthesis of MAs which has evolved in graminaceous plants. NAAT activity is dramatically induced by Fe deficiency and suppressed by Fe resupply. Based on the protein sequence of NAAT purified from Fe-deficient barley roots, two distinct cDNA clones encoding NAAT, *naat-A* and *naat-B*, were identified (Takahashi et al. 1999). Their deduced amino acid sequences were homologous to several aminotransferases, and shared consensus sequences for the pyridoxal phosphate-binding lysine residue and its surrounding residues. The expression of both *naat-A* and *naat-B* is increased in Fe-deficient roots, while *naat-B* has a low level of constitutive expression in Fe-sufficient roots. No detectable mRNA from either *naat-A* or *naat-B* is present in the leaves of either Fe-deficient or Fe-sufficient plants. One genomic clone with a tandem array of *naat-B* and *naat-A* in this order was identified. *naat-B* and *naat-A* each have six introns at the same locations. The isolation of NAAT genes paves the way for understanding the mechanism of the response to Fe in rye plants, and may lead to the development of cultivars tolerant to Fe deficiency which can grow in calcareous soils.

The secretion of MAs shows a distinct diurnal rhythm. MAs solubilize sparingly soluble inorganic iron by chelation and possess a high chelation affinity for iron, but not for other polyvalent ions such as Ca^{2+}, Mg^{2+}, and Al^{3+}. The iron uptake process is regulated by a specific uptake system that transports the MA–Fe(III) complex as an intact molecule. This system specifically recognizes the MA–Fe(III) complexes, but not other MA–metal or synthetic chelator–Fe(III) complexes, suggesting that binding sites with strict recognition for stereo structure of the complex are located on the plasma membrane. All these regulatory systems are considered to represent an efficient strategy to acquire adequate amounts of iron and to avoid factors unfavorable for iron acquisition such as high pH, high concentrations of bicarbonate, Ca^{2-} and Mg^{2+}, microbial degradation, and uptake of other metals that are common in calcareous soils.

3.4.2.2 Copper

Copper is necessary for carbohydrate and nitrogen metabolism and, inadequate copper results in stunting. Copper also is required for lignin synthesis which is needed for cell wall strength and prevention of wilting. Cu is taken up as Cu^{2+} ion. Deficiency symptoms of copper are dieback of stems, yellowing of leaves, stunted growth, and pale green leaves that wither easily. Copper deficiencies are mainly reported in sandy soils, which are low in organic matter. Copper uptake decreases as soil pH increases. Increased phosphorus and iron availability in soils decreases copper uptake by plants.

Copper efficiency in rye appears to be a dominant trait controlled at a single locus on the long arm of chromosome 5R (Graham 1984, Schlegel et al. 1993). It was found that rye shows the ability to take up iron and copper under deficient conditions (Podlesak et al. 1990). As shown by Marschner et al. (1989) and Treeby et al. (1989), the release of phytosiderophores is the main mechanism by which grasses acquire iron and copper in the rhizosphere. An initial study demonstrated that rye secretes mainly HMA under iron-deficient conditions, but also MA and 2'-DMA (Mori et al. 1990, Ma et al. 1999). Schlegel et al. (1993) found a clear correlation between high amounts of those chelators and the presence of chromosome 5R in wheat–rye addition lines. Later the genes for high copper and iron efficiencies were physically mapped on the distal region of chromosome arm 5RL by using a specific wheat–rye translocation lines (see Figures 3.3 and 5.4).

Copper efficiency has been transferred from rye to wheat by a translocated part of 5RL to a chromosome of wheat (see Section 4.12.4). The translocated 5RL

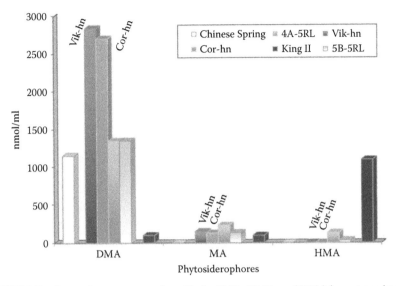

FIGURE 3.3 Approximate amounts (nmol/ml) of MA, DMA, and HMA in root exudates of Vik-hn, Cor-hn, 4A–5RL, and 5B–5RL wheat–rye translocations, Chinese Spring wheat, and King II rye after seedling culture in iron-deficient solution. DMA, 2′-deoxymugeneic acid; HMA, 3-hydroxymugeneic acid; MA, mugeneic acid.

chromosome segment carrying the copper-efficiency trait *Ce1* (see Figures 3.2 and 5.4) onto the 4BL chromosome confers on plants a much greater ability to mobilize and absorb copper ions tightly bound to the soil (Graham 1984). The translocation has been successfully incorporated into cultivars adapted to South Australia (Graham et al. 1987). The 5RL chromosome arm also confers copper efficiency in triticale, unless a copper-inefficient rye is used in the cross. Triticales show efficiency, are intermediate between wheat and rye, and have been used for this reason in many sandy or peaty copper-deficient soils. Additional experiments have shown that copper efficiency in rye is not clearly linked to zinc or manganese efficiency (see Table 3.4). Thus, independent and relatively specific genes are involved, and neither root system geometry nor size appears to be critical (Holloway 1996). Studies of rye addition lines suggest that 6R contributes a little to efficiency for all three elements, that is, Cu, Zn, and Mn, perhaps by way of a root geometry feature, but the major genes are elsewhere.

When copper is present in high concentration, it causes damage to plants, for example, in pots, in original sewage field soil, and in artificially polluted soil. The weakly polluted sewage field soil was contaminated with 2.2′,5.5′-tetrachlorobiphenyl, benzo[*a*]pyrene, cadmium, or copper as well as with combinations of these organic pollutants and heavy metals. These treatments were compared with those in an extremely contaminated sewage field soil. Rye showed typical symptoms of damage in artificially polluted soil, which was similar to effects on rye in extremely contaminated sewage field soil. Damage from single substance copper was sometimes stronger than that from combined metal availability (Dorn and Metz 1996).

TABLE 3.4

Comparison of Wheat and Rye Genotypes for Mn and Zn Efficiencies

Wheat Variety	Efficiency (%)		Rye Variety	Efficiency (%)	
	Mn	Zn		Mn	Zn
Avalon	54	49	Danae	64	84
Bezostaya 1	61	50	Graser	57	67
Bolal	68	30	Imperial	72	63
Borenos	56	64	Inbred	98*	61
Carola	39	42	King II	78	89*
Cheyenne	51	55	Pico	59	65
Chinese Spring	72	55	Pluto	86*	90*
Dagdas	74	48	Petka	78	81
Fakon	59	53			
Gerek 79	40	45			
Giza	61	74			
Holdfast	41	48			
Seri	29	36			
Trakia	91*	84*			
Trakia-alloploid	67	64			
Viking	65	60			
Vilmorin	62	73			
Mean	**51**	**55**		74	75

*Significant at $p = 5\%$.

3.4.2.3 Zinc

Zinc is an essential component of various enzyme systems for energy production, protein synthesis, and growth regulation. Rye takes it up as Zn^{2+} or $Zn(OH)_2$ ions. Zinc-deficient plants exhibit delayed maturity. Zinc is not mobile in plants so zinc-deficiency symptoms occur mainly in new growth. Poor mobility within the plant suggests the need for a constant supply of available zinc for optimum growth. The most visible zinc-deficiency symptoms are short internodes and a decrease in leaf size. Delayed maturity also is a symptom of zinc-deficient plants. Zinc deficiencies are mainly found in sandy soils low in organic matter and in organic soils. Zinc deficiencies occur more often during cold, wet, and spring weather and are related to reduced root growth and activity. Zinc uptake by plants decreases with increased soil pH. Uptake of zinc also is adversely affected by high levels of available phosphorus and iron in soils.

Although there was a considerable genotypic variation among cereals, rye showed the highest level of efficiency, particularly the German variety Pluto (see Table 3.4). However, some varieties of hexaploid wheat such as Trakia can compete with some of the rye varieties. Other genomes from *Avena sativa*, *T. turgidum*, *T. durum*, *T. persicum*, *H. villosa*, *A. elongatum*, *A. sharonense*, or *A. ventricosa* did not reach the level of efficiency of rye (Schlegel et al. 1998a).

TABLE 3.5

Comparison of Different Wheat–Rye Substitution and Translocation Lines for Cu and Zn Efficiencies

Chromosome Addition/Substitution/Translocation	Efficiency (%)		
	Fe	Cu	Zn
1R, Holdfast-King II	–	8	68
1RS/1A, Amigo	–	–	57
1RS/1B, Bovictus	19	21	68
1RS/1B, Salmon	–	–	71
1RS/1D, Chinese Spring strain	12	16	68
2R/4B, Transec	–	–	53
5R(5D), Chinese Spring strain	–	65	–
5R(4A), Chinese Spring strain	–	58	–
5RL/5A, strain	61	–	49
5RL/4B, Cornell	46	70	51
5RL/4B, Viking	56	71	53
5RL/5B, strain	37	–	51
7R, Holdfast-King II	–	12	76
Controls—18 wheats	14	26	54
Controls—11 ryes	63	69	77

Source: Schlegel, R., and Cakmak, I., *Plant Nutrition: For Sustainable Food Production and Environment*, Kluwer Academic Publishers, Tokyo, Japan, 1997. With permission.

Using aneuploid stocks, it has been established that the target genes of rye, determining Zn, Cu, and Fe efficiencies, are clearly expressed in the alien background of the wheat recipients (see Table 3.5). The results demonstrate a genetically dominant expression of the rye genes in the presence of the wheat genomes. This is true not only for Zn (Schlegel et al. 1998a) but also for Cu, Fe (Schlegel et al. 1991b, 1993), and Mn (Graham et al. 1987) efficiencies.

With genetic testers, it was possible to associate the effects of rye on Cu, Fe, and Zn efficiencies with individual chromosomes. Cu and Zn efficiencies seem to be determined by genes of independent linkage groups. While Cu efficiency is linked to chromosome 5R of rye, genes on chromosomes 1R and 7R might control Zn efficiency. These data are complemented by wheat–rye translocations of the homoeologous groups 1 and 5. The short arm of chromosome 1R carries genetic information for Zn efficiency and the long arm of 5R genes for Cu and Fe efficiencies, respectively.

Dosage effects of chromosome arm 1RS were not observed. Although genes of different linkage groups are involved in Cu, Fe, and Zn efficiencies, a common mechanism might be supposed for an efficient uptake of heavy metals. Mori et al. (1990) located genes encoding synthetases for MA and HMA on chromosome 5R of rye. Exudation of these phytosiderophores was found to be associated with Cu-, Fe-, and Zn-efficient cereal genotypes (Schlegel et al. 1993). In rye, there was a significantly positive correlation ($r = +0.97**$, $**P < 1\%$) between the amount of phytosiderophore exudation and the yield.

When different 5RL wheat–rye translocations were studied, it was clear that the amount of 2'-DMA, MA, and HMA was increased in different translocated 5RL lines compared to a nontranslocated wheat control (see Figure 3.2). By contrast, no traceable amounts of HMA could be found for Viking and Cornell selection translocations, even though these translocations produced about double as much DMA as the wheat control. The high DMA production of Viking and Cornell selection might be explained by a dosage effect. Cytological and molecular studies revealed two types of translocations (see Figure 5.4). A terminal 5RL segment in Viking-trans (cf. Schlegel et al. 1993) replaced a faint terminal segment of wheat. If the rye segment 5RL2.3 is homoeologous to group 4 chromosomes of wheat and chromosome 4B of wheat, there are genes coding for phytosiderophores; then by the rye segment, a second dose of those genes is supplemented, possibly giving rise to double the amount of DMA measured in the Viking and Cornell translocations.

In plant physiology, rye is recognized among cereals as a genotype that shows high ability of uptake of micronutrients, for example, under iron, copper, manganese, and zinc deficiencies (Erenoglu et al. 1999). When grown in zinc-deficient calcareous soil in the field, the rye cultivars had the highest, and the durum wheat the lowest zinc efficiency. A Turkish variety Aslim shows the best performance. Under zinc deficiency, rye had the highest rate of root-to-shoot translocation of zinc. The results indicated that high zinc efficiency of rye could be attributed to its greater zinc uptake capacity from soils. By utilization of wheat–rye addition lines, it was demonstrated that genes on chromosome arms 1RS and 7RS are associated with high zinc efficiency (Schlegel et al. 1999).

The different behavior of root uptake is not necessarily correlated with the concentration of micronutrients in the shoot (Schlegel et al. 1997). Nevertheless, the chromosomes 2R and 7R were associated with improved manganese and iron concentrations in the shoots, respectively, and chromosome 1R with zinc and 5R with copper concentration (see Figure 3.6).

When rye is stressed at the shooting stage with different zinc ($ZnCl_2$) concentrations of 0, 20, 200, and 400 mg dm^{-3} for 10 days at pH 4.5, the roots significantly decrease the fractions of surface acidic functional groups. It means that these groups have apparent surface dissociation. The same charge properties are found in the roots of control (Szatanik-Kloc et al. 2009). Root cation exchange capacity and specific surface area are mainly responsible for sorption and transport of water and nutrients by plants. The cation exchange capacity of roots governs the tolerance of rye to soil acidity. Moreover, zinc plays an important role in plant metabolism. It activates many enzymes such as carbonic anhydrase, alcohol dehydrogenase, dehydrogenase of reduced nicotinamide dinucleotide, and nicotinamide adenine dinucleotide phosphate. Zinc influences the processes of ribosome formation and cell membrane permeability, and controls the proportions of elements at the cellular level.

Zinc efficiency in rye does not appear to be clear-cut from studies on wheat–rye addition lines. Four or five chromosomes could be included: 2R, 3R, 4R, 7R, and, to a lesser extent, 5RL and 6R (Graham 1988a). Cakmak et al. (1997) have additionally linked Zn efficiency to 1R and 7R, both on the short arms (see Figures 3.4 and 3.5).

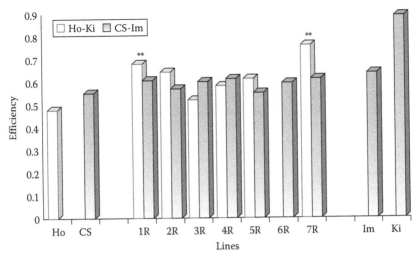

FIGURE 3.4 Zinc efficiency (−DW/+DW) of hexaploid wheat (Ho, CS), ryes (Im, Ki), and several disomic wheat–rye chromosome addition lines (Ho-Ki and CS-Im) under Zn-deficient growth (average of 2-year trials). CS, Chinese Spring; Ho, Holdfast; Im, Imperial; Ki, King II; **$P < 1\%$.

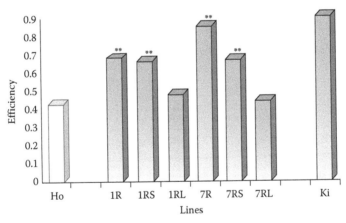

FIGURE 3.5 Zinc efficiency (−DW/+DW) of hexaploid wheat (Ho), rye (Ki), and several disomic wheat–rye chromosome addition lines (1R, 1RS, and 1RL) under Zn-deficient growth (average of 2-year trials). Ho, Holdfast; Ki, King II; **$P < 1\%$.

3.4.2.4 Manganese

Manganese is necessary for photosynthesis, nitrogen metabolism, and formation of other compounds required for plant metabolism. Interveinal chlorosis is a characteristic manganese-deficiency symptom. In very severe manganese-deficiency cases, brown necrotic spots appear on the leaves. Delayed maturity is another deficiency symptom. White-gray spots on leaves of rye are a sign of manganese deficiency. Rye takes Mn up as Mn^{2+} ion. Manganese deficiencies mainly occur in organic soils, high pH soils, sandy soils low in organic matter, and overlimed soils. Soil manganese may be less available in dry, well-aerated soils, but can become more available under wet soil conditions when manganese is reduced to the plant-available form. Conversely, manganese toxicity can result in some acidic, high-manganese soils. Uptake of manganese decreases with increased soil pH and is adversely affected by high levels of available iron in soils.

Manganese efficiency of rye is located on chromosome 2R, a conclusion supported by poor performance in manganese-deficient soils of triticale cv. Coorong- and Armadillo-type triticale lacking 2R. Generally, rye shows slightly better performance than wheat. Best results could be obtained from the advanced rye variety Pluto from Germany (see Table 3.4). Remarkable diversity for manganese efficiency exists also within wheat, especially in *T. aestivum* and *T. durum*. The old hexaploid Bulgarian wheat Trakia can be a source for better manganese efficiency (see Table 3.4). Manganese efficiency in barley appears to be simply inherited, taking the evidence of the cross of varieties Weeah (efficient) and Galleon (inefficient) (Graham 1988b). This major gene has recently been mapped to chromosome 4H. Studies have shown that two additive acting genes (Khabaz-Saberi et al. 1998) control Mn efficiency in durum wheat. It is likely that these are in homoeologous positions in the two genomes with barley (single gene), rye (single gene), and durum wheat (two genes). Three major loci can be predicted in bread wheat. Minor genes are almost certain to be involved in all species.

Moreover, manganese accumulation in seeds seems to be associated with the uptake, at least when chromosome 2R is considered. A study showed a significant increase of Mn concentration in wheat–rye addition lines 2R and 7R compared to wheat standards (see Figures 3.5 and 3.6). When copper was tested, chromosome 5R was critical (Schlegel, unpublished data).

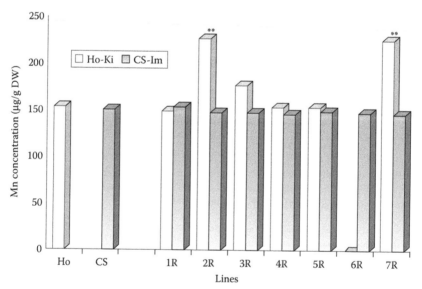

FIGURE 3.6 Manganese concentration (μg/g DW) in seeds of hexaploid wheats (Ho and CS) in comparison with disomic wheat–rye chromosome addition lines (Ho-Ki and CS-Im) under standard growth conditions. Ho-Ki, Holdfast-King II; CS-Im, Chinese Spring-Imperial; **$P < 1\%$.

3.4.2.5 Boron

A primary function of boron is related to cell wall formation, so boron-deficient plants may be stunted. Sugar transport, flower retention, and pollen formation and germination also are affected by boron. Seed and grain production are reduced in plants having low boron supply. Boron-deficiency symptoms first appear at the growth points. Boron is taken up as $B(OH)_3$ ion. This results in a stunted appearance (rosetting), barren ears due to poor pollination, hollow stems and fruit (hollow heart), brittle, discolored leaves, and loss of fruiting bodies. Boron deficiencies are found mainly in acidic, sandy soils in regions of high rainfall, and those with low soil organic matter. Borate ions are mobile in soil and can be leached from the root zone. Boron deficiencies are more pronounced during drought periods when root activity is restricted.

3.4.2.6 Molybdenum

Nitrogen metabolism, protein synthesis, and sulfur metabolism are affected by molybdenum. Molybdenum has a significant effect on pollen formation, so grain formation is affected in molybdenum-deficient plants. Because molybdenum requirements are very low, rye does not exhibit molybdenum-deficiency symptoms. Unlike the other micronutrients, molybdenum-deficiency symptoms are not confined mainly to the youngest leaves because molybdenum is mobile in the plant. Molybdenum deficiencies are found mainly in acidic, sandy soils in humid regions. Molybdenum uptake by the plant increases with increased soil pH, which is the opposite of the other micronutrients. Molybdenum deficiencies may be corrected by liming acidic soils rather than by molybdenum application. It is taken up as MoO_4^{2-} ion. However, seed treatment with molybdenum sources may be more economical than liming in some areas.

3.4.2.7 Chloride

Because chloride is a mobile anion, most of its functions relate to salt effects (stomatal opening) and electrical charge balance in physiological functions of the plant. Chloride also indirectly affects plant growth by stomatal regulation of water loss. Wilting and restricted, highly branched root systems are the main chloride-deficiency symptoms. Most soils contain sufficient levels of chloride for adequate plant nutrition. It is taken up as Cl^- ion. However, chloride deficiencies have been reported in sandy soils in high rainfall areas or in those derived from low-chloride parent materials. There are few areas of chloride deficiency so this micronutrient generally is not considered in fertilizer programs. In addition, chloride is applied to soils with KCl, the dominant potassium fertilizer. The role of chloride in decreasing the incidence of various diseases in rye is perhaps more important than its nutritional role from a practical viewpoint.

3.5 *IN VITRO* BEHAVIOR

Rye is known as one of the most recalcitrant cereal species for tissue culture. The lack of an efficient *in vitro* culture system limited the development of a reproducible genetic transformation protocol for this crop. The main factors determining the tissue culture response in rye include genotype, donor plant quality, and developmental stage of the explant at the time of culture initiation and culture medium composition, as well as in culture conditions.

3.5.1 EMBRYO CULTURE

The germination of cereals depends on the genotype and the age of caryopses. In rye, the optimum age of the embryo for better recovery of complete plantlets coincides with onset of the drying of hybrid caryopses in the field. From mature caryopses, on culture mediums prepared with different types of water, respectively, deuterium-depleted water (with 25 ppm D) or Pi water, as the substitute of distilled water from the recipe elaborated by Murashige-Skoog (1962), resulted in ~55% embryo germination and 50% plantlet recovery. Even from premature caryopses, embryo culture is technically easy, as it is for other cereal crops. Its utilization is limited because of the long time taken for vernalization. There is no real-time savings in order to accelerate the succession of generations during breeding. An *in vitro* long-term storage of embryos has not been much explored.

3.5.2 TISSUE CULTURE

Tissue culture response strongly varies between crop plants and within cereals. It has been shown to be highly genotype dependent. At the beginning, *S. cereale* did not provide well-responding genotypes, while genotypes with common genes from *S. vavilovii* have been described to be more suitable for anther cultures than for other rye species such as *S. cereale* (Friedt et al. 1983; Flehinghaus et al. 1991, Flehinghaus-Roux et al. 1995). However, *S. vavilovii* is of poor agronomic value mainly due to its fragile and brittle spikes.

Different tissues have been described for tissue culture initiation in rye. Apart from anthers, which are important for production of doubled haploid (DH) lines, immature embryos and immature inflorescences were the two most frequently used explants. Unexpectedly, Rakoczy-Trojanowska and Malepszy (1993) described immature embryos as less suitable as starting material than immature inflorescences (Rakoczy-Trojanowska and Malepszy 1995). The developmental stage of the explants has been found to be a crucial factor in the establishment of totipotent cultures (Vasil 1987).

To establish an efficient *in vitro* culture system for rye, a comparison of genotype-specific culture response and an optimization of culture media and culture conditions were performed. Eleven rye inbred lines, eleven single crosses, and the population variety Wrens Abruzzi were compared for their tissue culture response after short- and long-term callus culture, showing a high variation for regeneration potential. Genotypes with superior tissue culture suitability were identified. A multifactorial designed experiment allowed the development of genotype-specific *in vitro* culture protocols, maximizing the plant regeneration response from rye tissue culture. Precocious germination of explants, callus induction, and callus maintenance (formation of rhizogenic and caulogenic callus) as well as regeneration response (regenerating calli and number of regenerated plants per callus) were significantly influenced by genotype, carbohydrate and auxin sources, and their respective concentrations and interactions. Genotypes differed in the callus response to media sterilization procedure, basic salt composition, gelling agent, $CuSO_4$ complementation, and illumination. The selection of inbred lines with superior tissue culture response and the development of genotype-specific culture protocols provided a basis for the development of a genetic transformation protocol for rye.

Recently, Targonska et al. (2013) characterized the genotypes of ryes as "universally responding," that is, the level of regeneration *in vitro* is similar regardless of the culture or explant type, whereas the majority of genotypes were "differentially responding," that is, the level of regeneration *in vitro* depends on medium composition, explant type, other culture conditions, and interactions between these factors. Examples of universally responding forms are inbred line L318, which is classified as a positively responding genotype, and line L9, which is a nonresponding one. Most of the genotypes show a response to tissue culture only with one type of explant and/or under strictly defined culture conditions. For example, plants of inbred line H363 could be regenerated from immature inflorescences at an efficiency of over 75%, but completely failed to respond in the case of immature embryo or anther cultures. In contrast, no plants could be regenerated from immature inflorescences of line H316, but the regeneration efficiency of immature embryos was as high as 30%.

Genetic analysis performed in F_1, F_2, and F_3 generations obtained from crosses between selected inbred lines (DW28, H363, L318, D855, H32, Pw330, L9, L29, L299, and H316) that differ in their tissue culture response demonstrated that the *in vitro* response of immature embryos and immature inflorescences is controlled by a complex, polygenic system with various gene interactions, and that the plant regeneration ability is a recessive trait. For both explant types, embryogenic callus production and plant and root regeneration appear to be determined by recessive genes or suppressed by two dominant nonallelic complementary genes, whereas the reduced ability to produce nonembryogenic callus is most probably controlled by dominant genes. The lack of response was shown to be controlled by at least two interacting genes. The main difference between these two explant types is apparently caused by a heterosis effect, which positively influences embryogenic callus production and plant regeneration exclusively in immature embryos. Heterosis of donor plants was also found to promote androgenic plant regeneration from rye anthers. Heritability values of 0.8–0.93 and 0.36–0.5 were described for regeneration of embryogenic callus derived from immature embryos and immature inflorescences, respectively.

Genes on chromosomes 1R, 2R, 3R, 6R, and 7R positively stimulate tissue culture response using immature embryos, while 1R, 3R, and 4R do it when anthers are used.

3.5.3 Protoplast Culture

A protoplast is a plant cell that had its cell wall completely or partially removed using either mechanical or enzymatic means. More generally, protoplast refers to the unit that is composed of a cell nucleus and the surrounding protoplasmic materials. It can be used to study membrane biology, including the uptake of macromolecules and viruses, for DNA transformation, since the cell wall would otherwise block the passage of DNA into the cell, and for distant crosses in plant breeding, using a technique called protoplast fusion.

Early studies with rye protoplasts were reported by Evans et al. (1972) and Siminovitch et al. (1978). They used free protoplasts prepared from the epicotyls of nonhardened rye seedlings, which were subjected to fast and slow freezing on a microscope-adapted thermoelectric stage.

Protoplasts isolated from nonacclimated rye leaves, cooling to −10°C at a rate of 1°C/min results in extensive freeze-induced dehydration (osmotic contraction), and injury is manifested as the loss of osmotic responsiveness during warming. Under these conditions, several changes can be observed in the freeze-fracture morphology of the plasma membrane. These included lateral phase separations in the plasma membrane, aparticulate lamellae lying next to the plasma membrane, and regions of the plasma membrane and associated lamellae in various stages of lamellar-to-hexagonal II transition. Therefore, freeze-induced lamellar-to-hexagonal II phase transitions in the plasma membrane are a consequence of dehydration rather than subzero temperature per se (Gordon-Kamm and Steponkus 1984). When suspensions of protoplasts isolated from cold-acclimatized leaves were frozen to −10°C, no injury was incurred, and a hexagonal II phase transition was not observed.

Langis and Steponkus (1990) developed a procedure for the vitrification of mesophyll protoplasts isolated from leaves of nonacclimatized and cold-acclimatized winter rye seedlings (Puma). The procedure involves the following:

1. Equilibration of the protoplasts with an intermediate concentration of ethylene glycol at 20°C
2. Dehydration of the protoplasts in a concentrated vitrification solution made of 7 M ethylene glycol, 0.88 M sorbitol, and 6% bovine serum albumin (BSA) at 0°C
3. Placing the protoplasts into polypropylene straws and quenching in liquid nitrogen
4. Recovery of the protoplasts from liquid nitrogen and removal (unloading) of the vitrification solution

For nonacclimatized protoplasts, ~47% survival was obtained following recovery from liquid nitrogen if the protoplasts were first loaded with 1.75 M ethylene glycol prior to the dehydration step. However, to achieve this level of survival, nonacclimatized protoplasts had to be unloaded in a hypertonic (2.0 osm) sorbitol solution. If they were unloaded in an isotonic solution (0.53 osm), survival was ~3%. In contrast, survival of cold-acclimatized protoplasts following recovery from liquid nitrogen was ~34% when the protoplasts were loaded in a 2.0 M ethylene glycol solution and unloaded in an isotonic sorbitol solution (1.03 osm). If cold-acclimatized protoplasts were unloaded in a hypertonic sorbitol solution (1.5 osm), survival was ~51%.

The permeability of rye leaf protoplasts to glycerol was determined using 1,3-^{14}C glycerol and liquid scintillation spectrometry. The activation energy for glycerol permeability was 32.8 kJ/mol. The effect of electroporation on glycerol uptake was also explored. Treatments were performed with a field strength of 100 V/cm and an exponential decay constant of 5.8 ms. At 22°C, electroporation affected the rate and extent of glycerol permeation, causing an increase in the intercept of the glycerol uptake curve and a decrease in the slope (Pitt et al. 1997).

Over time, little success has been achieved using suspension and protoplast cultures of rye. Ma et al. (2003) developed a method for embryogenic callus induction and fertile plant regeneration from suspension cell-derived rye protoplasts, but only 7% of the embryogenic calli transferred to solid MS medium produced green shoots.

3.5.4   Anther Culture and DHs

3.5.4.1   Pollen Culture

Trinucleate pollen is usually difficult to germinate and grow *in vitro*. As such, rye pollen requires solid medium containing agar or agarose to obtain acceptable and consistent germination and pollen tube growth (Pfahler 1965, Heslop-Harrison 1979). The optimal medium contains 1% agarose or agar, 0.6 M sucrose, 10^{-3} M boric acid, and $1–2.5 \times 10^{-3}$ M calcium chloride or nitrate. However, after experiments with a number of additives, liquid media were found to give satisfactory germination and pollen tube growth. Germination of pollen begins in 60–70 s in this medium. Pollen tubes grow at ~10% of their normal *in vivo* rate. Tube lengths after half an hour are ~10% of final *in vivo* lengths. These media became an important factor for subsequent investigations into causes of incompatibility.

When anthers grow for 3 weeks on Blaydes medium with added indole-3-acetic acid (IAA), 6-benzylaminopurine(6-BAP), and gibberellic acid (GA) (each substance at 1 ppm), pollen grains of more than 20 forms can be obtained. Some of the pollen grains failed to develop typical vegetative cells and sperm nuclei. Differences between pollen grains result from various rates of nucleus divisions, disturbances of cytokinesis, polyploidization, and grade of differentiation of nuclei and cells. After a 4–5-week incubation, some anthers had multicellular embryoids (<60 cells).

Some experiments were carried out to diploidized rye pollen. Szakacs and Barnabas (2004) cultured spikelets. They were subjected to colchicine treatment at different stages of development and under differing *in vitro* conditions. Exposure to colchicine led to a drastic reduction in both the number of fertile pollen grains and the percentage seed setting, which was only observed in cultures inoculated in the early binuclear microspore stage. In the medium containing colchicine, the seed-setting percentage was 0.1%.

3.5.4.2   Doubled Haploids

Rye selection of parental lines is more effective in homozygotes than in heterozygotes, especially with respect to quantitative traits (see Table 3.6). Therefore, homozygous inbred lines are useful in testcrosses, particularly in hybrid rye breeding. The common production of inbred lines in allogamous rye is complicated by its self-incompatibility. At least, it requires five generations of controlled self-fertilization.

Remote hybridization techniques for haploid production, such as the *Hordeum bulbosum* and *Zea mays* pollination, do not really work in rye. Androgenesis for the haploid production seems to be the more applicable approach in rye. Following the culture of anthers or isolated microspores, growing embryos from immature

TABLE 3.6

Calculated Frequencies and Ratios of Completely Recessive Plants in F_2 Progeny of Doubled Haploids, Common Diploids, and Tetraploids

Number of Alleles	Frequency of Recessive Plants as Determined One of the Plants Given Below (1/×)		
	Doubled Haploid	Diploid	Tetraploid
n	2^n	2^{2n}	6^{2n}
1	2	4	36
2	4	16	1,296
3	8	64	46,656
4	16	256	1,679,616
5	32	1024	60,466,176

pollen and subsequent plantlet regeneration result in haploid plantlets. Either the formation of homozygous diploids occurs spontaneously, or it is induced by colchicine treatment. In triticale, androgenesis induction was associated with seven QTLs detected on chromosomes 5A, 4R, 5R, and 7R.

A DH is a diploid plant, which results from spontaneous or induced chromosome doubling of a haploid cell or plant, usually after anther or microspore culture by using different means. In rye, DH plants were produced using anther culture from populations, including breeders' strains and cultivars. DHs can also be achieved from F_1 plants. For breeding approaches, the DH technology is still difficult. Only 10%–40% of green regenerants produced via anther culture were suitable for research or breeding purposes because of low survival rate or low fertility (Tenhola-Roininen 2009). Spontaneously arising DH regenerants are more often fertile compared with the colchicine-treated ones. The fertility of spontaneous DHs varies from sterile to half of that found in a normal rye population, which has implications for the design of a crossing scheme and subsequent anther culture. Sometimes fertility can be low because of self-incompatibility factors. DHs for hybrid breeding and mapping populations were established.

Studies on anther culture and haploid production began in the 1970s (Wenzel and Thomas 1974, Thomas et al. 1975, Wenzel et al. 1977, Flehinghaus et al. 1991, Dainel 1993). Best results were first achieved with interspecific hybrid progenies of *S. cereale* and *S. vavilovii* (Flehinghaus-Roux et al. 1995). Deimling et al. (1994) completed these experiments with the successful regeneration of plantlets, derived from isolated microspores. Guo and Pulli (2000) and Ma et al. (2004a) were even able to increase the regeneration frequency of true rye microspore cultures. Immonen and Anntila (1996) as well as Rakoczy-Trojanowska et al. (1997) later described responsive genotypes of cereal rye. The so-called stress treatments such as cold or osmotic treatments contributed to the progress.

Embryo induction, plant regeneration, and albino-to-green plant ratio are independently inherited. Rye chromosomes 3RL and 5R are associated with embryo induction, whereas 1R, 4R, and 6RS are associated with regeneration rate

(Grosse et al. 1996, Schlegel et al. 2000). Nevertheless, rye haploid production *in vitro* is still less advanced compared to other crop species. Improved culture protocols led to 10%–36% of green regenerants produced via anther culture. Spontaneously arising DH regenerants are often more fertile compared with the colchicine-treated ones. The fertility of spontaneous DHs varied from sterile to half of that found in a normal rye population, which has implications for the design of a crossing scheme and subsequent anther culture. In the reciprocal crosses within one DH population, fertility was the lowest observed, probably because of self-incompatibility factors, whereas in the DH crosses with normal heterozygous cultivars, fertility was the highest (Tenhola-Roininen et al. 2006).

A direct induction of homozygous diploids was tried by spikelet culture. Szakacs and Barnabas (2004) subjected one rye and one barley variety to colchicine treatment in different stages of development and under differing *in vitro* conditions. Exposure to colchicine led to a drastic reduction in both the number of fertile pollen grains and the percentage seed setting, which was only observed in cultures inoculated in the early binuclear microspore stage. In the medium containing colchicine, the seed-setting percentage was 1.6% for barley and 0.1% for rye. Flow cytometry and root tip analysis revealed that all the progeny barley plants were diploid, while in the case of rye one was tetraploid, indicating that the egg cell may also be diploidized by colchicine treatment.

3.5.5 Somaclonal Variation

Somaclonal variation in rye was studied in populations derived from embryo culture. Induction media are usually supplemented with either 2,4-dichlorophenoxyacetic acid (2,4-D) or dicamba. In this way, several somaclones were produced. For instance, the sexual progenies of 40 R_0 regenerants—(1) somaclones derived in the medium with 2,4-D and (2) somaclones derived in the medium with dicamba—were analyzed according to the following traits: plant height, total number of tillers, number of productive tillers, spike length, number of spikelets per spike, spike compactness, number of normally developed grains per spike, weight of grains per spike, and weight of 1000-grain weight. The results for 22 R_1 plants surpassed the variability range for the control. The transmission of positive changes to the next generation was proved in the case of eight originally chosen R_1 plants: seven plants selected from the (1) somaclones and one plant from the (2) somaclones. Five out of the eight created somaclonal lines proved to be stable somaclonal variants. The absolute rate of the efficiency of positive somaclonal changes was calculated as ~0.6% (Rakoczy-Trojanowska 2002).

On random amplified polymorphic DNA (RAPD) analysis by Linacero et al. (2000), 40% of regenerants showed at least one variation. The number of mutations per plant was quite high, ranging from 1 up to 12. On some occasions (3% of the scored bands), the modified band was observed in only one plant or in several plants that originated from the same callus (variable band). In other cases (5%), the same band varied in several plants obtained from different calli. The authors called these as hypervariable bands. They could vary between plants belonging to different cultivars and/or with different origins, inflorescences, or embryos. Thus, they originated through independent mutational events. It was assumed that these bands represent hypervariable regions of the rye genome, that is, hot spots of DNA instability. Some of these bands proved to be unique sequences; others were present in a low copy number, while the remaining ones were moderately or highly repetitive.

In 2008, Puente et al. (2008) studied amplified fragment length polymorphisms (AFLPs) of 24 *in vitro* regenerated rye plants. Regenerants were obtained from cell lines derived from immature embryos and developed by somatic embryogenesis. Twenty-three regenerants showed variation when compared against sibling plants obtained from the same cell line. A total number of 887 AFLP markers were scored, and 8.8% identified the same polymorphism in plants obtained independently from different cell lines, revealing putative mutational hot spots. Genetic stability in the next generation was verified for only five of these polymorphisms, using controlled crossings and analysis of the corresponding progenies. The DNA sequence of marker A1-303 was identified as part of a tandemly repeated sequence, the 120-bp family, which is located at telomeric regions and is widely distributed among rye chromosomes. Another marker A5-375 showed high similarity with regions of *Angela* retrotransposons. In many species, there is a positive correlation between the copy number of retrotransposons and the genome size. Retrotransposons became excellent tools for detecting genetic diversity, as they are major generators of genomic changes. Moreover, sequences generated from retrotransposon-based molecular markers are often more polymorphic than those generated from RAPD, AFLP, or restriction fragment length polymorphism (RFLP) (see Chapter 5). By tissue culture, transposons can be activated. Two variable amplicons of rye were found as RYS1, a mobile element and foldback transposon (Alves et al. 2005). RYS1 is the first transposon described in rye and the first active plant foldback transposon as well. Preferential integration points in the rye genome exist, because the new insertions seem to be located in the same genome positions.

3.6   PREHARVEST SPROUTING

The germination of seeds on mature spikes (see Figure 2.9)—in wet years usually before harvest, or when harvest is delayed—is common in rye, wheat, barley, or oats. It can cause severe loss of quality and thus significant economic losses worldwide. The most widely used parameters for preharvest sprouting determination are Hagberg falling number and α-amylase activity, which are inversely correlated (Hagberg 1960, Perten 1964).

Seed dormancy of rye grains is partially associated with the embryo (embryo dormancy) and partially with the covering layers (coat dormancy). Rye coat dormancy is conferred by the endosperm including the testa/pericarp. Both determine the timing of grain germinability. Thus, sprouting-susceptible cultivars are those whose coat-imposed dormancy is terminated well before harvest maturity. Seed dormancy is lower in cultivated populations rather than in wilds (see Figure 2.9).

Dominant alleles promote expression of red pigments (phlobaphenes) within the testa/pericarp that tightly surrounds the embryo. Grain-color genes provide increased dormancy or are very closely linked to other dormancy-increasing loci. The role of the viviparous1 transcription factor (VP1) in controlling preharvest sprouting and abscisic acid sensitivity has been investigated. VP1 controls high-α-amylase promoter and alternative splicing. In addition, a protein kinase accumulates in mature embryos and is responsive to abscisic acid.

There are three classes of selection-responsive loci underlying preharvest sprouting in rye (Masojć et al. 2009): (1) Ten preharvest sprouting directional loci located on chromosomes 1RL (three), 3RS (two), 3RL (two), 5RL (two), and 7RS (one) responded significantly to both directions of the disruptive selection and were epistatic to the remaining two classes. (2) Nine preharvest sprouting resistance loci mapped on chromosomes 1RS (one), 1RL (one), 2RS (one), 3RS (one), 4RS (one), 5RS (one), 5RL (one), and 6RL (two) responded only to selection for sprouting resistance, being neutral for selection carried out in opposite direction. (3) Eight preharvest sprouting enhancing loci mapped on chromosomes 2RL (one), 3RL (one), 4RL (one), 5RL (one), 6RS (one), and 7RS (three) were affected by selection for sprouting susceptibility and did not respond to selection for sprouting resistance. Several QTLs for α-amylase activity as well as other QTLs involved in controlling sprouting resistance have been found in rye (Masojć et al. 1999; Masojć and Milczarski 2005; Twardowska et al. 2005; Masojć et al. 2007; Masojć and Milczarski 2007, 2009). One major QTL was found, which explained 16.1% of phenotypic variation. The QTL was localized on the long arm of chromosome 5R. Microsatellites SCM74, RMS1115, and SCM77, nearest to the QTL, were proposed by Tenhola-Roininen et al. (2011) for marker-assisted selection to decrease sprouting damage (see Chapter 5).

Masojć and Kosmala (2012) found qualitative and quantitative differences between two-dimensional electrophoretic spectra of 546 proteins from two bulked samples of mature rye grain representing (1) 20 recombinant inbred lines extremely resistant to preharvest sprouting and (2) 20 recombinant inbred lines extremely susceptible to preharvest sprouting. Mass spectrometry of resolved proteins showed that four spots specific for preharvest sprouting susceptibility represented high molecular weight glutenin subunit, glutathione transferase, 16.9 kDa heat shock protein, and monomeric α-amylase inhibitor. Two spots specific for preharvest sprouting resistance contained cytosolic malate dehydrogenase and functionally unrecognized protein with sequence homology to rubber elongation factor protein. Majority of 14 proteins with at least twofold higher accumulation level in preharvest sprouting-susceptible lines relative to that found in sprouting-resistant lines showed sequence homology to proteins involved in defense mechanisms against biotic and abiotic stresses including oxidative stress, and those that took part in energy supply. Two spots were identified as regulatory proteins from the 14-3-3 family with one molecular form prevailing in sprouting-susceptible lines and another form highly accumulated in sprouting-resistant lines.

3.7 VERNALIZATION

The treatment of germinating seeds with low temperatures to induce flowering at a particular preferred time is called vernalization or jarowization. In thale cress (*Arabidopsis thaliana*), a gene, *Flc*, suppresses the formation of flowers during cold periods. Another gene, *Vrn2*, triggers that suppression. However, in spring time, the suppression is raised in such a way that the gene "remembers" the previous cold period, because the gene *Vrn2* itself is cold insensitive. There is a similar gene in *Drosophila melanogaster*, which also serves as "chemical memory."

Different plants have different cycles of vegetative growth, flowering, and ripening. However, some biennials such as rye produce vegetative structures in one season and induce flowering in the following season, only after they are exposed to prolonged winter or cold treatment. In fact, farmers sometimes used to cultivate this winter rye by subjecting the water-imbibed grains to cold treatment and growing them in spring and harvesting the same in summer.

For normal growth and development, every plant requires optimum temperature. Nevertheless, for vernalization, the optimum temperature required is 3°C to 17°C, which varies depending upon the species involved. Even the duration of treatment varies from species to species. Individual requirements have to be determined independently by testing. In rye, var. Petkus, the most effective range of temperature is 3°C to 7°C. The efficiency of cold treatment in bringing about vernalization is determined by the number of days shortened between germination and flowering stage.

Among many plants, rye is a short-day plant. It was the first to be used for vernalization experiments (see Table 8.4). Vernalization through cold treatment is very effective at the seed stage or seedling stage. Even the embryos can be successfully vernalized. However, in many cold recurring species, vernalization is not effective until and unless the plant possesses at least a few leaves. The requirement of a few leaves for effective vernalization is called "ripeness to flowering." This suggests that plants need a certain degree of photosynthesis to respond to cold treatment. Additional support for this view comes from an experiment on the embryos of rye. The embryos were separated from the endosperm, and then if the embryos alone are subjected to cold treatment, they fail to be vernalized. However, if such embryos are subjected to cold treatment along with carbohydrates such as sucrose solution or endospermous tissue, vernalization is very effective. Probably the role of carbohydrates in vernalization is to supply some energy. Nevertheless, the most sensitive site, which acts as the perceptive organ, is the meristematic region of the shoot apex. Even leaves, which act as the sites, should have a certain amount of meristematic tissues in them.

Along with the cold treatment, plants also require water and oxygen for effective vernalization. The seeds or embryos should possess at least 40%–50% water in their cells, without which cold treatment has no effect. Similarly, oxygen is very essential for biological oxidation. The essentiality of carbohydrates for effective vernalization supports the view of requirement of oxygen.

Various experiments in the past have revealed that during cold treatment, either the meristematic cells found in stem apex or leaves are stimulated to produce some substance. The presence of such substance has been demonstrated by grafting a vernalized plant to another nonvernalized plant at normal temperatures. The above experiments clearly demonstrated that some substance is synthesized and such substance is now called vernalin that is capable of diffusion. Attempts to isolate and identify the components of vernalin have failed. Whether the vernalin is the same as florigen or a precursor of florigen is not known. Florigen is a universal hormone that supposedly causes plants to change from the vegetative to the reproductive state. In 2005, after 70 years of hypothesis, Huang et al. (2005) identified mRNA of the *Ft* gene of thale cress that is produced in leaves and induces flowering when transported to the apex tissue.

If vernalized seedlings or seeds of rye are subjected to higher temperature such as 35°C to 40°C, the plants that develop from such treatment fail to flower. Such a nullifying effect by higher temperatures is called devernalization. Nevertheless, if the vernalized plants are maintained at sufficiently low temperatures for a long period of time, which has to be determined for every species, devernalization is not possible. This may be because the putative vernalin would have already acted upon the genetic material and committed to its flower formation. However, devernalized plants can be revernalized by subjecting the same seedling or seed again for another period of cold treatment by repetition of vernalization and devernalization cycles. Prolonged vernalization decreases the viability of seedlings and flower formation.

Rye plants that require cold treatment also require proper photoperiodic treatment for the induction of flowers, without which vernalization does not have any effect. If such plants are treated with gibberellins, they produce flowers without subjecting the plants to cold and photoperiodic treatments. It means that gibberellins substitute not only vernalization but also photoperiodic treatment. However, in rye, these gibberellins were ineffective.

Recently, it can be stated that cold treatment induces the expression of CO, which in turn induces the Ft protein, which with the Fd protein induces the expression of floral meristem identity genes. These on expression act on floral part-inducing genes. Even if rye is not vernalized or grown under normal temperature and proper photoperiodic treatment is given to it, it still flowers but takes a long time. In contrast, if it is subjected to vernalization and then subjected to a long photoperiod, it produces flowers in a shorter time. This means that the synthesis of hypothetical A to B substances takes place all the time, but it is accelerated under cold treatment. Once substance B accumulates in sufficient amounts, B is converted to C and then to D, which is actually under the control of long-day conditions. Under short-day or day-neutral conditions, C is converted back to B and then to E that keeps the plant in vegetative conditions.

Actually, rye contains the Flc protein bound to specific loci that represses genes required for flowering. During cold treatment, the Flc dissociates from the loci and allows other factors to bind and activate gene expression. This makes the plants flower in response to vernalization. During cold treatment, the concentration of *Flc*, the repressor, goes down with time, which means that its binding to chromatin is reduced with time under cold treatment. The gene *Flc* acts to block flowering and *Fri*, in turn, acts to upregulate *Flc* levels. The autonomous pathway genes promote flowering by repressing *Flc* expression, that is, vernalization promotes flowering by causing an epigenetic shutoff of the *Flc* gene.

After Munoz-Florez et al. (2009), there are three vernalization genes in rye, *Vrn1* syn. *Sp1* [on chromosome 5RL, linked to the *ct2* locus by 11.7 cM, *Xpsr426* marker by 7 cM, *Est2* locus by 5 cM, and *Xwg644* marker by 0 cM; cf. Schlegel and Korzun (2013)], *Vrn2*, and *Vrt2*, available in the varieties Puma (winter type; *Vrn1*), Sangaste (winter type; *Vrn1*), or Gazelle (spring type; *ScVrn1*). Under long-day conditions, *Vrn2* is downregulated after 2 days of vernalization, but gene transcripts in Puma and Sangaste begin to accumulate only after 14 days of cold acclimation. In contrast, *Vrn2* gene expression is not detected under short-day conditions at any of the sampling times, but *Vrn1* transcripts begin to accumulate after 35 or 42 days of cold acclimation. The results suggest that *Vrn2* allele expression is influenced by photoperiod with the result that the short day delays the transition from vegetative to reproductive growth.

4 Cytology

4.1 GENOME STRUCTURE

In the late 1970s and 1980s, the cloning and sequencing of DNA segments of the rye genome were initiated in several countries around the world. This activity focused on repetitive DNA sequences because they were more accessible for establishing the technologies required for using these sequences as molecular markers in genetic mapping and characterizing breeding lines. Early studies in Australia showed the value of the repetitive DNA sequences for characterizing chromosome arms through visualizing their distribution by *in situ* hybridization to root tip squashes (Appels et al. 1978). By excluding the chromosomes that contain C^+/N^+ heterochromatin (Schlegel and Gill 1984) from the correlation analysis, the correlation between chromosome size and DNA content per chromosome is $r = +0.87$. Only 10%–20% of the genome can be assigned, biochemically, to the major part of the genome, which belongs to the repeated sequence category. The kinetics of genome organization has revealed that repeated sequences are generally interspersed among unrepeated sequences. The discovery of a very rapidly reannealing class of DNA constitutes 4%–10% of the genome, and although it is believed to be composed largely of sequences capable of renaturation, this class contains long tandem arrays of simple, repeated sequences (Appels 1982). Ranjekar et al. (1974) were the first to demonstrate several buoyant-density components in a fraction of DNA renaturing with a density of 0–0.1 (10%–12% of the genome). However, the predominant component is a well-defined species at 1.7 g/cc in a CsCl gradient. Smith and Flavell (1977) considered this class of DNA to consist mainly of palindromic sequences, which are distributed in clusters throughout at least 30% of the genome. DNA, with a mean fragment length of 500 bp, was fractionated to allow recovery of very rapidly renaturing fraction ($C_o t$ 0–0.2). This DNA was shown to contain several families of highly repeated sequence DNA. Two of them were purified which resulted in a fraction renaturing to a density of 1.7 g/cc comprising 0.1% of the total genome. The second fraction, the polypyrimidine tract of DNA, also comprised 0.1% of the genome. Other hybridization studies between cereals have shown that 22%–24% of the DNA is rye-specific repeated sequence (Rimpau et al. 1978).

The 1C DNA content of the rye autosomic genome with seven chromosome pairs ranges from 7.8 pg (Bashir et al. 1993) to 8.3 pg (Bennett 1976) to 9.5 pg (Flavell et al. 1974, Bennett and Smith 1976). Among the cultivated cereals, the DNA 1C-value of rye is one of the highest, ~34% greater than the largest haploid genome of *Aegilops* species and cultivated wheat. *Secale cereale* has ~33% more DNA than the largest of the diploid wheat genomes. Estimates of the 4C DNA content per nucleus in several *Secale* species range from 28.85 pg in *S. silvestre* to 33.14 pg in *S. cereale*, and 34.58 pg in *S. vavilovii*, respectively. It seems that during evolution a significant increase in the amount of nuclear DNA occurred in the genus. Variation in the size of

TABLE 4.1
DNA Amount and Percentage of Telomeric Heterochromatin in Some Species of the Genus *Secale* L.

Species	4C DNA Content (pg)	Telomeric Heterochromatin (%)
Secale cereale	34.14	12.2
S. vavilovii	34.58	7.5
S. montanum	33.49	8.9
S. africanum	29.68	6.1
S. silvestre	28.85	6.1

Source: After Bennett, M. D. et al., *Chromosoma*, 61, 149–176, 1977. With permission.

TABLE 4.2
DNA Content of Individual Rye Chromosomes

Chromosome	DNA Content (pg)
1R	0.9
2R	1.1
3R	0.8
4R	1.3
5R	1.0
6R	1.3
7R	1.4

Source: Bashir, A. et al., *Cytometry*, 4, 843–847, 1993. With permission.

rye chromosomes can mainly be attributed to different contents of heterochromatin located mainly at the telomeres (see Table 4.1). The 1C DNA content of individual rye chromosomes ranges from ~1.2 to 1.4 pg, considering that chromosome length and DNA content are directly proportional. The short arm of rye chromosome 1R (~400 Mbp) represents only 5% (see Table 4.2). However, flow cytometric studies showed a somewhat different pattern (Bashir et al. 1993).

The big amount of repetitive DNA could also have regulating functions. Data that are more detailed provided Bartos et al. (2008) with 7917 Mbp per 1C content (see Figure 4.1). In relation to some of the other small grains, the rye genome is estimated to be larger than barley (4.9 Gbp) but smaller than oat (11.3 Gbp) and wheat (16.5 Gbp). Repetitive sequences comprise 84% of the genome (see Table 4.3) and seem to be inert. Data from a recent human genetic project "ENCODE" revealed that parts of the genome do not code for proteins, which however represent an enormous regulation system. It contains millions of feedback units regulating the activity of genes. Disturbances of this noncoding and regulating system may cause numerous problems for the organism. Why should rye be an exception?

For the chromosome arm 1RS, a detailed study revealed at least 3121 gene loci and at least 1882 different gene functions on 1RS (see Figure 4.2). Excluding the

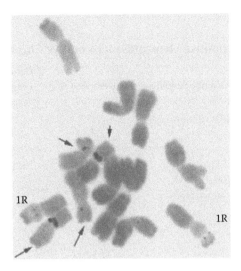

FIGURE 4.1 FISH localization of COPI repeat on mitotic metaphase chromosomes [COPI, coat protein that is an ADP ribosylation factor (ARF)-dependent adaptor protein involved in membrane traffic]; COPI repeat (originally red) is localized on three chromosome pairs of rye (arrows) including the short arm of chromosome; chromosomes were counterstained with 4′,6-diamidino-2-phenylindole (DAPI) (originally blue). (Courtesy of Dr. Dolezel, J., Olomouc, Czech Republic.)

ribosomal and secalin genes, 3.36% of all sequence reads (30,118 reads) identified the gene space of 1RS. Approximately three-quarters of the 1RS genome consist of repeated DNA. The quantity of sequence information resulted in 0.436× sequence coverage of the 1RS chromosome arm, permitting the identification of genes with estimated probability of 95%. More than 5% of the 1RS sequence consisted of gene space, identifying at least 3121 gene loci representing 1882 different gene functions. Repetitive elements comprised ~72% of the 1RS sequence, gypsy/sabrina (13.3%) being the most abundant (Kubalakova et al. 2003, Simkova et al. 2008). Six transposon classes dominate the 1RS repeat landscape. A ty3/gypsy retrotransposon is even specific for rye centromeres. The latter contain a 365-bp fragment consisting of a reverse transcriptase, RNaseH, and an integrase domain (Gonzalez-Garcia et al. 2012). The repetitive DNA probe pMD-CEN-3 labels also centromere-specific.

Revolver is a new class of transposon-like gene composing the Triticeae genome. An 89-bp segment of Revolver which is enriched in the genome of rye was isolated by deleting the DNA sequences common to rye and wheat. The entire structure of Revolver was determined by using rye genomic clones, which were screened by the 89-bp probe. The Revolver sequence consists of 2929–3041 bp with an inverted repeated sequence on each end and is dispersed through all seven chromosomes of the rye genome. Revolver is transcriptionally active, and the isolated full-length cDNA (726 bp) reveals that it harbors a single gene consisting of three exons (342, 88, and 296 bp) and two introns (750 and 1237 bp), and encodes 139 amino acid residues of protein, which shows similarity to some transcriptional regulators. Revolver variants ranging from 2665 to 4269 bp, in which 5′ regions were destructed, indicate

TABLE 4.3
Representation of Repetitive Element Groups on Rye Chromosome 1RS

Type	Genome Specificity	Described As	Cumulative Length (bp)	Fraction of 1RS (%)
Class I elements			1,306,781	64.3
	LTR retrotransposons		1,275,443	62.8
		Gypsy	989,195	48.7
		Copia	281,937	13.9
		TRIM	281,937	13.9
	Non-LTR retrotransposons		31,338	1.5
		LINE	23,400	1.2
		SINE	1,877	0.1
Class II elements			100,854	5.0
		CACTA	90,323	4.4
		Mutator	1,394	0.1
		MITE	4,681	0.2
		LITE	2,753	0.1
Unclassified elements			30,340	1.5
Other known repeats			106,376	5.2
	Ribosomal genes		94,665	4.7
	Simple repeats		5,049	0.2
	Tandem repeats		6,662	0.3
Known repeats			1,536,587	75.6
Unknown repeats			178,027	8.8
Total			**1,712,364**	**84.2**

Source: Bartos, J. et al., *BMC Plant Biol.*, 8, 95–102, 2008. With permission.
Note: LTR, long terminal repeat.

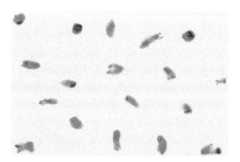

FIGURE 4.2 Flow-sorted chromosomes 1RS after double FISH with a fluorescein isothiocyanate (FITC)-labeled probe for 45S rDNA. The probe highlights the nucleolus organizer region (NOR) in originally yellow-green color at secondary constriction of the short arm (pale band); the Cy3-labeled probe shows the rye-specific repeat pSc200 [originally red color of the telomere (dark band)]; the chromosomes were counterstained with 4′,6-diamidino-2-phenyl-indole (DAPI) (originally blue color). (Courtesy of Dr. Dolezel, J., Olomouc, Czech Republic.)

structural diversities around the first exon. Revolver does not share identity with any known class I or II autonomous transposable elements of any living species (Tomita et al. 2008).

More than 4000 single-sequence repeat (SSR) sites mostly located in gene-related sequence reads were identified; the existence of chloroplast insertions in 1RS has been verified by identifying chimeric chloroplast-genomic sequence reads. Synteny analysis of 1RS to the full genomes of *Oryza sativa* and *Brachypodium distachyon* revealed that about half of the genes of 1RS correspond to the distal end of the short arm of rice chromosome 5 and the proximal region of the long arm of *B. distachyon* chromosome 2. Comparison of the gene content of 1RS to 1HS barley chromosome arm revealed high conservation of genes related to chromosome 5 of rice (Fluch et al. 2012).

The large 8-Gbp diploid genome of diploid rye remains a major challenge for genome analysis due to its size and complexity. The rye genome has undergone a series of rearrangements compared to other cereals, such as the wheat and barley genomes. Since 2011, a strategy for genome analysis was developed which involves chromosome sorting, genome fractionation, development of a dense marker map, next-generation sequencing, and syntenic integration (GenomeZipper). The dense marker scaffold allows delineating syntenic segments and chromosomal breaks with high resolution and positioning 22-k rye genes on the genome. This approach allows analysis and comparison of the rye genome with unprecedented depth, and generates fundamental insights into the evolution of the rye genome and the genomic consequences of an outbreeding lifestyle.

Until now, a virtual gene order for each of the seven individual rye chromosomes was obtained by using onefold 454 sequencing coverage of sorted rye chromosomes (Martis et al. 2013). As a result, a linear gene order map of 22,426 nonredundant gene loci of the rye genome has been obtained. Compared to barley, the rye genome exhibits multiple chromosomal rearrangements, and an evolutionary model involving five consecutive translocations can be proposed for the rye genome. Comparison of the degree of syntenic conservation, the sequence homology, and the synonymous substitution rate against related grass genomes demonstrates pronounced distortions among different segments. In line with phylogenetic observations, it is hypothesized that the modern genome of *S. cereale* is a mosaic of segments from several ancestral genomes rather than the result of a series of intragenomic rearrangements. It seems that the observed genomic mosaic involved a series of polyploidization and/or introgression events that might have been eased by the allogamous flowering of rye.

4.2 CHROMOSOME NUMBER

The chromosome number of rye (*S. cereale*) is $2n = 2x = 14$. The genome formula of diploid rye is given as $R^{cer}R^{cer}$. Haploids are occasionally observed among twin seedlings. Natural polyploids are not known. Induced autotetraploids were produced for breeding purposes (see Section 7.2.1). However, karyotypes with so-called B chromosomes are reported (see Section 4.5.2). The subtelomeric regions of standard A chromosomes are often cytologically heterochromatic. The ends of rye chromosomes are visibly condensed throughout most stages of the cell cycle (see Section 4.3).

In an embryological study, Bennett (1974) demonstrated that the presence of rye chromosomes in antipodal cells could be detected with the help of at least eight to nine heteropycnotic bodies found in addition to seven densely stained centromeric regions. They are grouped around one cell pole. These bodies are rather spherical or subspherical and Feulgen-positive. The number of rye bodies exceeded the number of seven rye chromosomes, because each rye chromosome has a single body except chromosome 1R, which shows two—one located at the end of each chromosome arm. In chromosomes 2R, 4R, 5R, 6R, and 7R, the body is found on the short arm, while in chromosome 3R it is found on the long arm. It was suggested that the bodies represented the telomeric heterochromatin, and each of the body constituted 4% of the total haploid genome.

Various genomic analyses in rice, sorghum, and maize have suggested an ancestral whole-genome duplication predating the divergence of the different cereal genomes ~77 million years ago (see Table 2.1), but the structure and basic number of chromosomes of the ancestral genome have remained uncertain. Recently, a model was proposed for the evolution of the cereal genomes from an ancestor with a basic number of 12 chromosomes whose structure is similar to the rice genome. Salse et al. (2008) compared 42,654 rice gene sequences to 6,426 mapped wheat expressed sequence tags (ESTs) using improved sequence alignment criteria and statistical analysis. This allowed the identification of duplications covering ~70% of both the rice and wheat genomes. The authors then conducted a detailed analysis of the length, composition, and divergence time of these duplications and made comparisons with sorghum and maize. This led to a model of grass genome evolution from a common ancestor with a basic number of five chromosomes. The authors suggest that the common ancestor with $x = 5$ chromosomes underwent a whole-genome duplication that resulted in an $x = 10$ intermediate, followed by two interchromosomal translocations and fusions that led to the construction of two new chromosomes, resulting in the $x = 12$ intermediate ancestor that has been described previously. As in the model, the proposed rice has retained this intermediate ancestral chromosome number, whereas it has been reduced in wheat and its relatives, maize, and sorghum genomes, following a pattern that may be typical of plant chromosome evolution in general.

Gene-based single-nucleotide polymorphisms for construction of a high-resolution chromosome map in *Aegilops tauschii* (Luo et al. 2009) and comparison of the map with the rice, sorghum, and *B. distachyon* genome sequences showed that the basic chromosome number x was reduced in the Triticeae lineage from $x = 12$ to $x = 7$. Insertions of entire chromosomes by their termini were translocated into the centromeric regions of other chromosomes. The same process of dysploid reduction was independently responsible for reductions of basic chromosome number from $x = 12$ to $x = 5$ in *B. distachyon*, to $x = 10$ in sorghum, and to $x = 9$ in finger millet. Of the 17 dysploid events in the Panicoideae, Chloridoideae, and Pooideae subfamilies, all took place via this mechanism, with a modification in three cases, which indicated that this is the dominant mechanism of dysploidy in the grass family, including rye. In the *B. distachyon* genome, all insertion points could be located on the genome sequence—in some chromosomes into regions less than 100 kb. The distribution of genes, and in some instances of repeated nucleotide sequences around the insertion points, is consistent with telomeric and pericentromeric heterochromatin participating in chromosome insertions.

4.3 KARYOTYPE AND HOMEOLOGY

4.3.1 KARYOTYPE

Until now, several attempts have been made to establish a common chromosome designation (see Table 4.4), a general karyological characterization, and a standard karyotype in diploid rye. All efforts, however, did not completely account for the natural variation of chromosome morphology and structural heterozygosity in allogamous rye populations (Heneen 1962). To overcome the difficulties at least partially, the participants of the first and second workshops on *Rye Chromosome Nomenclature and Homoeology Relationships* decided to consider the Imperial rye additions to the hexaploid wheat variety Chinese Spring as the standard rye chromosome set, although these chromosomes are not completely identical with those of the population of Imperial variety (Sybenga 1983).

Based on the agreements, a preliminary karyotype and its C-banding patterns, a homeologous designation of added rye chromosomes and a generalized C-banding pattern were proposed. The participants anticipated producing a standard diploid rye homozygous for genetic, biochemical, and/or molecular studies, which should, in addition, be available for cytogenetic testcrosses. Therefore, Schlegel et al. (1987) established from the spring rye variety Petka, a haploid, a dihaploid, and, finally, a tetrahaploid plant from the original haploid genome. The variety was chosen because of its spring growth habit, dominantly greenish grains, and traits of modern cultivars Petka. It was bred in Petkus (Germany) and released in 1961. Its genetic background is related to the gene pool of

TABLE 4.4
Comparison of Different Proposals for Designation of Rye Chromosomes

Author (year)	Chromosome						
Lewitzky (1931)	IV	V	I	III	VI	VII	II
Pathak (1940)	A	B	F	G	D	E	C
Tijo and Levan (1950)	I	II	V	III	IV	VI	VII
Oinuma (1953)	b	g	c	a	e	d	f
Lima de Faria (1952)	II	IV	III	I	VI	VII	V
Bhattacharyya and Jenkins (1960)	I	III	II	V	IV	VI	VII
Heneen (1962)	1	2	3	4	5	6	7
Vosa (1974)	I	IV	II	V	VI	III	VII
Verma and Rees (1974)	2	3	5	1	4	6	7
de Vries and Sybenga (1976)	II	III	I	IV	VI	V	VII
Schlegel and Mettin (1982)	2R	7R	3R	4R	5R	6R	1R

Source: Compiled from different sources.

FIGURE 4.3 Commemorative chalkboard of F. von Lochow near the entrance of the former Institute of Cereal Breeding, Petkus, Germany.

Petkus rye, which has been used worldwide in breeding and research. F. von Lochow (see Figure 4.3) at Petkus, south of Berlin, Germany, bred Petkus rye since 1882. The material was derived from the landrace Probsteier Roggen, and marketed by *Probsteier land- und forstwirtschaftlicher Verein* (see Section 7.4), Schönberg, near Kiel, Germany.

The data demonstrate remarkable differences in mean total length as well as relative arm length, which range from 124.24 units in chromosome 1R to 162.46 units in 2R, and from 45.03 units in chromosome arm 5RS to 93.25 units in 5RL, respectively. The arm ratios vary from 1.02 in chromosome 3R to 2.07 in 5R. The small sample standard deviations mean that most of the length differences or arm index variations are statistically significant (see Table 4.5). A further survey given in Table 4.6 shows that related chromosomes of other species may strongly deviate from the standard measures. Length variation and variation of arm index are clearly seen for the satellited chromosomes.

The karyogram drawn from the data has been used for detailed description using the C-banding pattern (see Figure 4.4). Prominent blocks of telomeric heterochromatin are stained, which is a common pattern of diploid rye, *S. cereale* L. (see Figure 4.5). Rye chromosomes 2R, 3R, and 6R show also prominent N bands near the centromeres (Schlegel and Gill 1984). Moreover, there is quite a good correlation between the distribution of heavy knobs of chromomeres described by Lima de Faria (1952) and C and/or N bands. Recently, tetrad-fluorescence *in situ* hybridization (FISH) analysis and linkage maps based on restriction fragment length polymorphism (RFLP) markers clearly indicated that heterochromatin strongly suppresses recombination of completely chromosomal regions (Kagawa et al. 2002).

The centromeres are inserted proximal to subterminal (5R, 6R). It has been known for decades that centromere size varies across species. In grasses, centromere

TABLE 4.5

Variation of Chromosome Length in Haploid Rye Rosta-7

Characteristics	Chromosome						
	1R	2R	3R	4R	5R	6R	7R
Satellite	15.24						
SE	*1.40*						
Short arm	39.32	75.44	63.73	65.25	45.03	50.51	58.78
SE	*0.40*	*0.63*	*0.65*	*1.26*	*1.30*	*0.99*	*0.76*
Long arm	69.68	87.02	65.63	88.21	93.25	89.67	71.54
SE	*1.26*	*1.15*	*1.26*	*4.70*	*1.18*	*0.82*	*4.10*
Total length	**124.24**	**162.46**	**129.36**	**153.47**	**138.29**	**140.19**	**130.33**
SE	*1.35*	*1.60*	*1.82*	*5.99*	*1.47*	*1.48*	*5.51*
Arm ratio	1.27[a]	1.15	1.03	1.35	2.07	1.78	1.22
SE	*0.06*	*0.02*	*0.02*	*0.03*	*0.08*	*0.04*	*0.01*

Source: Schlegel, R. et al., *Theor. Appl. Genet.*, 74, 820–826, 1987. With permission.

Note: Confidence limits at $p = 1\%$ of probability for the total chromosome length (6.30), the arm length (3.64), and the arm ratio (0.06).

[a] Satellite included in the short arm.

sizes can vary 40-fold in genome size. There is a clear linear relationship between total centromere size and genome size. Bennett and Rees (1967) investigated whether changes in chromosome size are accompanied by changes in genome size. They showed that in rye there was no change in genome size despite changes in chromosome volume of 50% depending on the phosphate level.

Species with large genomes and few chromosomes tend to have the largest centromeres (e.g., rye), while species with small genomes and many chromosomes have the smallest centromeres (e.g., rice). However, within a species, centromere size is surprisingly uniform. In general, the most abundant tandem repeat is the centromere DNA as a general property of genomes. High-copy centromere tandem repeats are highly variable in sequence composition and length. While centromere position seems to be epigenetically determined, tandem repeats are highly prevalent at centromeres. A functional role for such repeats can be assumed, perhaps in promoting concerted evolution of centromere DNA across chromosomes.

The band positions and band sizes are related to the relative arm length, so there are sufficient references for identifying each of the seven chromosomes individually. Applying the standard chromosome band nomenclature taken from Schlegel et al. (1986b), a specific reference karyogram was established excluding structural heterozygosity of the genome (see Figure 4.5). This standard karyotype and fully homozygous genotypes of the dihaploid and tetrahaploid progenies were proposed as general reference material in genetic and cytogenetic studies.

In recent times, the progress of molecular techniques contributes very much to chromosome classification. Regarding the abundance of repetitive sequences in rye genome (Smith and Flavell 1977), SSRs have been used first in taxonomic studies

TABLE 4.6
Mean Chromosome Length (Relative) and Chromosome Arm Ratios (Long/Short) in Diploid Rye Species (*Secale cereale, S. ancestrale, S. vavilovii, S. montanum, S. kuprijanovii, S. africanum,* and *S. silvestre*)

Chromosome Characteristics	Chromosome (Longest to Shortest)						
	2R	4R	6R	5R	7R	3R	1R
Secale cereale (standard; Schlegel et al. 1987)							
Total length	162.46	153.47	140.19	138.29	130.33	129.36	124.24
Arm ratio	1.1	1.3	1.8	2.0	1.2	1.0	1.3[a]
Secale cereale (Danae population)							
Total length	160.38	154.55	150.17	138.51	131.22	125.39	118.10
Arm ratio	1.1	1.0	1.8[a]	1.1	1.6	1.5	1.8
Secale ancestrale							
Total length	152.17	147.95	146.54	143.72	135.26	129.63	122.58
Arm ratio	1.6	1.6	1.6[a]	1.7	1.7	1.3	1.4
Secale vavilovii							
Total length	154.58	153.22	142.27	139.54	135.43	131.33	121.75
Arm ratio	1.1	1.0	1.7[a]	1.4	1.4	1.5	1.0
Secale montanum							
Total length	156.84	154.21	146.30	143.66	139.71	133.12	104.12
Arm ratio	1.1	1.7	1.4	1.4[a]	1.0	1.0	1.0
Secale kuprijanovii							
Total length	156.40	149.60	144.16	140.08	137.36	133.28	116.96
Arm ratio	1.4	1.4[a]	1.9	1.2	1.5	1.8	1.4
Secale africanum							
Total length	155.18	151.37	143.74	141.19	137.38	133.56	115.75
Arm ratio	1.4	1.1	1.4[a]	1.1	1.3	1.0	1.5
Secale silvestre							
Total length	159.18	151.09	137.60	141.65	136.25	129.50	122.76
Arm ratio	1.6	1.1	1.5[a]	1.5	1.7	1.4	1.6

Source: Taken from five mitotic metaphase spreads per species as compared to the standard karyotype according to Schlegel, R. et al., *Theor. Appl. Genet.* 74, 820–826, 1987. With permission.

[a] Satellite chromosome, satellite included in the short arm.

of *Secale*. Later it was demonstrated that the patterns of microsatellite sequence distribution enable the identification of individual rye chromosomes (Hackauf and Wehling 2003). Another microsatellite sequence-based marker system used in rye is inter-SSR (ISSR), also named SCIM (*S. cereale* intermicrosatellite; Camacho et al. 2005). ISSR markers are generated from nucleotide sequences located between two microsatellite priming sites inversely oriented on opposite sides. Primers used in ISSR analysis can be based on any of the SSR motifs (di-, tri-, tetra-, or pentanucleotides) anchored at 3′ or 5′ end by—two to four arbitrary nucleotides. Extensive diversity of distribution and copy number variation are the features that

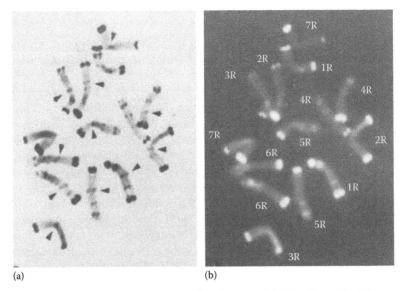

(a)   (b)

FIGURE 4.4 Somatic spread of diploid rye after (a) sequential C-banding and (b) fluorescence 4′,6-diamidino-2-phenylindole (DAPI) staining.

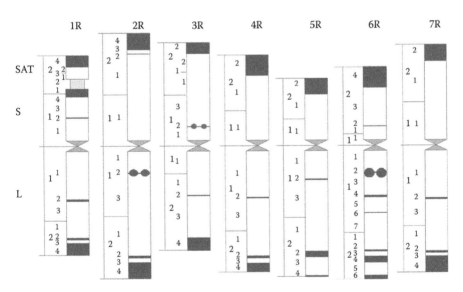

FIGURE 4.5 Standard karyogram of diploid cereal rye, established according to chromosomal measurements as well as C- and N-band distribution.

characterize both microsatellites and retrotransposons. These types of sequences are prone to rearrangements, which make them useful in genetic divergence and phylogenetic studies. In case of *Secale* genus, the analysis of diversified pattern of distribution of these sequences can contribute significantly to clarification of phylogenetic relationships between taxa (see Section 2.2).

4.3.2 HOMEOLOGY

It is supposed that cereals within the subtribe Triticeae have a common origin and, likewise, a partial structural homology (see Table 2.1). It was first concluded by more or less good ability of chromosomes to substitute each other in interspecific hybrids (Gupta 1971, Koller and Zeller 1976, Schlegel 1990). Later, it was supported by intergeneric hybridization, particularly with *Triticum* species, that rye chromosomes can even show chiasmatic pairing with wheat chromosomes (Schlegel and Weryszko 1979). The comparatively high degree of *Triticum–Secale* chromosome pairing and, thus, recombination became of interest for plant breeding utilizing the incorporation of useful characteristics of rye species for wheat and triticale improvement. The frequency of wheat–rye pairing ranged from 0% in crosses with *S. silvestre* to 2.4% per pollen mother cell (PMC) in *S. montanum*, respectively.

After comparative studies of Moore et al. (1995), it was suggested that cereal chromosomes have evolved from a single ancestral chromosome. Rice, wheat, barley, and rye share orthologous regions among their genomes. This has been demonstrated by numerous studies of comparative mapping of EST–SSR markers in barley, wheat, rye, and rice (Khlestkina et al. 2004). For example, during the evolution of rye and wheat chromosomes, the common ancestral Triticeae chromosomes were reorganized in such a way that different chromosomes contained different segments of the ancestral chromosomes (see Figure 4.6).

Molecular maps show remarkable conservation of gene order among wheat, rye, barley, millet, rice, and so on disrupted only by a few gross interchromosomal

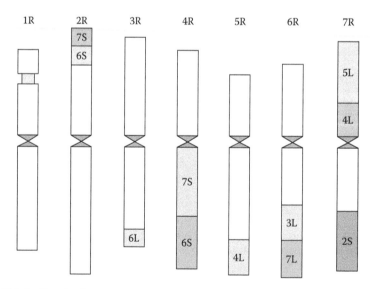

FIGURE 4.6 Standard karyogram established according to the karyological rearrangements during the cereal grass evolution revealed by molecular markers. (After Devos, K. M. et al., *Theor. Appl. Genet.* 85, 673–680, 1993.)

rearrangements and the emergence of genome-specific noncoding sequences, particularly at the physical ends of chromosomes. Rye has diverged from wheat by at least 7–13 translocation events (see Table 2.1; Figure 4.5) after only 6 million years of divergence, while, for example, barley appears to reflect precise synteny with the basic wheat genome (Devos et al. 1993). Other authors provide evidence that some genomes fix rearrangements more readily than others do. These different rates of species divergence through chromosomal rearrangement do not correlate with the breeding system, because high levels of evolutionary translocations are found both in rye, an outbreeder, and in, for example, *Ae. umbellulata*, a predominantly self-pollinated species.

The latest knowledge of homeologous relationships among cereal genomes benefits genetics and breeding. For example, when the distal ends of chromosome 5 of rye and wheat evolved from the interchromosomal interchange of Triticeae chromosomes 4 and 5, the recent rye chromosome 5R, in particular 5RL, has a distal end of former Triticeae chromosome 4 and the present chromosome 4R, in particular 4RL, of the former chromosome 5, respectively (see Section 2.3). Both chromosome ends were shown to be agronomically important for plant height and preharvest sprouting (Börner et al. 1998a, Masojć and Milczarski 2009).

4.3.3 Neocentric Activity

Neocentric activity is a newly derived kinetic activity outside the proper centromere. Kattermann (1939) first described the phenomenon in rye. Neocentromeres are rare in plants, but less infrequent in meiosis. Three types of neocentromeres have been described in rye:

1. The neocentromere shows a stable structural differentiation at one end of a given chromosome. It is inherited as a Mendelian gene locus.
2. The neocentromeres are located at terminal regions of some chromosomes. They are mostly associated with distal heterochromatin. They are variable in activity and number among individuals and cells within an individual. There is a polygenic control (Viinikka 1985). They occur in both inbred lines and allogamous populations.
3. Schlegel (1987) described an additional type in haploid rye. A proximal constriction present on the long arm of chromosome 5R is cooriented with the ordinary centromere. It behaves like a dicentric chromosome.

However, molecular studies show that the 5RL constriction lacks detectable quantities of two repetitive DNA sequences, CCS1 and the 180-bp knob repeat, present at cereal centromeres and neocentromeres (Manzanero et al. 2000). Nevertheless, it seems that meiotic transmission behavior and frequent chromosome breakage resulting in 5R deletions are due to the activation of the neocentric region (Dubovets 2005).

4.4   CHROMOSOME PAIRING

4.4.1   MEIOTIC CYCLE

The duration of meiosis has been estimated in more than 100 organisms, including prokaryotes and eukaryotes. The duration of female meiosis has been estimated in far fewer species than that of male meiosis. However, meiosis in the PMCs is concurrent with that of the egg cell in the ovule. Despite large variation in the duration of meiosis, three generalizations can be made: first, prophase is always very long compared with the remaining meiotic stages; second, the rate of meiotic development is very slow compared with the rate of development in dividing somatic meristem cells of the same organisms under the same conditions; and third, the duration of meiosis is characteristic of the genotype and species.

Four main factors that affect or determine the duration of meiosis have been recognized, namely:

1. Environmental factors (e.g., temperature)
2. Nuclear DNA content
3. Ploidy level of the organism
4. The genotype

Because nuclear DNA content plays a major role in determining the duration of meiosis, it has been suggested that DNA influences the rate of meiotic development in two ways: first, through its informational content (the genotype) and, second, indirectly by the physical and mechanical effects of its mass independently of its informational content (i.e., the nucleotype). Thus, the observed duration of meiosis is the result of a complex genotype–nucleotype–environment interaction. With the obvious exception of variation caused by developmental holds, changes in the duration of meiosis usually involve proportional changes in the duration of all its stages (Bennett 1977). By sampling methods and by timing the intervals between the premeiotic DNA synthesis and meiotic stages following the incorporation of tritiated thymidine, Bennett et al. (1971) estimated the duration of meiosis.

Meiosis takes ~24 h in wheat, 21 h in triticale, and ~51 h in diploid rye. The lengths of the meiotic stages relative to that of the division correspond reasonably well in the three forms studied, but zygotene and pachytene were much longer in rye than in wheat and triticale (see Table 4.7). The first prophase accounts for ~80% of the total length of meiosis, while the stages from metaphase I (MI) to telophase II (TII) together account for only 20%. The duration of individual stages of meiosis ranges from 20 h for leptotene while it is 50 to 40 min for diakinesis. The prophase stages during which chromosome pairing and chiasma formation occur are long, zygotene lasts for over 11 h and pachytene for 8 h. The second prophase is much briefer than the first prophase. Tetrads persist for over 8.5 h before they break up into pollen grains.

Scanning electron microscopy (SEM) proved that in zygotene transversion of chromatid strands to their homologous counterparts becomes evident. In pachytene, segments of synapsed and nonsynapsed homologs alternate. At synapsed regions, pairing is so intimate that homologous chromosomes form one filament of structural entity.

TABLE 4.7

Mean Duration of Meiosis (h) in Diploid Rye at 20°C in Comparison with Hexaploid Wheat and Triticale

Stage of Meiosis	Rye (*Secale cereale*, $R^{cer}R^{cer}$)			Wheat (*Triticum aestivum*, BBAADD)	Triticale (*Triticum turgidocereale*, BBAAR^{cer}R^{cer})
	15°C	20°C	25°C		
Leptotene		20.0		20.4	7.5
Zygotene		11.4		3.4	3.0
Pachytene		8.0		2.2	2.2
Diplotene		1.0		0.6	1.0
Diakinesis		0.6		0.4	0.5
Metaphase I		2.0		1.6	1.8
Anaphase I		1.0		0.5	0.5
Telophase I		1.0		0.5	0.5
Dyad stage		2.5		2.0	1.5
Metaphase II		1.7		1.4	1.2
Anaphase II		1.0		0.5	0.5
Telophase II		1.0		0.5	0.5
Meiosis (total)	**87.5**	**51.2**	**39.0**	**24.0**	**20.7**
First prophase		41.0		16.0	14.2
MI–TII		10.2		8.0	6.5
Tetrad formation		8.5		10.0	7.3

Source: Bennett, M. D. et al., *Proc. R. Soc. Lond.*, 178, 259–275, 1971. With permission. Bennett, M. D. et al., *Can. J. Genet. Cytol.*, 14, 615–624, 1972. With permission.

Chiasmata are characterized by chromatid strands, which traverse from one homolog to its counterpart. Bivalents are characteristically fused at their telomeric regions. In MI and MII, there is no structural evidence for primary and secondary constrictions (Zoller et al. 2004b).

A highly synchronized mitotic division in most cells of the tapetum occurs immediately before, at, or just after commencement of meiotic prophase—in rye, just after the start of leptotene, followed by a second tapetal division.

Meiosis, together with the premeiotic DNA synthesis (S phase), may be regarded as two cell cycles with only one S period. The durations of the two division cycles of meiosis are very different from the duration of the mitotic cycle time of ~12 h. The duration of the first meiotic division in rye exceeds 40 h, while the second division is completed in ~6 h. The differences between the durations of the two meiotic divisions and the duration of the mitotic cell cycle result presumably from the fact that chromosome pairing and formation of chiasmata prolong the first meiotic division, while the second division is shortened by its lack of DNA synthesis. Ayonoadu and Rees (1968) estimated the duration of S phase in root tip cells as ~6.75 h, which, when added to the duration of the second meiotic division (~6 h), gives a value equal to the mitotic cycle time, in both rye root tip and tapetal meristematic cells.

Both cell cycle time and meiotic duration are influenced by temperature. Over a range of 15°C to 25°C, meiotic duration and pollen maturation time decreased with increasing temperature (Bennett et al. 1972). In plants grown at 25°C, abnormal meiotic and pollen development resulting in male sterility can be seen in several anthers. The cause of male sterility appears to be thickening of tapetal cell walls concurrent with abnormal late meiotic development in PMCs.

Effective somatic cell synchronization in root tip meristems and improved chromosome spreading were achieved in rye by application of hydroxyurea and amiprophos-methyl or colchicine, combined with a pretreatment of ice water and modified fixative, as well as enzymatic digestion of the meristems. The protocol provides metaphase indices of >50%. The chromosomes and chromosomal DNA show minimum distortion, providing useful material for chromosome banding studies, *in situ* DNA–DNA hybridization, microdissection, and microcloning (Pan et al. 1993).

4.4.2 Prophase Pairing and Bouquet Stage

Homologous pairing and synapsis depend on the meiotic recombination machinery that repairs double-strand DNA breaks produced at the onset of meiosis. The culmination of recombination via crossover gives rise to chiasmata, which are located distally in rye. Although synapsis initiates close to the chromosome ends, a direct effect of regions with high crossover frequency on partner identification and synapsis initiation has not been demonstrated.

It is generally thought that if a trial-and-error process identifies homology, there must be a mechanism(s) in the early prophase cell that increases the number or efficiency of random contacts. One such mechanism could be a widespread phenomenon known as the bouquet stage: the clustering of telomeres to a small region of the nuclear envelope during zygotene. Bouquet formation is an active process in rye. Chromosome pairing is disrupted when colchicine is applied well before the bouquet stage, during the end of the preceding mitosis and during early premeiotic interphase. The bouquet is formed through a gradual, continuous tightening of telomeres over ~6 h. To determine whether the motion of chromosomes is random or directed, Carlton et al. (2003) developed a computer simulation of bouquet formation to compare with other observations. They found that the bouquet is formed in a manner comparable to that observed in cultured meiocytes only when the movement of telomeres is actively directed toward the bouquet site. Directed motion, as opposed to random diffusion, is thus required to reproduce the observations, implying that an active process moves chromosomes to cause telomere clustering.

Using three-dimensional light microscopy in rye, it was shown that telomeres are randomly distributed in the premeiotic interphase and early leptotene, and then transported to a small region of the nuclear envelope in prezygotene. The clustering of telomeres preceded the alignment of nontelomeric loci, suggesting that the bouquet is one of the first steps in the pairing process.

Modern SEM down to a resolution of a few nanometers in addition to light microscopy allowed structural investigation and morphometry of meiotic chromosomes at all stages of condensation of meiosis I + II. Remarkable differences during chromosome condensation in mitosis and meiosis I can be found with respect to initiation, mode, and degree of condensation. Mitotic chromosomes condense in a linear fashion, shorten in length, and increase moderately in diameter. In contrast, in meiosis I, condensation of chromosomes in length and diameter is a sigmoidal process with retardation in zygotene and pachytene and acceleration from diplotene to diakinesis. Controlled loosening with proteinase K (after fixation with glutaraldehyde) provides an enhanced insight into chromosome architecture even of highly condensed stages of meiosis. By selectively staining with platinum blue, DNA content and distribution can be visualized within compact chromosomes as well as in a complex arrangement of fibers. Chromatin interconnecting threads, which are typically observed in prophase I between homologous and nonhomologous chromosomes, stain clearly for DNA.

The basic structural components of mitotic chromosomes of rye are "parallel fibers" and "chromomeres," which become highly compacted in metaphase (Zoller et al. 2004a). Although chromosome architecture in early prophase of meiosis seems similar to mitosis in principle, there is no equivalent stage during transition to MI when chromosomes condense to a much higher degree and show a characteristic smooth surface. No indication is found for helical winding of chromosomes in either mitosis or meiosis. Based on measurements, a mechanism for chromosome dynamics in mitosis and meiosis can be proposed, which involves three individual processes: (1) aggregation of chromatin subdomains into a chromosome filament; (2) condensation in length, which involves a progressive increase in diameter; and (3) separation of chromatids.

Meiosis in rye can be summarized as happening in 10 phases. After DNA replication in the premeiotic interphase, meiotic chromosomes are first identifiable in leptotene as long threads with the sister chromatids tightly pressed together. The two sister chromatids of each leptotene chromosome are bound to a common protein core known as an axial element, which appears to hold the meiotic chromatin in a looped configuration. It is not known whether the axial element attachment regions correspond to the scaffold attachment regions of mitotic chromosomes. During zygotene, genetic recombination is probably initiated. The chromosomes begin to coil, and for a brief period the sister chromatids become visibly distinct. These specialized condensation patterns may be regulated in part by prophase I-specific chromatin proteins such as meiotin-1. The homologous zygotene chromosomes begin to synapse along their length via a ribbon-like structure called the synaptonemal complex (SC). The axial elements become the lateral elements of the SC, which are joined together by transverse elements and a central element. Synapsis proceeds to completion in rye even though large heterochromatic regions at the ends of the chromosomes interfere with the end-to-end associations (Gillies 1985).

Phillips et al. (2008) applied antibodies to two SC-associated proteins (Asyl and Zypl) and two recombination-related proteins (*Spo11* and *Rad51*) of *Arabidopsis thaliana*. They were bound to meiocytes throughout meiotic prophase of rye and visualized using conventional fluorescence microscopy and confocal laser scanning microscopy. Moreover, Western blot analysis (Schlegel 2009) was performed on proteins extracted from pooled prophase I anthers, as a prelude to more advanced proteomic investigations. The four antibodies of *Ar. thaliana* reliably detected their epitopes in rye. The expression profile of *Rad51* gene is consistent with its role in recombination. The protein encoded by this gene assists in repair of double-strand DNA breaks. Asy1 protein was shown for the first time to cap the ends of bivalents. Western blot analysis reveals structural variants of the transverse filament protein Zyp1. The putative structural variants of Zyp1 may indicate modification of the protein as bivalents are assembled. Asy1 cores are built by elongation of early foci. The persistence of foci of *Spo11* to late prophase does not fit the current model of molecular recombination. Valenzuela et al. (2012) were able to demonstrate that repair of double-strand DNA breaks via crossover is essential in both the search of the homologous partner and the consolidation of homologous synapsis.

Sosnikhina et al. (2005) created a genetic collection of meiotic mutants of rye. Mutations were detected in inbred F_2 generations after self-fertilization of F_1 hybrids, obtained by individual crossing of Vyatka rye or weedy rye with plants from autofertile lines. The mutations caused partial or complete plant sterility. They were maintained in a heterozygous state. Genetic analysis accompanied by cytogenetic study of meiosis revealed six mutant types:

1. Nonallelic asynaptic mutants *sy1* (on chromosome 7RL) and *sy9* (on chromosome 2RL) cause formation of only axial chromosome elements in prophase and anaphase. The SCs are absent, the formation of the chromosome "bouquet" was impaired, and all chromosomes are univalent in MI in 96.8% (*sy1*) and 67% (*sy2*) of cells. The *sy9* gene cosegregates with two SSR markers *Xscm43* and *Xgwm132*. The asynaptic gene *sy1* was mapped within the interval between the isozyme locus *Aat2* and the two cosegregating loci *Xrems1188* and *Xrems1135* that are located at a distance of 0.4 cM proximally and 0.1 cM distally with respect to the gene locus.
2. Weak asynaptic mutant *sy3* hinders complete termination of synapsis in prophase I. Subterminal asynaptic segments are always observed in the SC, and at least one pair of univalents is present in MI, but the number of cells with 14 univalents does not exceed 2%.
3. Mutants *sy2*, *sy6*, *sy7*, *sy8*, *sy10*, and *sy19* cause partially nonhomologous synapsis, that is, change in pairing partners and foldback chromosome synapsis in prophase I. In MI, the number of univalents varies and multivalents are present.
4. Mutant *mei6* causes the formation of ultrastructural protrusions on the lateral SC elements and causes gaps and branching of these elements.
5. Allelic mutants *mei8*, *mei9*, and *mei10* cause irregular chromatin condensation along chromosomes in prophase I and sticking and fragmentation of chromosomes in MI.
6. Allelic mutants *mei5* and *mei10* cause chromosome hypercondensation, defects of the division spindle formation, and random arrest of cells at different meiotic stages. However, these mutations do not affect the formation of microspore envelopes even around the cells, whose development is blocked at prophase I.

Meiotic analysis of double mutants reveals epistatic interaction in the mutation series: *sy9* > *sy1* > *sy3* > *sy19*. It reflects the order of switching these genes during meiosis. The expression of genes *sy2* and *sy19* is controlled by modifier genes.

4.4.3 INITIATION OF SYNAPSIS

In many plant species, synapsis starts at, or close to, the chromosome ends and this has been related to the distal location of chiasmata. In this regard, Naranjo et al. (2010) studied the meiotic behavior of rye chromosome pair 5R in a wheat background using FISH. The use of different DNA probes allowed the identification of the two rye homologs, their centromeres and subtelomeric heterochromatic chromomeres, and the telomeres of all chromosomes in prophase I and metaphase I. Three types of plants were analyzed: homozygotes for the standard chromosome 5R, homozygotes for a deficient chromosome 5R (del5R) with only the proximal 30% of its long arm (del5RL), and heterozygotes. Synapsis of the deficient chromosome arm pair 5RL was completed in most meiocytes at pachytene, but the number of chiasmata formed was much lower than that in the intact 5RL arm. Deletion facilitated the migration of the telomere of the accompanying chromosome arm 5RS during bouquet organization. This was followed by an increase of synapsis and chiasma frequency in this arm with regard to its counterpart in the intact chromosome. Results demonstrated that crossing-over formation depends on the DNA sequence or the chromatin organization of each chromosome region and that homologous alignment, synapsis, and chiasma formation may be conditioned by chromosome conformation.

An estimate of the number and location of successful homology identification events per chromosome can be obtained by direct observation of SC formation. SC initiation sites are correlated with the presence of at least one early so-called recombination nodule. Observations are consistent with the idea that telomeres frequently initiate pairing: SC is first formed in the subterminal regions of the chromosomes. Internal sites of pairing initiation are also very frequent. In rye, there are 9–20 per bivalent (Abirached-Darmency et al. 1983).

At pachytene, the chromosomes are fully synapsed and often dispersed in the nucleus so that they can be easily identified. Pachytene chromosomes are much longer than mitotic prophase chromosomes and have been used in several species to make rough cytological maps, such as Lima de Faria (1952). In diplotene, the homologous chromosomes separate but remain associated with chiasmata. In diakinesis, the chromosomes contract lengthwise by a spiraling process and prometaphase I (immediately before metaphase I) and are thickened and highly condensed. The spindle is formed in prometaphase to metaphase. In anaphase I, the chiasmata are released and sister chromatids segregate to the same pole. During the interphase between meiosis I and II (=interkinesis), there is no DNA replication. The chromosomes again become visible at prophase II, and after a mitotic-like division in meiosis II, the sister chromatids disjoin to form four haploid daughter cells.

4.4.4 HAPLOIDS AND CHROMOSOME BEHAVIOR

Haploid rye plants were first described by Müntzing (1937). He found them when spikes were treated with cold temperatures during anthesis prior to pollination. Usually they are selected among diploid offspring. They spontaneously arise, mostly by pathenogenetical development of haploid egg cells or synergid cells. In most cases they appear without external influence. Sengbusch (1968) selected twin plants of rye in the hope of finding polyploids among them. Out of 654 twins, 4 triplets and 1 quadruplet were found among 2 million germinating seeds. Just one seedling was identified as haploid. Often haploid development takes place by distant crosses.

In 1980, another source was observed by the author (unpublished). As compared to a diploid control, the progeny of trisomic 3R of the variety Danae frequently provided twin seedlings. Of them more than 60% were haploid. It seemed that the extra dosage of chromosome 3R contributed to this behavior. This coincides with more recent data of Pershina et al. (2007). They observed in hybrid combinations between *Hordeum vulgare–Triticum aestivum* alloplasmic recombinant lines and five wheat substitution lines [1R(1D), 2R(2D), 3R(3B), 5R(5A), and 6R(6A)]. Chromosomes 1R and 3R of rye cultivar Onokhoiskaya proved to affect the expression of polyembryony in the hybrid combinations that involved the alloplasmic recombinant lines of common wheat as maternal genotypes. Based on this finding, polyembryony was regarded as a phenotypic expression of nuclear–cytoplasmic interactions involving chromosomes 1R and 3R and the *H. vulgare* cytoplasm.

Several workers have reported haploids and meiotic studies on haploids in rye as well. Associations of two or more apparently nonhomologous chromosomes are often observed at first metaphase (MI). Levan (1942) was the first to demonstrate statistically that chiasma formation between the seven chromosomes was not random. He suggested that one particular chiasma is formed at quite a high frequency while the remaining arise at random.

Differentiation between true chiasma formation and secondary end-to-end attachments can be made by either chromosome coorientation or proving chromatid bridges and acentric fragments during anaphase I. No chiasma formation was observed in the heterochromatic telomeres. This indicates that this kind of repetitive DNA does not function as a homologous region contributing to crossing-over (Puertas and Giraldez 1979, Schlegel et al. 1987).

The mean chiasma frequency per PMC as given by different authors ranged from 0.03 (Nordenskiöld 1939), to 0.26 (Heneen 1965) and 0.34 (Neijzing 1982), and to 0.44 (Levan 1942).

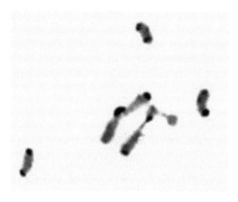

FIGURE 4.7 Metaphase I chromosome pairing in haploid rye (*Secale cereale* L.) after C-banding, showing three univalents and two rod bivalents.

Schlegel et al. (1987) revealed 6.38 univalents, 0.29 rod bivalents, and 0.01 chain trivalent per PMC considering 200 cells (see Figure 4.7). This resulted in 0.32 chiasmata per PMC. The frequency corresponded well with previous results.

Using Giemsa C-banding, it was possible to discriminate pairing behavior between chromosomes. Chromosomes 1R, 2R, 3R, and 7R, showing heavy telomeres on both chromosome arms, are more frequently involved in chiasmatic pairing than chromosomes 4R, 5R, and 6R. Apparently, chiasma formation in haploids of rye is nonrandom.

In addition, the chiasmatic chromosome pairing points to homologous chromosome regions on nonhomologous chromosomes. Obviously, there are DNA duplications of quite considerable size on several chromosomes (see Section 4.3).

4.4.5 Diploids

4.4.5.1 Variation of Chromosome Pairing

4.4.5.1.1 Intraindividual Variation

Studies on intraindividual variation of chromosome pairing and chiasma formation were carried out by numerous researchers (Müntzing and Akdik 1948; Rees 1955a, 1955b; Rees and Thompson 1955, 1956; Lawrence 1958; Sybenga 1958). Rees and Naylor (1960) as well as Jones and Rees (1964) demonstrated intraindividual pairing variation, mainly for the chiasma frequency between tillers, anthers, and PMCs. Chromosomal rearrangements, for example, reciprocal translocations, may prevent recombinational transfer of genes from a donor genotype to a recipient, especially when the gene is located in an interstitial segment. When genetic recombination was studied in translocation heterozygotes by using trisomy of chromosome arm 1RS, an alteration of the crossing-over pattern around the translocation breakpoint could be detected, with a special increase in the interstitial segment of 6RS and adjoining regions, normally hardly accessible to recombination (Sybenga et al. 2012). Obviously, heterogeneity widens the distribution of crossing-overs, including segments normally not accessible to recombination, but decreases average recombination in other segments. Sybenga (2012) even discusses epigenetic control of pronounced within-plant heterogeneity of meiosis. Meiosis and gametogenesis are associated with marked changes in chromatin organization and with the activity of genes regulating preferential chromosome transmission or responsible for the evaluation of environmental conditions (Caperta et al. 2002).

Environmental conditions may strongly modify the number of chiasmata per chromosome (Lamm 1936, Dowrick 1957, Lawrence 1958, Law 1963); for example, Bennett and Rees (1970) applied different concentrations of phosphate to two genotypes of rye and observed a significant increase in chiasma frequency correlated with an increasing availability of phosphate. In autotetraploid rye, a significant change in the pattern of pairing was found in response to different nitrogen treatments, but the chiasma frequency remained the same (Hossain 1978).

Despite studies on individual plants and inbred lines, the most comprehensive experiments were presented by Schlegel and Mettin (1975a, 1975b) using heterozygous clonal plants from diploid and tetraploid rye populations (see Figure 4.8). Under field conditions, they revealed quite low variability within genotypes and/or between clones, that is, without statistical significance (see Table 4.8). However, extreme temperature conditions

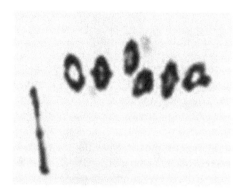

FIGURE 4.8 Metaphase I chromosome pairing in diploid rye (*Secale cereale* L.) after Feulgen staining, showing six ring and one rod bivalents.

during meiosis may result in significant differences of homologous chromosome pairing and chiasma frequency per PMC, particularly in reduced chromosome pairing, univalent formation, and partial sterility of florets. As in maize, there is a sort of compensation effect for chiasmata. Missing chiasmata on one chromosome may be compensated by more crossing-overs on one or more of the others. In diploids, the chiasma frequency is strongly correlated with the formation of rod and/or ring bivalents. In tetraploids, the tetrasomic pairing seems to be more complex, that is, the correlation between chiasma frequency and chromosome associations is less pronounced (see Table 4.9).

4.4.5.1.2 Interindividual Variation

The variability of chromosome pairing is much higher between genotypes, in both diploids and polyploids. Among ~115 genotypes of 5 diploid rye populations, the

TABLE 4.8
Meiotic Chromosome Pairing in 40 Spikes of a Clonal Genotype (V-43) of Diploid Rye Studied over a Period of 6 Weeks under Varying Field Conditions in May/June of Hohenthurm (Germany)

	Mean Metaphase I Configuration per PMC				
Number of PMCs		Bivalents			Number of Chiasmata
Scored	Univalents	Ring	Rod	Total	per PMC
20	0	6.65	0.35	7.00	14.25
20	0	6.65	0.35	7.00	14.10
20	0	6.85	0.15	7.00	14.35
20	0	6.60	0.40	7.00	13.90
20	0	6.50	0.50	7.00	14.40
20	0	6.80	0.20	7.00	13.90
20	0	6.35	0.65	7.00	13.85
20	0	6.65	0.35	7.00	13.95
20	0	6.85	0.15	7.00	14.21

TABLE 4.8
(Continued) Meiotic Chromosome Pairing in 40 Spikes of a Clonal Genotype (V-43) of Diploid Rye Studied over a Period of 6 Weeks under Varying Field Conditions in May/June of Hohenthurm (Germany)

| Number of PMCs Scored | | Mean Metaphase I Configuration per PMC | | | | Number of Chiasmata per PMC |
| | Univalents | Bivalents | | | | |
		Ring	Rod	Total		
20	**0**	6.70	0.30	7.00		14.11
20	0	6.76	0.24	7.00		14.36
20	0	6.56	0.44	7.00		14.16
20	0	6.76	0.24	7.00		14.36
20	0	6.71	0.29	7.00		14.21
20	0	6.51	0.49	7.00		14.00
20	0	6.71	0.29	7.00		14.01
20	0	6.61	0.39	7.00		14.26
20	0	6.71	0.29	7.00		14.06
20	0	6.66	0.34	7.00		14.66
20	0	6.56	0.44	7.00		14.31
20	0.10	6.50	0.45	6.95		14.85
20	0	6.75	0.25	7.00		14.70
20	0	6.50	0.50	7.00		14.35
20	0	6.60	0.40	7.00		14.50
20	0.20	6.05	0.85	6.90		14.20
20	0	6.90	0.10	7.00		14.50
20	0	6.40	0.60	7.00		14.30
20	0	6.55	0.45	7.00		14.30
20	0	6.55	0.45	7.00		13.90
20	0	6.45	0.55	7.00		14.15
20	0	6.60	0.40	7.00		14.15
20	0	6.65	0.35	7.00		14.45
20	0	6.65	0.35	7.00		14.50
20	0	6.70	0.30	7.00		14.24
20	0	6.65	0.35	7.00		13.90
20	0	6.55	0.45	7.00		13.95
20	0	6.45	0.55	7.00		14.05
20	0	6.30	0.70	7.00		14.20
20	0	6.65	0.35	7.00		14.10
20	0	6.60	0.40	7.00		14.05
Total (800)	**0.01**	**6.60**	**0.39**	**7.00**		**14.19 ± 0.01**

Source: Schlegel, R., and Mettin, D., *Biol. Zbl.*, 94, 539–555, 1975a. With permission.

Note: No statistically significant differences between the clones, spikes, and conditions. Bold text represents mean value over all entries.

TABLE 4.9

Meiotic Chromosome Pairing in 18 Spikes of Two Clonal Genotype (I-18 and No. 18888) of Tetraploid Rye Studied over a Period of 5 Days under Varying Field Conditions in May/June of Hohenthurm (Germany)

| Genotype | Number of PMCs Scored | Univalents | Bivalents | | | | Trivalents | | | | Quadrivalents | | | | Number of Chiasmata per PMC |
			Ring	Rod	Total	Chain	Pan	Others	Total	Ring	Chain	Others	Total	
I-18	30	0.47	6.63	1.87	8.50	0	0	0	0	1.56	1.17	0	2.73	25.20
	30	0.43	7.06	1.13	8.20	0.13	0.03	0	0.16	1.50	1.13	0.03	2.66	26.36
	30	0.43	6.33	1.10	7.45	0.13	0.13	0	0.26	1.90	1.00	0.03	2.93	26.46
	30	0.56	6.80	1.20	8.00	0.16	0	0	0.16	1.63	0.96	0.13	2.73	26.40
	30	0.43	7.66	1.20	8.86	0	0.03	0	0.03	1.10	1.26	0.07	2.43	26.10
	30	0.46	6.23	1.60	7.83	0.77	0.07	0	0.84	1.60	1.25	0.03	2.86	25.66
	30	0.70	6.20	1.40	7.60	0.13	0.03	0	0.16	1.36	1.53	0	2.90	25.35
	30	0.43	7.10	1.47	8.57	0.10	0	0	0.10	1.83	0.73	0.03	2.53	26.43
	30	0.37	7.25	1.50	8.73	0.10	0	0	0.10	1.83	0.60	0.03	2.47	26.60
	30	0.27	6.97	1.40	9.37	0.13	0.07	0	0.20	1.60	1.00	0	2.60	26.60
	300	**0.42**			**8.21**				**0.14**				**2.96**	**26.12**
No. 18888	30	0.33	6.03	1.07	7.10	0.13	0	0	0.13	2.30	0.90	0.07	3.27	26.30
	30	0.27	6.70	1.30	8.00	0.10	0	0.03	0.13	1.90	0.83	0.10	2.83	26.03
	30	0.17	6.00	1.19	7.13	0.10	0.03	0.03	0.16	2.57	0.60	0.10	3.27	26.70
	30	0.43	6.90	1.73	8.63	0.23	0	0	0.23	1.73	0.60	0.07	2.40	25.53
	30	0.23	6.60	1.47	8.07	0.10	0	0	0.10	2.13	0.60	0.10	2.83	26.13
	30	0.30	7.77	1.43	9.20	0.03	0	0	0.03	1.60	0.67	0.03	2.30	25.90
	30	0.17	7.53	1.13	6.67	0.03	0	0	0.03	1.87	0.63	0.10	2.60	26.47
	30	0.53	4.80	1.13	5.93	0.50	0.07	0	0.57	2.40	0.67	0	3.07	25.50
	240	**0.30**			**8.04**				**0.11**				**2.82**	**26.07**

Source: Schlegel, R., and Mettin, D., *Biol. Zbl.*, 94, 539–555, 1975a. With permission.

Note: Bold text represents mean value over all entries of two genotypes.

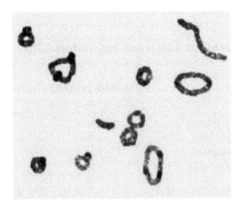

FIGURE 4.9 Metaphase I chromosome pairing in autotetraploid rye (*Secale cereale* L.) after Feulgen staining, showing one univalent, six ring bivalents, and four quadrivalents (three rings + one chain).

pairing configurations and the chiasma frequencies ranged between 0 and 0.05 univalents per PMC, 0.26 and 0.78 rod bivalents per PMC, and 13.70 and 14.47 Xta per PMC, respectively. The variety Carsten's Roggen showed the highest chiasma frequency. The ratio of rod bivalents varied between 3% and 11% of the total number of bivalents (Schlegel and Mettin 1975b).

4.4.6 TETRAPLOIDS

Among four tetraploid rye populations, the mean frequencies of univalents and trivalents per PMC were not statistically significant, although a wide range of pairing configurations between PMCs can be observed (see Figure 4.9; Table 4.2). Corresponding to other induced tetraploids (Ahloowalia 1967, Growley and Rees 1968, Swami and Thomas 1968), Schlegel and Mettin (1975c) demonstrated that in rye the frequencies of different chromosome configurations might vary independently from the chiasma frequency. The number of multivalents is not just a function of chiasma frequency, as Hazarika and Rees (1967) suggested. There is a genetic control for both chiasma frequency and type of chromosome association. The karyotype structure can influence the type of pairing association (see Section 7.2).

4.4.6.1 Chromosome Pairing in Egg Mother Cells

Inbred lines of rye, which have been used extensively for pairing studies, show significant differences in both cell chiasma frequency and bivalent variance. However, while chiasma frequency is affected by inbreeding, there is little effect on chiasma distribution and a majority of chiasmata is distal localized in one or both arms of each bivalent of PMCs and embryo-sac mother cells (EMCs), and within a given line, no significant differences exist between the two sexes (see Table 4.10). It was concluded that chiasma formation in PMCs and EMCs of rye is governed and regulated by a single controlling system of genes, and that variation in this genetic system is expressed identically in the two sexes.

TABLE 4.10

Mean Chiasma Frequency in PMCs and Egg Mother Cells of Diploid Rye (*Secale cereale* L.)

Genotype	Chiasmata per PMC	Chiasmata per EMC
Stalrag (inbred)	12.88	12.60
J33 (hybrid from interspecific cross)	10.84	10.10
J93 (hybrid from interspecific cross)	12.49	12.60
J113 (hybrid from interspecific cross)	9.04	9.50
J115 (hybrid from interspecific cross)	8.99	9.00

Source: Davies, E. D. G., and Jones, G. H., *Genet. Res.*, 23, 185–190, 1974. With permission.

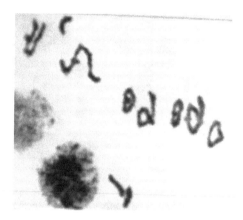

FIGURE 4.10 Metaphase I chromosome pairing in triploid rye (*Secale cereale* L.) after Feulgen staining, showing three univalents, three ring bivalents, and four trivalents (two chains + two frying pans).

4.4.7 TRIPLOIDS

Chromosome pairing in rye triploids was studied in a few cases. Kostoff (1939) first reported two to three trivalents in triploid PMCs, Lamm (1944) 3.75, and Balkandschiewa (1971) 4.30 in average, respectively (see Figure 4.10). The data of Balkandschiewa (1971; see Table 4.11) and former authors demonstrated incomplete homologous chromosome pairing, resulting in additional univalents and bivalents as well as reduced chiasma numbers per chromosome. Compared to the diploid and tetraploid parents with 1.15 and 0.95 chiasmata per chromosome, the triploids form only 0.77.

4.4.7.1 Chromosome Pairing in Different Genetic Backgrounds

The transfer of the rye genome into an alien genetic backgrounds, for example, into wheat, leads to striking modification of the chromosomal pairing pattern. In both hexaploid and octoploid triticales, the chiasma frequency is significantly reduced

TABLE 4.11

Meiotic Chromosome Pairing in Seven Triploid Genotypes of Rye
(*Secale cereale* L.) Resulting from Crosses of Autotetraploid Heines
Hellkorn × Diploid Danae

Number of PMCs Scored		Mean Metaphase I Configuration per PMC								Number of Chiasmata per PMC
			Bivalent			Trivalent				
	Univalent	Ring	Rod	Total	Chain	Pan	Others	Total		
30	2.90	2.90	0	2.90	2.30	1.80	0	4.10		16.53
30	2.10	2.10	0	2.10	2.10	2.80	0	4.90		17.40
30	3.03	3.00	0.03	3.03	2.53	1.47	0	3.97		15.80
30	3.43	3.27	0.17	3.43	3.00	0.57	0	3.57		14.57
30	2.33	2.20	0.13	2.33	2.43	2.20	0.03	4.67		17.17
30	2.09	2.09	0	2.09	2.75	2.18	0	4.91		16.27
30	2.63	2.57	0.07	2.63	2.33	2.03	0	4.37		16.10
Total (210)	**2.70**	**2.64**	**0.06**	**2.70**	**2.47**	**1.83**	**0.05**	**4.30**		**16.26**

Source: Balkandschiewa, J., Morphologische und cytologische Untersuchungen an triploidem Roggen und seinen Nachkommen zum Aufbau einer Trisomenserie, PhD thesis, 1971.

(see Table 4.12). Lowest chromosome pairing is found in primary hybrids with hexaploid wheat and highest chromosome pairing in hexaploid triticale. The pairing reduction is stronger when meiosis happens under higher temperatures. It could be shown that univalents and rod bivalents of rye predominantly involve chromosome arms carrying heterochromatic telomeres. Even if the rye genome is studied in its own cytoplasm (RRBAD), rye chromosome pairing is disturbed in the presence of the wheat chromosomes.

The *Ph1* locus on chromosome 5B that enforces strictly bivalent pairing in polyploid wheat also controls chromosome pairing in autotetraploid rye by apparently restricting chiasma formation between dissimilar homologs. Unlike in wheat, the effect appears to be dosage dependent, which may be a reflection of an interaction between *Ph1* and the rye chromosome pairing control system. With two doses of *Ph1* present, chiasmatic pairing is severely restricted, resulting in a significantly higher number of univalents and bivalents per cell than those in the controls. The restrictions imposed by *Ph1* virtually eliminated MI pairing of chromosome arms polymorphic for their C-banding patterns and did not appear to affect arms with similar patterns. The fact that *Ph1* operates in rye in the same fashion as in polyploid wheat suggests that it controls some basic mechanism of chromosome recognition (Lukaszewski and Kopecky 2010). On the diploid genome level, the dominant *Ph1* allele affects the homologous rye chromosome pairing as well (Schlegel et al. 1991a), even in the hemizygous situation. By chromosome arm addition of 5BL to diploid rye, an increase of univalent and rod-bivalent formation is observed, that is, a significant reduction of chiasma frequency from approximately 13.74 to 11.21 Xta per PMC, respectively.

TABLE 4.12
The Chiasma Frequency of Rye Chromosomes in Combination with Different Wheats

		Mean Number of Chiasmata per PMC		
Genotype	Origin	15°C	20°C	25°C
RR+5BL	Diploid rye plus chromosome 5BL of wheat		11.21	
RRBAD	F₁ from cross of tetraploid rye × hexaploid wheat		5.81	
BARR	F₁ from cross tetraploid wheat × tetraploid rye		7.01	
BADRR (N5A/D5B)	F₁ from cross of hexaploid wheat × tetraploid rye		7.53	
BBAARR	Hexaploid triticale	11.70	8.30	6.85
BBAADDRR	Octoploid triticale	9.43	8.13	7.22
RR	Control (diploid inbred rye)	13.74	11.79	7.95

Source: Schlegel, R., and Schrader, O., *Proc. 2nd Int. Triticale Symp.*, 359–367, 1991. With permission.
Schlegel, R. et al., *Plant Breed.*, 107, 226–234, 1991b. With permission.

4.5 SPOROGENESIS

4.5.1 MICROSPOROGENESIS AND POLLEN DEVELOPMENT

Microsporogenesis comprises events, which lead to the development of the unicellular microspores into mature microgametophtyes containing two male gametes. The first mitosis in microspores results in two structurally and functionally different nuclei: one generative and one vegetative. This process is accompanied by an asymmetric cell division that partitions the mother cell into two daughter cells of equal size. The smaller generative cell undergoes a second division. It results in the formation of two reproductive sperm cells. They fuse during the fertilization with the egg and central cell. The functional differentiation of pollen nuclei is accompanied by contrasting nuclear chromatin configurations. The generative nuclei are highly condensed. The vegetative nucleus is characterized by diffuse chromatin, including higher transcriptive activity. The nuclear differentiation occurs during the first pollen mitosis or in the early bicellular pollen. Soon after the asymmetric division of the microspore, the generative cell enters the S phase and then passes into the G2 phase, while the vegetative cell is arrested in G1 phase (see Section 4.4.1).

Rye differs from other small grains in that it is cross-pollinated. Each large spike consists of many two-flowered spikelets with long awns. Although pollen can and does fall into flowers of the same plant, little self-pollination takes place even when the spikes of the same plant are bagged. Self-sterility ranges between 96% and 99%. Together with the gametophytic incompatibility, self-fertilization is effectively prevented, as a typical characteristic of allogamous rye.

Meiotic divisions begin in the middle part of the spike and gradually proceed upward and downward in the spike, usually a few days before emergence of the spike from the leaf sheath. Male meiosis precedes female meiosis by ~15 h in the same floret (Bennett et al. 1973). There is only a single megaspore mother cell per floret compared to ~16,000 microspore mother cells.

Anthesis occurs ~3–10 days after the ear emerges from the flag leaf sheath, when a number of closely correlated events occur in a very short time. The first flowers to open are usually two-thirds of the way up the spike. They then open in both directions, up and down the spike, over a period of several days. The flowers remain open for 15 to 45 min. Unfertilized flowers stay open for as long as 20 days. Rye stigmas can be receptive up to 14 days. There are two phases of receptivity, which are dependent on temperature and humidity. In the first phase that lasts ~6 days at 20°C and 60% humidity, stigmas are receptive and result in good seed setting. The second phase (~5 days) is characterized by sudden drop in seed set. Higher temperatures and lower humidity reduce both phases.

The lodicules of each floret swell up forcing the lemma and the palea apart. The filaments of the stamens elongate and can eventually attain a length of ~15 mm. As the filament grows, the anther dehisces, each chamber developing a longitudinal split, starting at the tip of the anther, through which ~42,000 pollen grains are released. The peak of pollen release comes at different times of day in different locations. It is influenced by temperature, humidity, and sunlight. Pollen production is ~2.5 times greater in diploid than in tetraploid plants. Wodehouse (1935) reported rye pollen as being rather large and ellipsoidal of ~62 × 40 μm with the germ pore on the side near one end. This characteristic is unique among grasses and serves to distinguish this species from others.

The stigma lobes, which are pressed together before anthesis, move apart, and the receptive branches are spread widely giving a large area for pollen interception. In rye, there is dichogamy, that is, pollen and stigma ripen at different times. The whole process is complete within ~5 min. Anthesis occurs first in floret 1 of the spikelets of the upper two-thirds of the ear. The next day it progresses to the first floret of the basal spikelet and to the second floret of the upper spikelets. Within the crop, anthesis occurs first in the main shoot, but all shoots commence anthesis within 3 or 4 days. The whole process is usually complete in 10 days or less, depending on the weather.

Microgametogenesis in rye results in two structurally and functionally different cells, one generative cell, which subsequently forms the sperm cells, and the vegetative cell. The chromatin properties of both types of nuclei are different after first and second pollen mitoses in rye. The condensed chromatin of generative nuclei is marked by an enhanced level of histone H3K4/K9 dimethylation and H3K9 acetylation. The less-condensed vegetative nuclei are RNA polymerase II-positive. Trimethylation of H3K27 is not involved in transcriptional downregulation of genes located in generative nuclei as H3K27me3 was exclusively detected in the vegetative nuclei (Houben et al. 2011). The global level of DNA methylation does not differ between both types of pollen nuclei. In rye, centromeric histone H3 is not excluded from the chromatin of the vegetative nucleus and the condensation degree of centromeric and subtelomeric regions does not differ between the generative and vegetative nuclei.

4.5.1.1 Unreduced Gametes

Doubling of chromosomes in rye can be achieved by either colchicine and other means or utilization of spontaneous formation of unreduced gametes (see Section 7.2). By using a genetic seed marker (xenia), the frequency of unreduced gametes formed in diploid rye can be estimated (Mettin et al. 1972b). In crosses of a pale-grained tetraploid rye as female and 18 different diploid green-grained rye varieties/populations, the frequency of green-grained tetraploid F_1 caryopses (a measure for the frequency of unreduced gametes) ranged between 0% and 7.6% (see Table 4.13). On average, 0.6% of diploid pollen grains led to tetraploid xenia seeds in rye—a percentage that can successfully be applied for meiotic production of new tetraploids, without the adverse effects of colchicine treatment. Lelley et al. (1987) also describe an inbred

---

TABLE 4.13

Frequency of Tetraploid Xenia (Unreduced Gametes) in Valence Crosses of a Tetraploid (4×) Pale-Grained Rye (Strain H1/67) with Diploid (2×) Green-Grained Varieties in 1968–1972

Cross 4× (St. H1/67) × 2× Variety	Number of Caryopses Harvested	Number of Green Xenia Caryopses	% 3× Seedlings	% 4× Seedlings
× Danae (1968)	12,262	182	0.27	0.59
× Danae (1969)	1,826	16	0.22	0.06
× Danae (1970)	5,415	19	0.13	0.07
× Danae (1971)	5,769	766	0.33	7.60
× Danae (1972)	4,084	17	0.37	0.02
Petkuser kurz (1968)	8,062	125	0.12	0.89
Petkuser kurz (1969)	4,882	10	0.02	0.04
Petkuser kurz (1970)	6,997	32	0.07	0.06
Rye strain, mildew resistance (1968)	6,938	68	0.25	0.55
Rye strain, mildew resistance (1969)	3,225	15	0.22	0.03
Rye strain, mildew resistance (1970)	5,390	14	0.08	0.06
Rye strain, brown rust resistance (1968)	7,975	58	0.19	0.34
Rye strain, brown rust resistance (1969)	3,189	16	0.40	0
Rye strain, brown rust resistance (1970)	7,372	10	0.01	0.07
Rye strain, brown rust resistance (1971)	3,578	310	0.87	2.91
Rye strain, brown rust resistance (1972)	8,233	23	0.16	0.05
Edelhofer (1968)	6,679	80	0.15	0.43
Edelhofer (1969)	4,416	16	0.18	0
Edelhofer (1970)	4,047	16	0.10	0.03
Rye strain, sprouting resistance (1968)	6,789	36	0.15	0.22
Rye strain, sprouting resistance (1969)	4,589	9	0.04	0
Rye strain, sprouting resistance (1970)	5,023	24	0.04	0
Oberkärtner (1969)	3,990	24	0.23	0.10
Oberkärtner (1970)	5,499	23	0.11	0.02
Harrach 15/63 (1969)	4,708	66	0.32	0.28

TABLE 4.13

(Continued) Frequency of Tetraploid Xenia (Unreduced Gametes) in Valence Crosses of a Tetraploid (4×) Pale-Grained Rye (Strain H1/67) with Diploid (2×) Green-Grained Varieties in 1968–1972

Cross 4× (St. H1/67) × 2× Variety	Number of Caryopses Harvested	Number of Green Xenia Caryopses	% 3× Seedlings	% 4× Seedlings
Harrach 15/63 (1970)	6,029	38	0.05	0.02
Otterbacher (1969)	5,588	44	0.64	0.39
Otterbacher (1970)	5,185	31	0.21	0
Dominant (1969)	4,233	26	0.30	0.07
Dominant (1970)	6,492	37	0.15	0.02
Carsten kurz (1969)	6,470	10	0.06	0
Carsten kurz (1970)	6,173	31	0	0
Carsten kurz (1971)	3,524	365	0.35	5.67
Kungs II (1969)	5,305	10	0.06	0
Kungs II (1970)	8,236	31	0	0
Kungs II (1971)	1,823	172	0.26	4.09
Värne (1969)	5,397	29	0.13	0.22
Värne (1970)	7,324	21	0.04	0.01
Tellö (1969)	5,181	5	0.02	0
Dańkowskie Zł. (1969)	5,148	42	0.14	0
Gawczyna (1970)	6,051	59	0.12	0.20
Total	**229,096**	**2,926**	**0.18**	**0.61**

Source: Mettin, D. et al., *Tag. Ber. AdL (DDR)*, 119, 135–152, 1972b. With permission; Schlegel, R., Experimentelle Untersuchungen zur chromosomalen Stabilität und deren Bedeutung für die Fertiltät des autotetraploiden Roggens (*Secale cereale* L.), PhD thesis, 1973.

genotype of rye that quite regularly produces unreduced gametes. A crossing study revealed a polygenic and recessive control of this feature.

Nevertheless, there is a wide range of variability for both the formation of xenia and the percentage of unreduced gametes. Both characteristics are influenced by the environmental conditions and the genotype. The German variety Danae shows higher percentage of unreduced gametes as compared to the variety Tello or Dańkowskie Złote.

4.5.1.2 Gametocides

Natrova and Hlavac (1975) investigated the effect of gametocides FW-450, maleic hydrazide, and Dalapone during the course of microsporogenesis, gametogenesis, and starch synthesis in pollen grains of winter rye var. Esto. Solutions containing various concentrations of tested substances were applied on leaves in amounts of 2.5 and 5.0 ml per plant at the beginning of leaf emergence, at full emergence of leaves, and at the end of shooting. The effect of gametocides is detectable already in the phase of microspore formation when a decrease of anther weight, pollen grain size, and starch content in grains, and an increase of the number of microspores with unfinished development were observed. The manner and the extent of injuries are dependent on the applied substance, its concentration, and the period of its application.

4.5.1.3 Pollination and Fertilization

The diameter of the microspores is ~40 nm, the thickness of tectum is 0.2 μm, and the thickness of nexine and lumen is 0.15 μm (Heslop-Harrison 1979). The starch (amyloplasts) in the fully mature but nongerminated pollen is massed mostly in the hemisphere of the vegetative cell opposite to the aperture, leaving a starch-free zone adjacent to the aperture itself. This distribution is characteristic for germinable grains. After the contact of pollen with the stigma, it takes ~22 s for the pollen meniscus to appear, ~36 s for the exudate, and 75 s for the germination to start in compatible pollination. In incompatible pollination, it takes 24, 44, and 76 s, respectively. It is fast as compared to other grasses. The germinability of rye pollen held under laboratory atmosphere (23°C, 50% relative humidity) is ~2 h; afterward, it significantly drops, and after 9 h, the grains are incapable of germinating.

Approximately 80% of pollen shed is viable for at least 10 h. Collins et al. (1973) reported that pollen stored in liquid nitrogen (−192°C) germinated *in vitro* and produced seed set *in vivo*. Generally, lower temperatures and humidity are detrimental to pollen.

The pollen grain, when settled on a stigma, germinates in about one and a half hours to produce a pollen tube. Germination already starts after 3 min. The pollen tube grows down the style, between the cells, and eventually reaches the embryo sac via the micropyle. Two sperm nuclei move down to the tip of the pollen tube; a tube nucleus is also present, but this may not leave the pollen grain. At normal temperatures, the pollen tube reaches the embryo sac in ~40 min (Bennett et al. 1972). On reaching the embryo sac, the sperm nuclei are discharged and fuse with the egg nucleus (which develops to become the embryo) and the polar nuclei (which form the endosperm).

4.5.1.4 Embryo Development

Division of the fertilized egg nucleus commences later than that of the endosperm (Bennett et al. 1973). The early divisions produce a five-celled embryo with a basal cell, although variation in the pattern of development has been observed. Continuing cell division produces at first a club-shaped structure, which ultimately differentiates to form a mature embryo in the ripe seed. The growth of the embryo is supported by nutrients derived from the antipodal cells and the hydrolysis of parenchyma cells of the nucellus and neighboring endosperm cells. Later, while in development, transfer cells appear in the nucellar endosperm epidermis near the base of the embryo and the provascular tissue.

4.5.1.5 Endosperm Development

Following the fusion of the sperm nucleus and the polar nuclei, cell division is, for a time, synchronous, the number of endosperm cells doubling every 4–5 h. At first, the endosperm is coenocytic, but after ~3 days, cell walls are formed. Cell wall growth commences at the edge of the embryo sac and furrows inward to the central vacuole.

The cells at the periphery of the endosperm divide, and eventually the entire embryo sac is cellular. After cell formation is complete, the subcellular structures, which will synthesize the protein bodies, and the other cell components are formed. Amyloplast division ceases before cell division, and starch grains differ in growth rate in different cell layers of the endosperm.

4.5.2 MACROSPOROGENESIS

The female gametophyte can be studied by light microscope analysis of semithin sections of embryo sacs containing egg cells developing *in planta*. The morphological and ultrastructural details of egg cells isolated 3–18 days after emasculation can be examined to determine the function of the female gametophyte and its suitability, for example, for micromanipulation. A sufficient number of gametoplasts in the right stage of development is required for the successful microinjection and *in vitro* fusion of egg cells and male gametes. Various *in vitro* fertilization and microinjection techniques could be of service in gamete fusion experiments aimed at the creation of interspecific and intergeneric hybrids, which do not occur in nature due to sporophytic incompatibility (Timar et al. 2002).

The female gametophyte reaches its maximal dimension 2 or 3 days before the ripening of the male gametophyte. Five developmental stages can be assessed: tetrad >>> macrospore >>> bicellular >>> tetracellular >>> heptacellular (= eight-nucleate). The development of the seven-cellular female gametophyte is also faster than that of male gametophyte. The embryo sac is formed 3–4 days from the macrospore as a result of three consecutive rapid mitoses, compared to the 7 days of the trinucleate pollen stage out of two microspore mitoses. Thus, the development of the female and male gametophytes is asynchronous, demonstrating protogyny in rye.

Ova can be studied either *in planta* or when excised and/or isolated. When *in planta* preparation is applied, sections are made from pistils. When excised from the tissue, ova take a spherical shape. The size varies with their age. The diameters of 6- and 9-day-old cells are ~50 and 65 μm, respectively. In the cytoplasm of 3-day-old ovum prior to anthesis, lipid bodies, mitochondria, amiloplasts, and starch granules can be observed in the peripheral vesicles.

The receptive ovum closest to anthesis is ~9 days old. It shows dense cytoplasm, large number of mitochondria, endoplasmic reticulum, and polarization. In the old ovum, the nucleus is degraded and chromatin residues adhere to the membrane of the nucleus forming the so-called blebbings. Lipids, starch, and proteins are accumulated in the cytoplasm. Lysis is observed in the vesicle. Autophagic vacuoles appear, just following the features of programmed cell death.

4.6 PRIMARY ANEUPLOIDS

4.6.1 PRIMARY TRISOMICS AND TELOTRISOMICS

Primary trisomics of spontaneous origin in rye have been described as early as in 1935 (Takagi 1935, Sybenga 1965). A complete series of primary trisomics was established in 1962 (Kamanoi and Jenkins 1962); however, it got lost later on. Two new

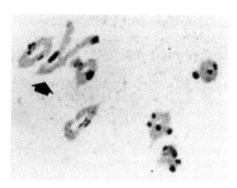

FIGURE 4.11   Metaphase I configuration of primary rye trisomic 5R ($2n = 2x = 15$) with $6^{II} + 1^{III}$ (arrow) after C-banding.

sets of trisomics were prepared from the winter rye varieties Danae by Mettin et al. (1972a) and Heines Hellkorn by Zeller et al. (1977). The first set was derived from triploid into diploid crosses. The trisomics were identified based on cytological and morphological observations (see Figure 4.11). Between the trisomics, there are striking differences in growth habit. They are suitable for morphological differentiations.

The extra chromosome in primary trisomics transmitted ~15% through the egg cell and almost 0% through the pollen. Consequently, backcrosses to the female plants are needed in order to maintain primary trisomics.

Although the extra chromosome paired quite regularly with its homologs (a positive correlation between the relative chromosome length and the number of chiasmata per chromosome was evident), in the progeny monotelocentric and other aberrations could be observed for almost all of the 14 chromosome arms (Schlegel and Sturm 1982, Sturm and Melz 1982, Melz and Schlegel 1985, Benito et al. 1996).

Recent studies on the short arm of chromosome 1R showed great differences in pairing and recombination patterns not only between plants but also among spikes from different tillers and clones of the same plant. However, anthers within spikes were rarely different. This confirms earlier studies of Schlegel and Mettin (1975a, 1975b).

It is suggested that there is an epigenetic system that rigidly maintains the pattern of chromosome pairing throughout generative differentiation. In competitive situations, the genotypes most competent for pairing will pair preferentially, forming specific meiotic configurations with different frequencies for different spikes of the same plant. This would explain the heterogeneity between spikes and the homogeneity within spikes. The epigenetic system could involve chromatin conformation or DNA methylation. However, there were no signs of heterochromatinization (Sybenga et al. 2008).

Until molecular studies became standard in gene mapping during the 1990s, primary trisomics and telotrisomics have been the most powerful tool for gene localization in rye (Sturm 1978, Melz et al. 1984). Depending on the backcross or selfing progeny, different segregation patterns in F_2 were used to assign a morphological or physiological character to a particular chromosome (see Table 4.14).

TABLE 4.14

Expected F$_2$ Segregations of Trisomic F$_1$ Plants from a Critical Cross of Trisomic by Disomic, Excluding Any Selection, Male Transmission of $n + 1$ Gametes, and Abnormal Chromosome Segregation

F$_1$ Genotype	Selfing or Backcrossing of Disomic Recessive	Expected Segregation Depending on the Transmission of $n + 1$ Gametes through Female	
		50%	30%
Aa	Selfing	3:1	3.0:1.0
Aa	Backcrossing	1:1	1.0:1.0
Aaa	Selfing	17:1	11.9:1.0
AAa	Backcrossing	5:1	3.3:1
Aaa	Selfing	2:1	1.6:1.0
Aaa	Backcrossing	1:1	1.0:1.3

Source: Schlegel, R., and Sturm, W. *Tag. Ber. Akad. Landwirtsch. Wiss.*, 198, 225–243, 1982. With permission.

4.6.2 B Chromosomes

B chromosomes (also called supernumerary chromosomes, accessory chromosomes, accessory fragments, etc.) usually have a normal structure, are somewhat smaller than the autosomes, and can be predominantly heterochromatic (e.g., maize) or euchromatic as in rye. These chromosomes of rye are very invasive. They undergo strong drive, which is counteracted by harmful effects on fertility and instabilities at meiosis. These result in lower seed set.

Morphological traits, for example, plant length, can also be influenced (see Table 4.15), and even quantitative trait loci are modified when B chromosomes are present. Diploids and tetraploids respond almost in the same way. The number of B chromosomes is correlated with the strength of change. Therefore, in advanced rye varieties, B chromosomes are rare. Both nondisjunction and meiotic behavior, and consequently the establishment of B polymorphisms, mainly depend on the B chromosomes themselves. Nevertheless, the adaptive nature of B chromosomes remains controversial.

The autonomous nature of drive in rye B chromosomes was clearly demonstrated by Lindström (1965), who showed that this enigmatic chromosome behaved in just the same way in the pollen of wheat as it does in rye, although its pairing properties at meiosis in this alien environment are compromised. The same thing happens when the B chromosome of *S. cereale* is transferred to *S. vavilovii* (Puertas et al. 1985). The B chromosome of rye is relatively large. A rye plant, each with 1C DNA content of ~580 Mbp, can tolerate 0–8 B chromosomes. It amounts to ~10% of the DNA of the haploid A genome (see Figure 4.12). B chromosomes are unexpectedly rich in A-derived genic sequences.

TABLE 4.15

Influence of B Chromosomes on Relative Seed Set and Plant Length in Diploid (2×) and Tetraploid (4×) Rye

Number of B Chromosomes	Plant Height		Seed Set	
	2×	4×	2×	4×
1	99.2	99.3	92.9	94.8
2	100.4	97.5	82.3	87.1
3	99.9	89.4	56.2	67.1
4	94.8	95.7	39.7	72.7
5		86.2		59.0
6		80.0		58.2
>6		89.8		40.8
Mean	**98.6**	**91.1**	**67.8**	**68.5**
0 B chromosomes, control	**100**	**100**	**100**	**100**

Source: Hoffman, W. et al., *Lehrbuch der Züchtung landwirtschaftlicher Kulturpflanzen*, Verl. P. Parey, Berlin, 1970. With permission.

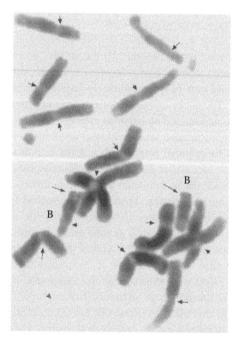

FIGURE 4.12 FISH of a *Secale cereale* plant with two B chromosomes; the rye-specific centromere DNA sequence Bilby is shown in originally green (arrowheads) and the B-specific DNA sequence E3900 in originally red on the telomere of the long arm of B chromosomes (arrows). (Courtesy of Houben, A., Gatersleben, Germany.)

The origin of rye B chromosomes was estimated to ~1.1–1.3 million years ago, thus overlapping in time with the onset of the genus *Secale* (1.7 million years ago). It has a widespread global distribution and exhibits little structural variation. B chromosomes are generally considered to be nonfunctional and without any essential genes. Nevertheless, recent studies demonstrated B sequences with a big homology to genes of sequenced plant genomes (Martis et al. 2012).

Rye B chromosomes apparently contain several prominent blocks of conserved genes corresponding to barley chromosomal regions 2H, 3H, 4H, and 5H, along with thousands of short genic sequences scattered all over the A chromosomes (Martis et al. 2013). It is presumed to be derived largely or entirely from the A chromosome set based on similar DNA composition, most recently shown by using *in situ* hybridization with genomic DNA or the rye-specific R173 repeat. By the so-called next-generation DNA sequencing of flow-sorted A and B chromosomes, it was shown that the B chromosomes are rich in gene-derived sequences, which allowed tracing back their origin. Some fragments of rye B chromosomes correspond to chromosomes 3R and 7R of the A complement. However, compared to A chromosomes, B chromosomes were also found to accumulate large amounts of specific repeats and insertions of organellar DNA!

B chromosomes appear to be transcriptionally inactive, implying either loss of coding sequences or selective amplification of repetitive DNA in its evolution. By amplified fragment length polymorphism (AFLP) studies, it was demonstrated that DNA products specific to B-containing plants were seen at a frequency of ~1%, significantly less than that predicted on a mass basis. Most of the fragments were found to be highly repetitive and were shown by FISH to be distributed over both A and B genomes. Dispersed fragments fell into two groups: the first showed an equal density of distribution across both A and B genomes, consistent with earlier observations, while the second showed an increased concentration at a pericentric site on the long arm of the B chromosome. The redundancy of members of the second group suggests that this region is less complex than the average for the B chromosome, and may represent a hot spot of sequence amplification.

The terminal heterochromatin of B chromosome contains both B-specific sequences and sequences present on the A chromosomes (Houben et al. 1996). The B-specific D1100 family is the major repeat species located in the terminal heterochromatin. The most distinctive region of the B chromosome is a subtelomeric domain that contains an exceptional concentration of B-specific DNA sequences. At metaphase, this domain appears to be the physical counterpart of the subtelomeric heterochromatic regions of the standard B chromosome (Langdon et al. 2000). This Giemsa banding-positive heterochromatic subterminal domain undergoes decondensation during interphase. In contrast to the heterochromatic regions of A chromosomes, this domain is simultaneously marked by trimethylated H3K4 and trimethylated H3K27, an unusual combination of apparently conflicting histone modifications. Both types of B-specific high copy number repeat families (E3900 and D1100) of the subterminal domain are transcriptionally active, although with different tissue type-dependent activity. The lack of any significant open reading frame and the highly heterogeneous size of mainly polyadenylated transcripts indicate that the noncoding RNA may function as structural or catalytic RNA (Carchilan et al. 2007).

Rye is the only plant where the B chromosomes undergo nondisjunction at the postmeiotic mitosis of both female and male gametophytes. It is controlled by a gene or gene complex on the B chromosome itself. In the first pollen mitosis, sister chromatids of rye B chromosomes are attached to each other at the chromatid adhesion site, leading to nondisjunction. In addition, a rye B chromosome with a partial deletion of the long arm increases in number by nondisjunction when it is present together with a complete rye B chromosome in the same plant (Lima de Faria 1962).

Nondisjunction is controlled by two elements: pericentromeric sticking sites and a *trans*-acting element at the distal region of the long arm of the standard B chromosome. The distal region of the long arm provides this function and carries the B-specific repetitive DNA sequences (Endo et al. 2008). Through reduced fertility and persistent univalent formation a high degree of chromosomal instability is found, which results in a highly polymorphic population. The long-arm distal region (so-called transacting control element) and pericentromeric region (so-called chromatid adhesion site) of the rye B chromosome are critical to nondisjunction. The distal region that carries the element controlling nondisjunction comprises a concentration of B-specific sequences from two families, E3900 and D1100, assembled from a variety of repetitive elements, some of which are also represented in the A genome (Sandery et al. 1990, Blunden et al. 1993, Houben et al. 1996, Langdon et al. 2000). No genes have been found in the region, which begs the question about what genetic process controls nondisjunction, since its occurrence happens with very high frequency.

It was speculated from cytological studies that B centromeres act in the normal way in the anaphase of the first pollen mitosis and can be seen to be separated and pulling at opposite poles. The B chromatids appear to be transiently held together at sensitive sticking sites (receptors) on either side of the centromere, and since the spindle is asymmetrical, the equator is closer to the pole, which will passively include the B chromatids in the generative nucleus. Kousaka and Endo (2012) concluded that nondisjunction is gametophytically controlled by the transacting control element. Recent studies prove that nondisjunction of B chromosomes is accompanied by centromere activity and is likely caused by extended cohesion of the B sister chromatids. The B centromere originated from an A centromere, which accumulated B-specific repeats and rearrangements. Because of unequal spindle formation at the first pollen mitosis, nondisjoined B chromatids get preferentially located toward the generative pole. The failure to resolve pericentromeric cohesion is under the control of the B-specific nondisjunction control region. Hence, a combination of nondisjunction and unequal spindle formation at first pollen mitosis results in the accumulation of B chromosomes in the generative nucleus, and therefore ensures their transmission at a higher than expected rate to the next generation (Banaei-Moghaddam et al. 2012).

Despite the general similarity, the B chromosome does not pair with any A chromosomes. However, translocations between B and A chromosomes were observed (Pohler and Schlegel 1990). For example, the autosome 3R was involved in one of the rearrangements. Schlegel and Pohler (1994) demonstrated that a terminal segment, including a large terminal heterochromatic block, was transferred to the terminal region of the long arm of B chromosome. Although several authors assumed a monophyletic origin of rye B chromosomes in the distant past and a perpetuation in various populations, the results imply similar events also nowadays, which is partially supported by stable fragments occasionally revealed as well as the presence of B chromosomes in numerous plant species.

Defined A–B translocations were thought to be useful tools for gene amplification, chromosome identification, genetic investigations, and gene transfer experiments.

4.6.2.1 Midget Chromosomes

In derivatives of wheat–rye crosses, sometimes midget chromosomes are observed. Kota et al. (1994) described one that is necessary for plump, viable seed development, fertility restoration in the alloplasmic line with rye cytoplasm, and a hexaploid wheat nucleus. This midget chromosome of rye represents one-fifteenth of the physical length of the chromosome 1R of rye from which it derived. C-banding analysis indicated that the centromeric and pericentric regions (~30% physical length) of the midget chromosome are heterochromatic and the distant 70% physical length is euchromatic. These data suggest that the midget chromosome may represent the pericentric region of the long arm of chromosome 1R. In contrast to other reports, an array of rye-specific repeated sequences (both dispersed and tandem) is present on the midget chromosome.

4.7 RECIPROCAL TRANSLOCATIONS

Rye is known as a species that maintains chromosomal interchanges (see Figure 4.13) within populations in a more or less high frequency and complexity (Candela et al. 1979). The maintenance of a constant frequency (~20%) of structural heterozygotes for several generations suggests the existence of an equilibrium. The breakpoints for an interchange in rye are such that all three combinations, the normal structural homozygote, the interchange heterozygote, and homozygote, were readily identifiable at both mitosis and meiosis. Most of the interchange homozygotes are inviable. However, in populations subjected to increasing selection pressures imposed by, for example, high sowing density, the proportion of interchange heterozygotes among survivors surprisingly increased. In terms of survival (seed-to-adult viability), the

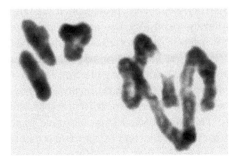

FIGURE 4.13 Heterozygous translocation including four chromosomes of diploid rye, showing three bivalents + one octovalent, after Feulgen staining in metaphase I of meiosis.

interchange heterozygotes at high sowing densities have a 102% advantage over the normal homozygotes, more than sufficient to generate a stable polymorphism for the interchange (Bailey et al. 1978).

Chromosomal interchanges are of particular importance for both evolutionary studies of rye and several genetic and/or breeding applications. Until the 1980s, the nomenclature of rye chromosomes was confusing (Schlegel and Mettin 1982). Therefore, several efforts were made in order to designate the chromosomes according to the homeologous relationships within the Triticeae (Sybenga 1983). An experimental translocation tester set of rye was established and used for crossing with wheat–rye chromosome addition lines. They originated from irradiated pollen grains of Petkus rye crossed with several inbred rye lines. In this way, for the first time a complete series of reciprocal translocations was described by Sybenga et al. (1985). Although several other interchanges have been established before and after those experiments (Sybenga and Wolters 1972, de Vries and Sybenga 1976, Augustin and Schlegel 1983), this series became the key tester set of rye involving all seven chromosomes by interchanged chromosomes as follows: 1RS–4RL, 1RL–5RS (T273), 1RS–5RL, 1RS–6RS (T248), 2RS–5RS, 2RL–5RS (T305), 2RL–6RL (T242), 3RS–5RS (T240), 3RS–5RL, 4RL–5RL (T501), and 5RL–7RS (T282).

Despite the intensive utilization for gene mapping in rye (de Vries and Sybenga 1976), reciprocal translocations were used for diploidization experiments in tetraploid rye (see Section 7.2). However, the complex structural heterozygosity of reconstructed tetraploid karyotypes not only increased preferential bivalent pairing but also decreased fertility.

Another type of complex genome rearrangements is pericentromeric regions of eukaryotic chromosomes that are commonly heterochromatic and the most rapidly evolving regions of complex genomes. The closely related genomes of Triticeae taxa share large conserved chromosome segments and provide a good model for the study of the evolution of pericentromeric regions. Five different pericentric inversions were detected in rye chromosomes 3R and 4R. This indicates that pericentric regions in the Triticeae including rye, especially those of group 4 chromosomes, are undergoing rapid and recurrent rearrangements (Lili et al. 2006).

4.8   GENETIC DONOR FOR OTHER CROPS

The numerous crossing experiments and chromosomal manipulation between cereals are exceptional within the plant kingdom and among crop plants. In 1975, the Scottish botanist A. C. Wilson (1876) reported the first wheat into rye cross; in 1888, Rimpau (1891) produced the first intergeneric and partially fertile hybrid between hexaploid wheat and diploid rye. It was during the crop season of 1918 at the Saratov Experimental Station in Russia that thousands of natural hybrids between wheat and rye appeared in many wheat fields. For the next 16 years, Meister and his colleagues exploited these hybrids. The name triticale first appeared in scientific literature in 1935 and is attributed to Tschermak, one of the rediscoverers of Mendelian law. It was also during the same year that Arne Müntzing at Svalöv, Sweden, initiated his lifelong dedication to triticale, mainly octoploids. Since the first success, a wide range of various combinations of genomes, individual chromosomes, and even chromosome segments were established (Schlegel 1990). Rye species contain genes associated with high protein content, high lysine, resistance to many diseases, sprouting, drought (see Section 5.1), winter hardiness, and many other morphological and biochemical features that can be transferred to related cereals (see Table 8.4).

4.8.1 α-AMYLASE INHIBITOR

Even the latest biotechnological means include rye as a genetic donor for other crops, for example, tobacco. Innumerable proteinaceous α-amylase inhibitors have been iso-lated and identified from different plant species. Among them, an α-amylase inhibitor gene with bioinsecticidal potential toward *Anthonomus grandis* (cotton boll weevil) was identified in diploid rye seeds (Campos-Dias et al. 2010). This cereal inhibitor was expressed in tobacco plants under the control of phytohemaglutinin promoter by using *Agrobacterium tumefaciens*-mediated transformation. The presence of *aBIII*-rye gene and further protein expression were confirmed by polymerase chain reaction (PCR) and Western blot analysis, respectively. Immunological assays indicated that the recombinant inhibitor was expressed in a concentration range from 0.1% to 0.28% of the total protein in tobacco seeds of R0 plants. From 14 independent transformants, 5 plants with expression levels between 0.20% and 0.28% in seeds were assayed *in vitro* against *An. grandis* amylolytic enzymes causing clear inhibition. Moreover, bio-assays using transgenic seed flour mixture for artificial diet produced 74% mortality in *An. grandis* first larval instar. These data suggest that rye inhibitor can be a promising biotechnological tool for producing transgenic cotton plants with an increased resis-tance to cotton boll weevil. Moreover, *aBIII*-rye gene should be considered a potential compound for a pyramiding strategy aiming at delaying insect resistance.

4.8.2 ALUMINUM TOLERANCE

The next molecular utilization of rye genes could be aluminum (Al) tolerance (Gustafson and Ross 2010). Rye is the most Al-tolerant cereal. Al-induced secretion of organic acids from the roots has been considered as a mechanism of Al tolerance. The Al response can be examined by measuring root elongation and cytoplasmic-free activity of calcium ($[Ca^{2+}]$cyt) in intact root apical cells. Measurement of $[Ca^{2+}]$cyt is achieved by loading a Ca^{2+}-sensitive fluorescent probe. Increases in $[Ca^{2+}]$cyt are correlated with inhibi-tion of root growth, generally measurable after 2 h. Addition of 400 µM malic acid (pH 4.2) largely ameliorated the effect of 100 µM Al on $[Ca^{2+}]$cyt in root apical cells and protected root growth from Al toxicity. An increase in $[Ca^{2+}]$cyt in root apical cells in rye is an early effect of Al toxicity and is followed by secondary effect on root elonga-tion (Qifu et al. 2002).

Aluminum induces rapid secretion of malate in the wheat, but a lag is observed in rye. The secretion of malate is not suppressed at low temperature in the wheat, but that of citrate is stopped in rye. Aluminum does not affect the activities of isoci-trate dehydrogenase, phosphoenolpyruvate carboxylase, and malate dehydrogenase in either plant. However, the activity of citrate synthase is increased by the exposure to Al in rye, but not in wheat (Li et al. 2000). The inhibitors of a citrate carrier, which transports citrate from the mitochondria to the cytoplasm (see Table 6.1), inhibit

Al-induced secretion of citrate from roots. All of these results suggest that alteration in the metabolism of organic acids is involved in the Al-induced secretion of organic acids in rye.

A wheat malate transporter [aluminum-activated malate transporter (ALMT)] was cloned and characterized as a constitutively expressed gene associated with aluminum tolerance in wheat (see Section 3.4.2). Then a rye bacterial artificial chromosome (BAC) library was screened for the presence of rye malate transporter (ALMT) gene complexes. Three rye BAC clones were identified which contained several ALMT copies and a transcription factor upstream from the ALMT region. The BAC sequences clearly showed the presence of more than one copy of the ALMT gene complex that could be used for transformation into wheat. Candidate aluminum tolerance genes encoding organic acid transporter (ALMT) and multidrug and toxic compound extrusion (MATE) families have been characterized in several plant species (Yokosho et al. 2010).

The primary mechanism of Al resistance described in plants is the chelation of Al^{3+} cations by release of organic acids into the rhizosphere. Rye possesses a complex structure at the *Alt4* locus for aluminum tolerance not found at a corresponding locus in wheat. Former genetic analysis attributed the Al tolerance in Blanco variety (see Table 7.12) to the *Alt4* locus on the short arm of chromosome 7R (7RS) and revealed the presence of multiple allelic variants (haplotypes) of the *Alt4* locus in the BAC library. Moreover, the Diversity Arrays Technology (Yarralumla, Australia) allowed the identification of additional quantitative trait locus markers associated with Al tolerance. Genetic maps suggested the putative location of markers on chromosomes 3R, 4R, 6R, and 7R (Niedziela et al. 2012). BAC clones containing clusters of the *Almt* genes were identified. They provided useful starting points for exploring the basis for the structural variability and functional specialization of *Almt1* genes at this locus (Shi et al. 2009).

During recent studies, in five different cultivars of rye, an aluminum-activated citrate transporter (*Aact1*) gene was isolated. It is homologous to aluminum-activated citrate transporter of barley. The *Aact1* gene was mapped to the 7RS, 25 cM away from the aluminum tolerance gene (*Alt4*). The gene consists of 13 exons and 12 introns and encodes a predicted membrane protein that contains the MatE domain and at least seven putative transmembrane regions. Expression of the *Aact1* gene is Al induced, but there were differences in the levels of expression among the cultivars analyzed (Silva-Navas et al. 2012).

4.8.3 THAUMATIN

Rye genes can even be utilized for dicotyledonous plants. The pathogenesis-related proteins have the potential for enhancing resistance against fungal pathogens. In particular, thaumatin-like proteins (TLPs) have been shown to have antifungal activity. They could be used for rapeseed improvement as an important oilseed crop. A serious problem in cultivation of this crop is yield loss due to fungal stem rot caused by *Sclerotinia sclerotiorum*. Therefore, a *tlp* gene was isolated from rye and introduced into the recipient plants. Transformation of cotyledonary petioles was achieved via *Ag. tumefaciens* (Zamani et al. 2012).

The amplified DNA fragment (~500 bp) was analyzed and confirmed by restriction pattern, cloned into pUC19, and designated as pUCNG1. Comparison of the cloned fragment with the DNA sequence indicated that this gene contains no intron. The *tlp* gene was predicted to encode a protein of 173 amino acids with an estimated molecular mass of 17.7 kDa. The deduced amino acid sequence of TLPs showed a significant sequence identity with TLP from rye. By a transgenic overexpression approach, the antifungal activity of expressed TLP on *S. sclerotiorum* was investigated. Antifungal activity was confirmed in transgenic rapeseed lines using detached leaf assay. The size of the lesions induced by *S. sclerotiorum* in the leaves of transgenic rapeseed was significantly retarded as compared to that of control plants.

4.8.3.1 Genome Additions

The basic aim of rye genome transfer into wheat was to combine the quality characteristics of wheat by the agronomic unpretentiousness of rye. From the many intergeneric hybridizations made between wheat and *Aegilops, Triticum, Haynaldia, Hordeum,* and *Agropyron* species, only the wheat–rye combination became valuable for breeding and agronomy. Less octoploid (BBAADDRR) and tetraploid triticale (AARR or DDRR or BBRR) but hexaploid triticale (BBAARR) became a new, worldwide-grown, man-made crop plant (Schlegel 1996). After a long period of cytogenetic research and breeding efforts, hexaploid triticale now enlarges the spectrum of cereal crops with its high yield, good resistance against diseases, and improved protein content. Hexaploid triticales are well established in agronomy, particularly for feeding animals.

New triticales are still produced. The *Imperata cylindrica*-mediated chromosome elimination approach has been demonstrated as a new system of chromosome elimination. It opens unique opportunities to shorten the breeding cycle, accelerate the varietal development programs, and generate gene mapping populations instantly in triticale and wheat. This system is used in combination with genomic *in situ* hybridization (GISH) and FISH tools for introgression, for example, of Himalayan rye genes into bread wheat genome (Chaudhary 2013).

4.8.3.2 Chromosome Additions

Among the more than 200 different wheat–alien chromosome addition lines, there are complete series of disomic wheat–rye additions, including the adequate disomic telocentric addition lines (Shepherd and Islam 1988). Individually added chromosomes are available from the rye varieties Imperial (see Figure 4.14) and King II. Incomplete sets exist from the varieties Dakold and Petkus rye (Riley and Chapman 1958). These addition lines were used not only for first genetic mapping studies in rye but also as a source for targeted gene transfer experiment from rye into wheat.

4.8.3.3 Chromosome Substitutions

In the past, wheat–rye or wheat–*Aegilops* chromosome substitutions were used for studies of homeologous relationships between cereal genomes. The better the wheat chromosome compensation by a given alien chromosome, the closer the genetic relationship of the chromosome and/or genome can be suggested. As measures for the close homoeologous ties, the plant vitality was used and the successful transferability pollen with the alien chromosome. First reports of spontaneous wheat–rye

FIGURE 4.14 Spike samples of disomic wheat–rye addition lines (1R–7R) of the series Chinese Spring–Imperial, including control Chinese Spring wheat (far left).

chromosome substitutions 5R(5A) were given already in 1937 (Kattermann 1937) and 1947 (O'Mara 1947). Driscoll and Anderson (1967) reported the substitution of the wheat chromosomes 3A, 3B, 3D, and 1D by rye chromosome 3R. Approximately 20 years later, 1R(1D) (Müller et al. 1989) and 1R(1A) substitutions (Koebner and Singh 1984) were produced in order to improve wheat for baking quality and resistance against diseases. Schlegel (1997) described other wheat–rye chromosome substitutions with breeding relevance.

4.8.3.4 Chromosome Translocations

Although a first experimental wheat–rye translocation (4B–2R) had already been produced in 1967 (Driscoll and Anderson 1967), introgression of rye genetic information into wheat became most famous by the spontaneous 1RS.1BL wheat–rye translocation (Mettin et al. 1973, Zeller 1973). Since 1973, this particular type of translocation was described in more than 250 cultivars of wheat from all over the world (Schlegel 1997, 2013). The most important phenotypic deviation from common wheat cultivars is the so-called wheat–rye resistance (Hsam and Zeller 1996), that is, the presence of wide-range resistance to races of powdery mildew and rusts, which is linked to decreased bread-making quality, good ecological adaptability, yield performance (Schlegel and Meinel 1994), and better nutritional efficiency. The origin of the alien chromosome was extensively discussed for genetic and historical reasons. It turned out that basically four sources exist: two in Germany (most likely one source, Schlegel and Korzun 1997), one in the United States, and one in Japan. The variety Salmon (1RS.1BL) is representative of the latter and the variety Amigo (1RS.1AL) is representative of the penultimate group, while almost all remaining cultivars can be traced back to one or the other German origin. Another wheat–rye translocation with breeding importance was found in the Danish variety Viking. It carries a 4B–5R interchange (Schlegel et al. 1993), causing high iron, copper, and zinc efficiencies compared to those in common wheat. By using a 2BS–2RL translocation, transfer of Hessian fly (*Mayetiola destructor*) resistance from Chaupon rye to hexaploid wheat was feasible (Friebe et al. 1990).

In some triticales, additional wheat–rye translocations are also reported. Taketa et al. (1997) found Bronco 90, both a 2D(2R) substitution and a translocated chromosome 5RS.5RL–4DL in the hexaploid variety.

By using *S. africanum*, even genes for yellow and leaf rust were successfully transferred to wheat. Two amphiploids, AF-1 (*T. aestivum* cv. Anyuepaideng–*S. africanum*) and BF-1 (*T. turgidum* ssp. *carthlicum–S. africanum*), were evaluated by chromosomal banding and *in situ* hybridization. Extensive wheat–*S. montanum* non-Robertsonian translocations were observed in both AF-1 and BF-1 plants, suggesting a frequent occurrence of chromosomal recombination between wheat and *S. africanum*. Moreover, introgression lines selected from the progeny of wheat and AF-1 crosses were resistant when field-tested with widely virulent strains of *Puccinia striiformis* f. sp. *tritici*. Three highly resistant lines were selected. GISH and C-banding revealed that the resistant line L9-15 carried a pair of 1BL.1RS-translocated chromosomes. This new type of *S. africanum*-derived wheat–*Secale* translocation line with resistance to *Yr9*-virulent strains will broaden the genetic diversity of 1BL.1RS for wheat breeding (Lei et al. 2012).

In several experiments, diploid barley was used as a recipient of rye introgression. An intergeneric *H. vulgare* (CM 67) into *S. cereale* (dwarf cultivar Snoopy) cross and progeny through F_5 have been studied. The cross was made by the chemical suppression technique of Bates et al. (1974) and subsequently involved chromosome elimination and spontaneous somatic doubling of the resultant haploid. The introgression that has required no backcrossing must be explained as a function of the chemicals (Bates et al. 1975).

5 Genetics

5.1 NOMENCLATURE AND DESIGNATION OF GENES

The rules of gene symbolization were adapted from the International Rules of Genetic Nomenclature, the recommended rules in wheat (McIntosh et al. 2008), the *Recommendation of the First International Workshop on Rye Chromosome Nomenclature and Homoeology Relationships* (Sybenga 1983), the *Third Workshop of Rye Genetics and Cytogenetics: Revision and Completion of the Genetic Map of Rye* (Melz and Sybenga 1994), and the *International Symposium on Rye Breeding and Genetics* (Schlegel and Korzun 2007):

1. In naming hereditary factors, the use of languages of higher internationality is given preference.
2. Symbols of hereditary factors, derived from their original names, are written in italics.
3. Whenever unambiguous, the name and symbol of a dominant allele begin with a capital letter and those of a recessive with a small letter.
4. Two-letter or three-letter symbols are given preference.
5. All letters and numbers used in symbolization are written on one line.
6. Two or more genes having phenotypically similar effects are designated by a common basic symbol. Nonallelic loci (mimics, polymeric genes, etc.) are designated in accordance with three procedures:
   a. In sequential polymeric series, an Arabic numeral immediately follows the gene symbol, for example, *ct2*.
   b. In homologous chromosomes, the basic symbol is followed by a hyphen (-) and by the locus designation taking the form of the accepted genome symbol of rye *R* and a homeologous set number represented by an Arabic numeral; for example, *Adh-R1* designates the *R*-genome member of the first *Adh* set. Different alleles, or alleles of independent mutational origin, are designated by a lowercase Roman letter following the locus number designation; for example, *Pm1a*.
   c. Temporary symbol designation. Where linkage data are not available, provision has been made for temporary symbols. These symbols shall consist of the basic symbol followed by an abbreviation of the genotype or stock and an Arabic number referring to the gene; for example, *YrKi2* and *YrKi3* refer to two genes for reaction with yellow rust in genotype King II.

   It is recommended that official records of temporary designations be kept, but it is not essential that subsequent numbers from other laboratories be checked against earlier numbers either phenotypically or genetically.

7. Inhibitors, suppressors, transporters, and enhancers are designated by the symbols *I*, *Su*, *Ta*, and *En* if they are dominant alleles or *i*, *su*, and *en* if they are recessive alleles, followed by the symbol of the allele affected (in brackets).

8. Whenever convenient, lethals are designated by the letter *l* or *L*, and sterility and incompatibility genes by *s* or *S*.

9. In rye and related wilds, linkage groups and corresponding chromosomes are designated by an Arabic numeral (*1–7*) followed by a genome designated by a capital Roman "R." In whichever context it appears, it is necessary to indicate from which *Secale* species a certain chromosome (or chromosome arm) is derived, thus a superscript of three letters is added. Thus, 1RL^{mon} indicates the long arm of chromosome 1 of the species *S. montanum*. Telocentric chromosomes are designated S (short) and L (long), respectively.

10. Genetic formulae are written as fractions, with the maternal alleles given first or as numerator. Each fraction corresponds to a single linkage group.

11. Chromosomal aberrations are indicated by the abbreviations *Tr* for translocation and *Tp* for transposition.

12. The zygotic number of chromosomes is indicated by $2n$, the gametic number by n, and the basic number by x.

13. Symbols for extrachromosomal factors are enclosed within brackets and precede the genetic formula.

14. All molecular markers including single-sequence repeats (SSRs), random amplified polymorphic DNAs (RAPDs), and amplified fragment length polymorphisms (AFLPs) are described with the basic symbol *X*.

   a. The *X* is followed by a laboratory designator written in small letters, except that a capital letter is specifically needed, for example, *Xlprm2-4R*. The laboratory designator is given in a separate list of marker abbreviations.

   b. Anonymous loci. The *X* is followed by a laboratory designator, a probe number, a hyphen (-), and, in case of doubt, the symbol of the chromosome in which the locus is located.

   c. For example, restriction fragment length polymorphism (RFLP) loci detected in different chromosomes with the same probe are assigned the same symbol except for the chromosome designation (*Xpsr161-R1* and *Xpsr161-R5*).

   d. Duplicate DNA loci located in the same chromosome are assigned the same symbol except for the addition of an Arabic numeral in parentheses immediately after the chromosome designation [*Xpsr158-R1(1)* and *Xpsr158-R1(2)*].

   e. DNA loci for which no product is described, the locus symbol consists of the basic symbol *X* followed by an abbreviation of the function, a hyphen, and a symbol of the chromosome in which the locus is located (*XNar-R2*).

   f. DNA loci that are "allelic" with named protein loci are designated with the protein locus symbol.

g. Clone designations should identify the type of vector, the species from which the cloned DNA was obtained, and the cloned DNA and source laboratory in the order (p = plasmid, l = lambda, c = cosmid, and m = M13 should be used to identify the vector). Initials of the species name (i.e., *Sc = Secale cereale, Ta = Triticum aestivum*, etc.) should be applied to designate the source of the cloned DNA and a unique number–letter combination chosen by the source laboratory should be used to designate the cloned DNA and the laboratory.

15. Defined regions and more precisely localized genes are given together with the appropriate nomenclature of the C and N bands of the chromosome (Schlegel and Gill 1984).

16. Naming quantitative trait loci (QTLs). A QTL name should reflect the trait, the cross, and the chromosome, instead the standardized wheat QTL naming scheme also contains an acronym for the institution that described the QTL. The basic symbol is *QTL* written in capital letters. Since trait or trait complex descriptions can be extensive, in the present compilation the acronym only includes a number and the chromosome or chromosome region designation, for example, *QTL14-1RL*. The details are given in a separate list of marker abbreviations.

In Table 5.1, a selection of gene symbols and abbreviations of rye genes and markers is compiled. Unfortunately, not all authors consider an adapted symbolization of genes and markers, which leads to numerous confusions and double or multiple naming.

TABLE 5.1
Symbolization and Abbreviations of Genes and Markers in Rye

Gene/Marker	Descriptor
al	Vertical arrangement of leaves
Aadh >>> Adh	Aromatic alcohol dehydrogenase
Aat syn. *Got*	Aspartate aminotransferase
AawR173-3 Petkus rye	Rye-specific PCR DNA primers
AawR173-3 King II	Rye-specific PCR DNA primers
Ac	Color of anthers
Acc2	Acetyl-CoA carboxylase
Acl	Acyl carrier protein
Acl1.2 syn. *Xwayc4*	
Acl1.3 syn. *Xwayc2*	
Aco	Aconitase
Acp	Acid phosphatase
Acph >>> Acp	
Adgp2 syn. *Xwye838*	
Adgp3 syn. *Xwye1858*	
Alb	Albinism of seedling

(Continued)

TABLE 5.1

(Continued) Symbolization and Abbreviations of Genes and Markers in Rye

Gene/Marker	Descriptor
Adh	Alcohol dehydrogenase
Adh syn. *Xcsd19*	
Adk	Adenylate kinase isomerase
Al	Length of anthers
al syn. *El*	Absent ligule
Almt1	Malate transporter gene likely responsible for aluminum tolerance
Alp	Alkine phosphatase
alpha-Amy3 syn. *Xpsr14*	
Alt1-6RS	Tolerance to excess of aluminum
Alt2-3RS	Tolerance to excess of aluminum, major locus; secretion of organic acids malate and citrate is associated with the presence of *Alt2*
Alt3-4RL	Tolerance to excess of aluminum
Alt4-7RS	Tolerance to excess of aluminum; encodes an aluminum-activated organic acid transporter gene that could be utilized to increase Al tolerance in Al-sensitive plant species
Amp syn. *Lap*	Aminopeptidase
Amy...	Amylase
an1 syn. *vi1*	Anthocyaninless
an2	Anthocyaninless leaf base
An1	Anthocyanin (green seeds)
An1a	Anthocyanin (green seeds)
An1b	Anthocyanin (green seeds)
An3,4,6	Anthocyanin (purple seeds)
An5 syn. *R, R1*	Anthocyanin (purple leaf base)
An7	Anthocyanin (purple leaf base)
Ans58	Anthocyanidin synthase of anthocyanin biosynthesis pathway
Anu...	Antinutritive component
Apr >>> Xapr	
APase	Root secretion of acid phosphatase
asc...	Asynaptic chromosome pairing
Asi >>> I(as) >>> Si	
Atp	Mitochondrial genes
Au	Aurea
AW15 >>> Xaw15	cDNA clone
Axc	Arabinoxylan content
B... >>> Xb...	Primers
BIII >>> I(a-BIII)	
BCD... >>> Xbcd...	Primers
Be syn. *Firm*	Brittle ear
Bg	Brown glumes

TABLE 5.1

(Continued) Symbolization and Abbreviations of Genes and Markers in Rye

Gene/Marker	Descriptor
beta-Glu1	β-Glucosidase
Bn	Bent lowest internode
br...	Brittle stem
bs syn. *Br*	Branched stem
br >>> bs	
Bt	Tolerance to excess of boron
Byd	Tolerance to barley yellow dwarf virus
Cat	Catalase
Cat syn. *Xpsr484*	
Cb	Brown stem
Cbt	Resistance to karnal bunt (*Neovossia indica*)
Ce	Copper efficiency
Centr	Centromere
Chi	Chalcone-flavanone isomerase of anthocyanin biosynthesis pathway
Chl	Chlorophyll deficiency
Cl	Shorter coleoptile length that seems to be partially dominant to long coleoptile; no correlation exists between seed weight and coleoptile length
cl >>> lu	
Clr1 >>> Ner	Hybrid chlorosis in wheat–rye hybrids
Cm	Chloroform–methanol proteins
Cm16 syn. *Xmtd862*	
Cnr syn. *Cre*	Resistance to cereal cyst nematode (*Heterodera avenae*)
Co	Corroded
cob	A mitochondrial gene responsible for an abnormal transcript causing male-sterile cytoplasm in rye (Pampa type)
Cxp1 syn. *Xwia483*	
Cre >>> Cnr	
CreR >>> Cnr and *Cre*	
cs	Shortened stamens and coalesced
ct...	Short straw mutants
ct2	Short straw mutant (compact spike)
ct3	Compactum growth habit
CTer	Terminal DNA sequence
Cut	Tolerance to excess of copper
Cxp	Carboxypeptidase
Cxp3 syn. *Xpsr8*	
Ddw... syn. *Dw...*	Dominant dwarfness
de	Downward directed spikes

(Continued)

TABLE 5.1
(Continued) Symbolization and Abbreviations of Genes and Markers in Rye

Gene/Marker	Descriptor
De syn. *N*	Dark ear
Dec	Dietary fibers content
Dia	Diaphorase
Dn7 syn. *Dnr* syn. *Gbr*	Resistance to the Russian wheat aphid
Dnr syn. *Gbr*	Resistance to (green bug) aphid, dominant (*Diuraphis noxia*)
dr syn. *sr*	Secondary root system defective
drr	Resistance to drought
ds1	Recessive dwarfness
dsc…	Desynaptic chromosome pairing
dw	Dwarf plant
dw8 syn. *Dw*	Dwarf plant, recessive
Dw1 syn. *Ddw1* syn. *Hl*	
Ec	Color of spike
eh	Early heading
el >>> *al*	Eligulatum
Embp syn. *Xrsq805*	
Embp1 syn. *Xrsq805*	
Embp2 syn. *Xrsq805*	
Eml-1Ra	Embryo lethality in wheat–rye hybrids, two alleles, compatible
Eml-1Rb	Embryo lethality in wheat–rye hybrids, incompatible
En(ae)	Enhancer of androgenous embryoids of wheat–rye substitutions *in vitro*
En(ai)	Enhancer of albino induction in anther culture
En(gi)	Enhancer of genome instability of wheat
En(hi)	Enhancer of anther culture ability and haploid induction
En(cg)	Enhancer of callus growth *in vitro*
En(edu)	Enhancer of equational division of univalents
En(em)	Enhancer of embryogenesis *in vitro*
Ep…	Endopeptidase
epr >>> *wa*	
epr1 >>> *wat*	
Er…	Early ripening
es >>> *wg*	
Est…	Esterase
Fa	Flag leaf area
Fbp syn. *Xpsr39*	
F3h	Flavanone 3-hydroxylase of anthocyanin biosynthesis pathway
firm >>> *Be*	
Fl	Length of flag leaf
Fr	Resistance to *Fusarium* ssp.

TABLE 5.1
(Continued) Symbolization and Abbreviations of Genes and Markers in Rye

Gene/Marker	Descriptor
fs	Fine stripe
fv	Flavovirescent
Fw	Width of flag leaf
(GACA)$_4$	Oligomer with tetranucleotide motif after fluorescence *in situ* hybridization (FISH); in the karyogram given as black dots
Gai	Gibberellic acid insensitivity
Gal	Galactosidase
Gb… >>> *Dnr* syn. *Gbr*	Green bug resistance, dominant
Gbr >>> *Dnr* >>> *Gb*	
Gbs	Primer derived from granule-bound starch synthetase I (GBSSI), also known as *waxy* gene
gd syn. *mn*	Grass dwarfness
Gerl	Gene encodes a bifunctional 3,5-epimerase-4-reductase in L-fucose synthesis
Gdh	Glutamate dehydrogenase
Glb3 syn. *Xwia807*	
Gli	Gliadin >>> secalin
Gli-R1 >>> *Sec-1*	
Gli-R3 >>> *Sec-4*	
Glob syn. *Xrsq808*	
GloB	Three-beta-globin
Glu…	Glutenin >>> secalin
Glu-R1 >>> *Sec-3*	
Glu syn. *Glu1*	
Glu1 >>> *Glu*	
Glu3 syn. *Xpsr11* >>> *Sec-3*	
Got syn. *Aat*	Aspartate aminotransferase
Gpd	Glucose-6-phosphatase dehydrogenase
Gpi syn. *Pgi*	Glucose phosphate isomerase
gr	Grassy plant habit
H syn. *Hfr*	Resistance to Hessian fly (*Mayetiola destructor*)
Hl syn. *Dwl*	Dominant short-strawed mutant
Ha syn. *Hp* syn. *Hs*	Hairy leaf sheath/peduncle
HemA	*HemA* gene encodes glutamyl-tRNA reductase
Hfr >>> *H*	
Hg syn. *V*	Hairy glume
Hl syn. *Dwl* syn. *Ddwl*	
Hml	Resistance to lethal leaf blight and ear mold disease caused by *Cochliobolus carbonum* race 1 (CCR1); common in all grasses; detected in maize and barley

(Continued)

TABLE 5.1

(Continued) Symbolization and Abbreviations of Genes and Markers in Rye

Gene/Marker	Descriptor
Hma	3-Hydroxymugineic acid synthetase
Hp >>> Hs >>> Ha	
Hs >>> Ha and *Hp*	Hairy leaf sheath
Hsp17.3 syn. *Xttu1935*	
Hsps >>> I(hsps)	
Ia >>> I(a)	
I(a…)	Alpha-amylase inhibitor
I(a-BIII)	Alpha-amylase inhibitor gene containing 354 nucleotides that encode amino acids
I(Amy) >>> I(a)	Amylase inhibitor
I(as) syn. *Si* syn. *Asi* syn. *Isa*	Alpha-amylase/subtilisin inhibitor
I(ae)	Inhibitor of androgenous embryoids wheat–rye substitutions *in vitro*
I(hsps)	Suppressor effect on accumulation of HSP18 and HSP70 transcripts
I(et)	Major endosperm trypsin inhibitor
Ibf	Iodine-binding factor
I(dha)	Inhibitor of insect alpha-amylase [rye dimeric alpha-amylase inhibitor (RDAI)]
I(dsc1)	Dominant inhibitor suppressing desynaptic chromosome pairing
I(ncw)	Inhibitor of novel cell wall formation in wheat
I(scx)	*Secale cereale* xylanase inhibitor
Il	Internode length
Isa syn. *I(as)*	
Kn	Number of seeds per spike
Kw…, kw…	Thousand-kernel weight
Lec syn. *Xmsu488*	
ln	Light nodes
Ln	Length of second internode
Iph >>> Per >>> Prx	
La	Anthocyanin in ligule
la	Leafy awn
Lap >>> Amp	
Lc	Color of leaf
Ldh	Lactate dehydrogenase
lg	Light-green leaf color
ln	Light node
Ln	Number of leaves of stem
lo	Onion-like leaves
Lr136–Lr syn. *Pr1*	Resistance to leaf (brown) rust (*Puccinia recondita, P. dispersa*) 6RL
Lr-a	Resistance to leaf rust (*P. recondita*) 1R

TABLE 5.1
(Continued) Symbolization and Abbreviations of Genes and Markers in Rye

Gene/Marker	Descriptor
Lr-c	Resistance to leaf rust (*P. recondita*) 1R
Lr-g	Resistance to leaf rust (*P. recondita*)
Lra(7)	Resistance to leaf rust (*P. recondita*), isolate 7
Lra(12)	Resistance to leaf rust (*P. recondita*), isolate 12
Lra(25)	Resistance to leaf rust (*P. recondita*), isolate25
Lra(81)	Resistance to leaf rust (*P. recondita*), isolate 81
Lra(108)	Resistance to leaf rust (*P. recondita*), isolate 108
lu syn. *cl*	Lutescent
Lw	Leaf waxiness
Lys	Lysine
Mal	Malic enzyme
Mas	Mugineic acid synthetase
mc	Monoculm growth habit
Mdh	Malate dehydrogenase
Me	Manganese efficiency
mn >>> gd	Multinodosum, dwarf habit
mn2	Dwarfness with double increased number of internodes
mo	Monstrous growth habit; shows additional spikelets per spike
mp	Multiple pistils
mrs1 >>> mo1 >>> mo	
ms	Male sterility
msh syn. *Xc, Xs, Xr*	Mismatch repair gene homologs *msh2* syn. *Xc11*; *msh3* syn. *Xs2*; *msh6* syn. *Xr2*
mu1 >>> mo	Multiflowered spike
N >>> De	
Nar	Nitrate reductase
Nc	Color of nodes
Nca	Neocentric activity
Ndh	NADH dehydrogenase
Ner	Hybrid necrosis
Ngc	Noncellulosic glucose content
Nl	Neck length
Nor	Nucleolar organizer region
np	Nana prostratum, short-stem inch-gitl
ol	Onion-accreted leaves
or	Orange
pAWRC.1	3.4-kbp repetitive segment of retrotransposon-like elements localized within the centromere of rye chromosomes without FISH signal in wheat chromosomes

(Continued)

TABLE 5.1
(Continued) Symbolization and Abbreviations of Genes and Markers in Rye

Gene/Marker	Descriptor
P(cp) >>> En(cp)	Homeologous pairing promoter
Pc	Purple culm
pe	Perennial growth habit
P(Edu) >>> En(edu)	Promoter of equational division of univalents
Per syn. *Prx*	Peroxidase
Per syn. *Xpsr833*	
Pdk1 syn. *Xhhu1*	
pfs	Partial floret sterility
Pgd	6-Phosphogluconate dehydrogenase
Pgi >>> Gpi	
Pgm	Phosphoglucomutase
Pgt	Parthenogenesis induction
Ph >>> QTL3-5RL	Plant height
Pi	Tolerance to inorganic orthophosphate (Pi) starvation stress
Pl >>> QTL4-5RL	Peduncle length
Pm	Resistance to powdery mildew
Pm20	Powdery mildew from rye prolific; confers a high level of resistance to the Chinese races of *Blumeria graminis*
pol-r	Mitochondrial gene
Pr3, Pr4, Pr5	Dominant resistance to *P. recondita* f. sp. *secalis*
Pr3 >>> Sec-4	55-kDa-seed protein
Pro	Proline
Prx >>> Per	Peroxidase
Ps syn. *Vs*	Purple seeds
psbD	Coding for components of the reaction center of photosystem II and D2 protein
psbI	Coding for a polypeptide
pSc34 >>> Xpsc34	350–480-bp family from *S. cereale*
pSc74 >>> Xpsc74	610-bp family from *S. cereale*
pSc119.2 >>> Xpsc119.2	120-bp family from *S. cereale*
pSc200 >>> Xpsc200	Highly repetitive sequence from *S. cereale*
pScJNK1 >>> XpscJNK1	Novel highly repetitive sequence from *S. cereale*
pScT7 >>> Xpsc77	5S rDNA form *S. cereale*
PSR154	cDNA clone, by HindIII
pTa71	25S–5.8S–18S rDNA from *Triticum aestivum*
Pur-R1 >>> Xpur-R1	
Py	Yield per plant
q	Speltoid spike
QTL	See Table 5.4a

TABLE 5.1
(Continued) Symbolization and Abbreviations of Genes and Markers in Rye

Gene/Marker	Descriptor
R, R1 >>> An5	
Rc	Red coleoptile
rd	Reed growth habit
re	Reduced ear
Reg	Red grain
REV1	Revolver; a transposon-like gene; consists of 2929–3041 bp with an inverted repeated sequence on each end and is dispersed through all seven chromosomes
Rf syn. *Rfc*	Male fertility restorer
Rfc >>> Rf	Male fertility restorer of the Pampa cytoplasm
Rfg	Male fertility restorer of the Guelzow cytoplasm
Rfp1	Male fertility restorer from Iranian primitive population
Rfp2	Male fertility restorer from Argentinean landrace
Rga	Enhanced root growth ability in wheat
rlt1412	cDNA clone from rye puma highest expressed in root tissue conferring information for cold tolerance
rlt1421	cDNA clone from rye puma highest expressed in leaf tissue conferring information for cold tolerance
Rm	Rye mildew
RNase	RNase
Rog	Round grain
rps	Mitochondrial genes, encoding the mitochondrial 18S ribosomal RNA
rps19	5′-Truncated mitochondrial pseudogene cotranscribed with a downstream *nad4L* gene
rr	Red rachis
rrn	Gene encoding the mitochondrial 18S ribosomal RNA
rs	Reduced seed set
3Rt	Anthocyanidin-3-glucoside rhamnosyltransferase of anthocyanin biosynthesis pathway
RYS1	A mobile element activated during tissue culture; a foldback (FB) transposon first described in rye and also the first active plant FB transposon reported
X5s...	Sequence-tagged site (STS) marker
S1 >>> al	Vertical arrangement of leaves
S syn. *Z*	Self-incompatibility
S5 syn. *Sf5*	Self-incompatibility
Sbp syn. *Xpsr804*	
Sc	Color of spike
SCX1 >>> I(scx)	
Sdf	Soluble dietary fiber content

(Continued)

TABLE 5.1
(Continued) Symbolization and Abbreviations of Genes and Markers in Rye

Gene/Marker	Descriptor
Sec…	Secalin
Sec-1	40–55-kDa ω-secalin showing 15 copies of the gene
Sec-1a >>> Sec-1	
Sec-1b >>> Sec-1	
Sec-2	40–75-kDa gamma-secalin
Sec-3	
Sec-3a >>> Sec-3	
Sec-3b >>> Sec-3	
Sec-4 syn. *Gli-R3* syn. *Pr*	
Sec-5	40–75-kDa gamma-secalin
Sf	Self-fertility
Si >>> Asi >>> I(as)	
Skd	Shikimate dehydrogenase
sl	Plants without spikes and spikelets
Sl	Length of spike
sl syn. *sp*	Spreading leaf
Sn	Number of spikelets per spike
Sod	Superoxide dismutase
sp >>> sl	
Sp syn. *Sp1* syn. *Vrn*	Spring growth habit
Spf >>> QTL13-1RS	Spikelet fertility
Sr	Resistance to stem rust (*P. graminis*)
Sr31	Resistance to stem rust (*P. graminis* f. sp. *tritici*) in 1AL.1RS translocated wheat Amigo = Insave rye
sr >>> dr	
ss1 syn. *Xpsr490*	
Ss2 syn. *Xpsr489*	
SsI(Ta)	Soluble starch synthetase I (SSI) transporter gene
Ssp…	Salt-soluble protein
Su(cp)	Homeologous pairing suppressor
Su(hsp)	Suppressor of heat shock proteins of wheat
Sut(Ta)	Sucrose transporter gene wheat, homeologous group 4
Sw	Stem waxiness
sy >>> syn	
sy1 = syn1 = asc1	It affects recombination events
sy2 = dsc1	
sy3 = asc4	Weak asynaptic mutant
sy2a = dsc2	
sy2b = dsc3	

TABLE 5.1
(Continued) Symbolization and Abbreviations of Genes and Markers in Rye

Gene/Marker	Descriptor
sy6	Causes nonhomologous synapsis
sy7	Causes nonhomologous synapsis
sy8	Causes nonhomologous synapsis
sy9 = asc2	Connected with defects of the assembly of synaptonemal complex axial cores
sy10 = asc3	Causing nonhomologous synapsis; coupled with the presence of protein Zyp1 in the core region. The assembly of proteins Asy1 and Zyp1 on the axes of meiotic chromosomes was shown to occur separately, which is a specific feature of rye, as compared to *Arabidopsis*
syn >>> asc or *dsc*	
T	Self-incompatibility
Ta(Almt1)	Aluminum-activated malate transporter renamed by the authors from TaALMT1; identical or closely linked with the *Alt4* locus
TAM… >>> Xtam…	
TBP5L/4L	Evolutionary translocation break point 5RL/4RL in rye
TBP6L/3L	Evolutionary translocation break point 6RL/3RL in rye
Tc	Color of tillers
td	Disturbed tetrad formation
tg	Tulip growth habit
Thi	Thionin
ti	Tigrina
Tp	Transposition
Tpi	Triosephosphate isomerase
trnS	Coding for tRNA^{Ser} (GCU)
Tr	Translocation break point
Tw	Thousand-grain weight
Tw1	Thousand-grain weight on chromosome 5R
Tw225	Thousand-grain weight on chromosome 7R
Tyr	Tyrosinase
tys	Transversal yellow stripes on leaf; lethal chlorophyll mutant
Uba syn. *Xpsr860*	
ug	Production of unreduced gametes in diploid rye; named by the author
V >>> Hg	
va	Viridoalbina
vb >>> vi	Viridis
Vdac1 syn. *Xtav1961*	
Vdac2 syn. *Xtav1960*	
vi syn. *an*	Viridis, virescent, anthocyaninless
vi1 >>> an1	

(Continued)

TABLE 5.1

(Continued) Symbolization and Abbreviations of Genes and Markers in Rye

Gene/Marker	Descriptor
Vrn >>> Sp	
Vs >>> Ps	Violet seeds
w >>> wal	Waxless plant with waxy nodes
wa syn. *epr*	
wal	Waxless plant
wg syn. *es*	Waxless glume
WG… >>> Xwg…	
Wh	White
Whg	White glume
Wsm…	Resistance to wheat streak mosaic virus
Wsp…	Water-soluble proteins
wx >>> wal	Waxy endosperm
xa	Xantha
Xabc…	cDNA clones of barley
X…	Molecular/DNA markers
yns	Yellow necrotic spots on leaves
Yr	Resistance to yellow rust (stripe rust) (*Puccinia striiformis*)
ys	Yellow stripe
Z >>> S	Self-incompatibility
1RS2.3	Indicates a region (band) along a chromosome according to the karyogram (see Figure 4.5)
1RS1.1-1RS2.1	Indicates a region between two bands (1RS1.1 and 1RS2.1) according to the karyogram (see Figure 4.5)

Source: Schlegel, R., and Korzun, V., Genes, markers and linkage data of rye (*Secale cereale* L.), 9th updated inventory, V. 03.13, 2013. With permission.

5.2 GENERAL

Because of the allogamous growth habit of rye, genetic studies appeared reluctant. First monogenic inherited characters were published as follows (see Table 5.2).

Polygenic inheritance was early observed for culm length with prevalence of long to short stem, for grain shape with prevalence of long to short seeds, for spike shape with prevalence of long to short, wide to narrow, dense to loose, as well as nodding and upright spike, or with prevalence of perennial to annual growth habit, respectively (Tschermak 1906). Despite its allogamous flowering, recessive traits (albinos, waxless, etc.) appear as off-types even in population varieties with low frequency— between 0.01% and 0.02% (Roemer and Rudorf 1950).

Since the first description of a mutant with forked spikes in 1757 (see Section 7.1.5), there is a tremendous progress in rye genetics. Thus, one can estimate more than 250 years' genetic studies in rye.

TABLE 5.2

Compilation of First Genetic Studies in Rye (*Secale cereale*), Demonstrating 3:1 Segregation in F₂

Trait	Allele		Author (Year)
	Dominant	Recessive	
Grain color	Green	Yellow	Rümker and Leidner (1914)
	Black	Yellow	Steglich and Pieper (1922)
Coleoptile color	Red	Green	Treboux (1925)
Seedling color	Green	White	Nilsson-Ehle (1913)
	Green	Green-white	Brewbaker (1926)
	Green	Aurea variegated	Brewbaker (1926)
Leaf type	Waxy	Waxless	Nilsson-Ehle (1928)
Ligula	With	Without	Krasnyuk (1936)
Culm	Hairy	Hairless	Tschermak (1906)
	Normal	Brittle	Brewbaker (1926)
Rachis	Brittle	Tough	Tschermak (1906)
Glume	Green	White	Sirks (1929)
Glume color	Brown	Yellow	Berkner and Meyer (1927)
Fertility	Self-sterile	Self-fertile	Heribert-Nilsson (1916b)
	Self-fertile	Self-sterile	Peterson (1934)
Seed set	Normal	Jagged	Pammer (1905)
Plant ripening	Early	Late	Tschermak (1906)
Growth habit	Spring-type	Winter-type	Tschermak (1906)

Source: Roemer, T., and Rudorf, W., *Handbuch der Pflanzenzüchtung*, Bd. 2, Grundlagen der Pflanzenzüchtung, P. Parey Verl., Berlin, 1950. With permission.

The development and utilization of new techniques in genetic analysis, karyotype identification and chromosome manipulation, and the elucidation of homeologous relationships in the Triticeae have greatly contributed to the present understanding of genetics and cytogenetics of rye. Knowledge of rye genetics, for a long period, was less developed compared to other diploid crops (maize, barley, tomato, or pea), is progressing, and becomes useful in marker-aided selection. While Jain (1960) in his review article stated that the genetics of rye had not received so much attention and those simple markers were hardly available, genetic and cytological markers for all rye chromosomes are now available in reasonable amount.

Based on compilation of the available data of genetics in rye, wheat, and triticale as well as wheat–rye addition and substitution lines from altogether 400 literature references, a comprehensive presentation of gene designation, localization, and linkage relationships has been established (Schlegel and Korzun 2013). Since 1960 (Jain 1960), along with the updates in 1982 (Schlegel and Mettin 1982), 1986 (Schlegel et al. 1986b), 1996 (Schlegel and Melz 1996), and 1998 (Schlegel et al. 1998b), recently there are ~1750 biochemical, molecular (see Figure 5.1), and morphological markers available. Of this number, ~86% are molecular markers (Rogowsky et al. 1992,

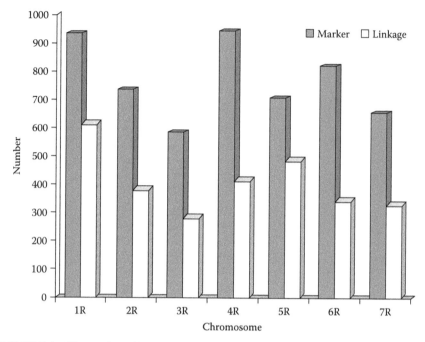

FIGURE 5.1 The number of localized genes and markers as well as linkages between genes and markers of individual chromosomes: Recent status. (After Schlegel, R., and Korzun, V., Genes, markers and linkage data of rye (*Secale cereale* L.), 9th updated inventory, V. 03.13, 2013.)

Quarrie et al. 1994) gathered during the last two decades, 11% are biochemical markers (Fra-Mon et al. 1984, Hart and Gale 1988, McIntosh 1988, Hull et al. 1991, 1992), and ~3% are mapped morphological features. First molecular linkage maps of rye were published in the early 1990s. Meanwhile suitable markers exist for important resistance and fertility traits applied in hybrid breeding of rye (Dreyer et al. 1996). The length of rye linkage maps varies between 340 and 3145 cM. First QTLs were mapped on several rye chromosomes (Schlegel and Meinel 1994, Börner et al. 2000) (see Section 5.5). The latest situation displays ~5400 markers, such as isozymes, RFLPs, RAPDs, SSRs, and amplified fragment length polymorphisms (AFLPs) (see Table 5.3). Molecular markers in rye are particularly suitable for the following:

1. Gene localization and characterization of loci controlling complex traits (yield, preharvest, etc.) or traits with low heritability
2. Marker-assisted selection in breeding (i.e., traits selected based on a marker)
3. Translocation studies within the rye genome or between homeologous cereal genomes
4. Genetic association mapping, that is, for out-crossing rye that requires very large number of markers or for whole-genome association;
5. High-resolution mapping, that is, searching for genetic linkages <1 cM

TABLE 5.3
Types of Molecular Markers So Far Applied in Rye

Type	Subtype	Notes
Hybridization-based markers	Restriction fragment length polymorphism (RFLP)	A common technique during the 1990s; it is laborious and reliable, and needs large amount of DNA; restriction enzymes (e.g., PstI, EcoRI) are used to cut DNA into fragments; fragments are hybridized on the membrane; the specific DNA is recognized with a stained probe (cf. Wang et al. 1991)
	Amplified fragment length polymorphism (AFLP)	DNA is cut by restriction enzymes (EcoRI, MseI, PstI); double-stranded adapters are ligated to the end of DNA fragments; DNA is then amplified between the known adapters; the method is rather complicated and needs large amount of DNA (cf. Bednarek et al. 2003)
	Diversity arrays technology (DArT)	Whole-genome fingerprints are generated by scoring the presence or absence of DNA fragments from genomic DNA sample; it provides a high number of markers; the technology is licensed (cf. Bolibok-Brągoszewska et al. 2009)
PCR- and DNA sequence-based markers	Random amplified polymorphic DNA (RAPD)	Arbitrary primers (10 bp) are used to amplify DNA; the method is quick and simple; reproducibility is sometimes difficult (cf. Iqbal and Rayburn 1995)
	Single-sequence repeats (SSRs) or micosatellites	A fragment with one to six nucleotides is repeated many times; the primers are outside the microsatellites; the markers are rather expensive to develop; however, the method is simple and transferable between populations; recently, it is the most widely used marker type in cereals (cf. Resch 2009)
	Intersingle-sequence repeat (ISSR)	DNA between two closely spaced inversely oriented microsatellites is amplified; there are *Secale cereale*-specific intermicrosatellites (SCIMs) (cf. Bolibok et al. 2005)
	Selective amplification of microsatellite polymorphic loci (SAMPLs)	Resembles the AFLP procedure (cf. Bolibok et al. 2005)
	Sequence-related amplified polymorphism (SRAP)	One of the primers anneals to an intron and the other to an exon (cf. Tenhola-Roininen et al. 2011)
	Sequence-characterized amplified region (SCAR) or sequence-tagged site (STS)	Developed from random markers by sequencing the original markers; then constructing a new pair of PCR primers that detect only one specific locus (cf. Stracke et al. 2003)

(Continued)

TABLE 5.3
(Continued) Types of Molecular Markers So Far Applied in Rye

Type	Subtype	Notes
	Single-nucleotide polymorphism (SNP)	Distinguishes one nucleotide difference between alleles in a single locus; DNA sequence data are needed; it yields sufficient polymorphism for rye mapping (cf. Varshney et al. 2007)
Retrotransposon-based markers	Interretrotransposon amplified polymorphism (IRAP)	DNA sequence products between two retrotransposons are amplified (cf. Tenhola-Roininen et al. 2011)
	Retrotransposon-microsatellite amplified polymorphism (REMAP)	DNA sequence products between a retrotransposon and a microsatellite are amplified (cf. Tenhola-Roininen et al. 2011)
	Sequence-specific amplified polymorphism (SSAP)	Resembles AFLP; the DNA fragment between a retrotransposon and a restriction site is amplified (cf. Nagy and Lelley 2003)

The best investigated is chromosome 4R with ~947 mapped loci, followed by chromosomes 1R (936 loci), 6R (822 loci), 2R (735 loci), 5R (703 loci), 7R (653 loci), and 3R (583 loci). The most DNA markers exploited in rye linkage maps are SSR, RAPD, and RFLP.

The difficulties in mapping experiments are the allogamous growth habit (i.e., heterozygous individuals when doubled haploids are not available) and the changing populations (varieties) of rye (i.e., allele numbers and allele frequencies are modified after each multiplication). The detection of rare alleles depends on the sample size used for the marker analyses. In addition, only a small part of the genome is sequenced. Thus, whole-genome sequencing becomes a difficult task. For example, the short arm of chromosome 1R (1RS) contains as much DNA base pairs as the complete rice genome! It represents approximately 5.6% of the rye genome with a genome size of 441 Mbp. Therefore, most rye maps are not saturated yet with molecular markers. Most of the maps contain less than 500 markers.

Implementation of molecular breeding in rye improvement programs depends on the availability of high-density molecular linkage maps. Several genetic maps of rye have been published so far, but the possibilities of their practical application are rather limited, mainly due to an insufficient saturation (Korzun et al. 1998). The average density of the most saturated rye linkage map published to date is 2.9 cM. However, there are gaps greater than 20 cM on maps. The average map interval length on the remaining rye maps exceeds 4.0 cM; in the majority of maps it is greater than 7 cM. On the other hand, as the number of sequence-specific, polymerase chain reaction (PCR)-based markers [SSRs and sequence-characterized amplified region (SCARs)] available for the species was until recently below 300, rye genetic maps were constructed predominantly with the use of usually poorly transferable AFLP and RAPD markers, or labor-intensive, time-consuming, and low-throughput RFLP markers. The highest number of sequence-specific, PCR-based markers on a rye linkage map equals 58. Lately, with the creation of bacterial artificial chromosome (BAC) library specific for 1RS and the development of 74 SSR markers for 1RS, a significant advance took place in rye genomics.

Using the *PstI/TaqI* method on the DNA of 16 rye varieties and 15 rye inbred lines, the mapping chances were significantly broadened. Microarray-based marker technologies, such as the so-called Diversity Arrays Technology (DArT) (Bolibok-Bragoszewska et al. 2009) or Rye5K-SNP Genotyping Array (Haseneyer et al. 2011) are highly suitable. They can be efficiently and effectively used for genome analyses—assessment of genetic similarity and linkage mapping, that is, association mapping.

Bolibok-Bragoszewska et al. (2009) produced DArT markers derived from more than 1022 clones that were polymorphic in the genotyped isogeneic lines and varieties and 1965 clones that differentiated the parental lines L318 and L9. Hierarchical clustering and ordination analysis were performed based on the 1022 DArT markers to reveal genetic relationships between varieties and inbred lines. Chromosomal location of 1872 DArT markers was determined using wheat–rye addition lines and 1818 DArT markers (among them 1181 unique, noncosegregating) were placed on a genetic linkage map of the cross of L318 into L9, providing an average density of one unique marker every ~2.7 cM. This was the most saturated rye linkage map based solely on transferable markers.

It allows comparing genetic relationships among varieties and genetic distances between inbred lines—an invaluable relief in the selection of partners for a crossing. Genetic similarity coefficients for all 600 possible pairs of genotypes calculated based on the Jaccard's coefficient ranged from 0.26 in the pair of inbred lines Ot1-3 to 0.98 in each pair of variety Dańkowskie Złote (see Figure 5.2).

Genotyping of wheat–rye addition lines on the rye genotyping array revealed 1872 (16.3%) DArT markers that were present in genomic representation of only one of the addition lines, and therefore, their chromosomal location could be determined. Proportions of DArT markers localized on individual chromosomes were not

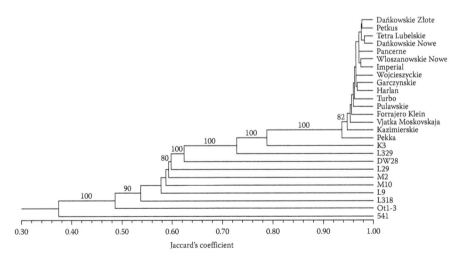

FIGURE 5.2 Unweighted Pair Group Method with arithmetic mean (UPGMA) dendrogram based on Jaccard's similarity matrix data obtained with 1022 DArT markers comparing rye varieties as well as inbred lines. (After Bolibok-Brągoszewska, H. et al., *BMC Genomics*, 10, 578–586, 2009.)

uniform and varied from 9.4% for the chromosome 1R (175 DArT markers) to 18.0% for the chromosome 6R (337 DArT markers). The result of the Spearman's rank correlation analysis of these proportions with physical lengths of rye chromosomes was not statistically significant; that is, there is a clear association of number of DArT markers and physical length of the individual rye chromosomes (see Figure 5.3). From the DArT marker with determined chromosomal location, 367 (19.6%) displayed polymorphism in parental lines of mapping population and thus can be used as anchor loci during map construction.

Musmann et al. (2012) reported ~3117 informative markers distributed across the genome. On elite breds, linkages to plant characteristics, such as plant height and 1000-grain weight, became evident.

It is a population-based survey of molecular marker analysis in order to identify trait–marker relationships based on linkage disequilibrium. The association between a pair of linked markers is also called linkage disequilibrium or, less frequently, gametic disequilibrium. However, association has a broader meaning that includes combinations of three or more linked markers, at least some of which are in linkage disequilibrium. These combinations are called haplotypes if specified for a single chromosome. Haplotypes occur in pairs, called diplotypes, consisting of one haplotype from each parent. The two haplotypes of a diplotype cannot be ascertained with certainty if two or more markers are heterozygous, except in special cases that include family studies, physical separation of chromosomes, and zero frequency of alternative haplotypes. Association mapping depends on the choice of map taken to represent linkage disequilibrium. Physical maps specify the distance in the DNA sequence, ideally measured in base pairs. The closest approach to this ideal is by the DNA sequence nominally finished, although errors in many relatively small areas remain and, of course, polymorphisms affecting the DNA sequence are represented

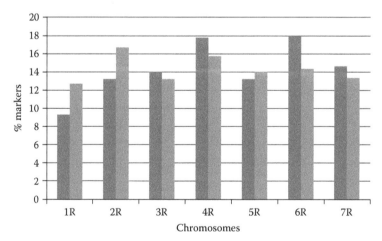

FIGURE 5.3 Relationship between the proportion of DArT markers localized on individual chromosomes using wheat–rye addition lines (dark bars) and the proportion of the physical length of the rye genome represented by individual chromosomes (pale bars). (After Schlegel, R. et al., *Theor. Appl. Genet.*, 74, 820–826, 1987; Bolibok-Brągoszewska, H. et al., *BMC Genomics*, 10, 578–586, 2009.)

by one arbitrary allele. For association mapping, it is convenient to represent location in kilobases to three decimal places, retaining full precision in the finished maps.

The 11,520-clone rye genotyping panel with several thousand markers with determined chromosomal location and accessible through an inexpensive genotyping service is a valuable resource for studies on rye genome organization and in molecular breeding (see Section 7.1.3). The association mapping method is an alternative to the QTL mapping. It has the advantage that it can be applied directly on elite strains, and thus immediately be used for the development of commercial varieties. Furthermore, it allows evaluating a potentially large number of alleles per locus. The low linkage disequilibrium in rye (Li et al. 2011b) offers a high resolution in genome-wide assays. Due to the large size of rye genome (8.1 Gbp), great efforts in preceding genotyping are required. Microarray-based marker technologies are advantageous. They consider multiparallel analysis of many loci, high degree of polymorphism, and high reproducibility of DNA fingerprints.

After detecting the high degree of collinearity within the grasses, rye became a candidate for comparative mapping as well. Based on the evolutionary conservation of cereal chromosomes, gene location in one species can be carried out rapidly, if information about the chromosomal location or mapping position in related species is already known and transferable. However, existing translocation differences between the species have to be considered. Microdissection and microcloning using rye chromosome 1R were first performed by Zhou et al. (1999).

5.2.1 MICRODISSECTION

Chromosome 1R was microdissected and collected from mitotic metaphase spreads by using glass needles. The isolated chromosomes were amplified *in vitro* by Sau3A linker adaptor-mediated PCR. After amplification, the presence of rye-specific DNA was verified by Southern hybridization. The second-round PCR products from five 1R chromosomes were cloned into a plasmid vector to create a chromosome-specific library, which produced approximately 220,000 recombinant clones. Characterization of the microclone library showed that the 172 clones evaluated ranged in size from 300 to 1800 bp with an average size of 950 bp, of which ~42% were medium/high copy and ~58% were low/unique copy clones. Chromosome *in situ* hybridization confirmed that the PCR products from microdissected chromosomes originated from chromosome 1R, indicating that many 1R-specific sequences were present in the library.

5.3 CHROMOSOMAL AND/OR REGIONAL LOCALIZATION OF GENES AND MARKERS

Among the first genes localized, three secalin loci have been reported. *Sec-1* and *Sec-3* loci are located on the short and the long arms of chromosome 1R, respectively. The *Sec-1* locus on 1RS harbors the γ-secalin (40 kDa) and ω-secalin (~45 kDa) genes, while the *Sec-3* locus on 1RL contains the high molecular weight (HMW) secalin genes (~100 kDa). The *Sec-2* locus located on the short arm of chromosome 2R codes for the 75-kDa γ-secalin genes. All 400 sequence reads were found to be related to the secalin genes. The majority (316 of the identified reads) represented ω-secalin genes, which were reported to be present in ~15 copies in 1RS, while 45 of them were related to secalin. In addition, 39 sequence reads could not be assigned to secalin or ω-secalin genes because of the high level of sequence homology of the two secalins in certain regions.

The 45S and the 5S rDNA (ribosomal DNA) loci were found in close vicinity to the distal portion of the chromosome arm 1R as shown by *in situ* hybridization and were reported to be present in ~2000 and 5000 copies comprising ~18 Mbp (3.9%) and 2.3 Mbp (0.5%) of the 1RS genome, respectively. This suggested a 7.8-fold difference in the spatial requirement of the two loci. For the 45S rDNA locus, 15,962 sequence reads (~1.8%) and for the 5S rDNA locus 2361 sequence reads (~0.3%) were identified, yielding a 6.75-fold difference.

In the text that follows, the localization of markers along the chromosomes and chromosomal regions is given (see Tables 5.4a–g). Since different authors published different results, often for the same marker, all available data have to be considered.

Because of limited space in this book, numerous authors, most of the mapping and linkage data, gene order along the chromosomes, and varying data of localization

---

TABLE 5.4A

A Selection of Genes and Markers Associated with Long and/or Short Arm of Chromosome 1R

1RS	AawR173-3P	Sr+	Xem1.18	Xora4	Xs17d-10a	Xtsm132	Xtsm552
	AawR173-3K	Sr1	Xem1.19	Xora6	Xs17d-12a	Xtsm143	Xtsm553
	Cl-1	Sr31	Xem1.20	Xora7	Xs17d-12b	Xtsm145	Xtsm556
	Cm1	5S-rDNA1	Xem1.21	Xora8	Xs17d-12c	Xtsm149	Xtsm559
	Cm3	18S-26S-rDNA	Xem1.22	Xora11	Xs17d-14a	Xtsm162	Xtsm565
	Cm4	Tr(T273W)	Xem1.24	Xora12	Xs17d-14c	Xtsm179	Xtsm575
	Dn7	Tri	Xem1.25	Xplug1017	Xs17u-1	Xtsm181	Xtsm587
	Dnr1	Wsm	Xem1.27	Xp6m12-P	Xscg75	Xtsm191	Xtsm592.2
	Dnr2	Xabc156	Xem1.28	Xp12/m53	Xscm9	Xtsm197	Xtsm593
	Dnr6	Xabc160	Xem1.29	Xp12/m59	Xs17u-9	Xtsm200	Xtsm596
	En(cg)1	Xad16-850	Xem1.30	Xp12/m61	Xs20-2a	Xtsm211	Xtsm598
	En(em)1	Xapr1.2	Xem1.31	Xp15/m53	Xs20-2b	Xtsm213	Xtsm604
	En(em)2	Xapr1.3	Xem1.32	Xp16/m56	Xs20-5	Xtsm221	Xtsm608
	En(hi)1	Xapr1.5	Xem1.33	Xp17/m57	Xs20-9	Xtsm228	Xtsm609
	Gpi1	Xapr1.6	Xem1.34	Xp17/m60	Xs20-16	Xtsm230	Xtsm621
	I(as)2	Xapr1.8	Xem1.35	Xp17/m60	Xscb190	Xtsm235	Xtsm623
	L1Per1	Xapr1.9	Xem1.36	Xpic31	Xscb241	Xtsm239	Xtsm625
	Lr1	Xapr1.10	Xgpi	Xpsc200	Xscb258b	Xtsm241	Xtsm626
	Lr3	Xapr1.21	Xglu3	Xpsr11	X5sdna	Xtsm264	Xtsm634
	Lr(7)	Xapr1.22	Xia215	Xpsr109	Xscg75	Xtsm266	Xtsm638
	Lr(12)	Xapr1.23	Xia294	Xpsr158	Xscm1	Xtsm268	Xtsm641
	Lr(25)	Xapr1.24	Xia299	Xpsr161	Xscm9	Xtsm275	Xtsm642
	Lr26, Sr31, Yr9	Xba1	Xiag79	Xpsr168	Xscm21	Xtsm279	Xtsm645
	Lr(81)	Xbsd12	Xiag95	Xpsr300	Xscm32	Xtsm282	Xtsm651
	Lr(108)	Xbcd22	Xiag229	Xpsr391	Xscm36	Xtsm294	Xtsm652
	msh2 (Xcl1)	Xbcd98	Xiag249	Xpsr393	Xscm39	Xtsm303	Xtsm655
	Nor1	Xbcd115	Xib159	Xpsr544	Xscm127	Xtsm306	Xtsm656

TABLE 5.4A

(Continued) A Selection of Genes and Markers Associated with Long and/or Short Arm of Chromosome 1R

pAWRC.1	Xbcd200	Xib262	Xpsr596	Xscm269	Xtsm312	Xtsm661
Per2	Xbcd921	Xib267	Xpsr601	Xtc6051	Xtsm314	Xtsm662
Per7	Xbcd927	Xib544	Xpsr634	Xtc6893	Xtsm315	Xtsm665
Per13/14	Xbcd1072	Ximc1	Xpsr937	Xter	Xtsm319	Xtsm676
Pgi	Xbcd1434	Ximc2	Xpsr941	Xtri	Xtsm322	Xtsm680
Pgt	Xbm2	Xir15	Xpsr949	Xtrx1100	Xtsm325	Xtsm683
Pm	Xbmac213	Xir26re7	Xpsr957	Xtsm10	Xtsm326	Xtsm685
Pm8	Xcdo99	Xis1.1	Xpsr960	Xtsm12	Xtsm329	Xtsm690
Pm111	Xcdo344	Xissr5	Xpsr688	Xtsm16	Xtsm332	Xtsm700
Pr3	Xcdo580	XksuD14	Xr173-1	Xtsm21	Xtsm347	Xtsm704
Pr-i	Xcdo658	XksuE18	Xr173-2	Xtsm25	Xtsm350	Xtsm706
pSc200	Xcdo1173	XksuE19	Xr173-3	Xtsm27	Xtsm355	Xtsm708
pSchet1	Xcdo1188	XksuG	Xrems1135	Xtsm29	Xtsm364	Xtsm714
pScT7	Xcsih69	XksuG9	Xrga39	Xtsm39	Xtsm366	Xtsm716
pSec2B	Xcslh69	Xlprm61	Xrms13	Xtsm54	Xtsm387	Xtsm718
pTa71	Xcslh6910	Xlprm62	X5s	Xtsm61	Xtsm411	Xtsm719
pTa794	Xcslh6913	Xlrk10	Xs10-1	Xtsm81	Xtsm422	Xuah30
QTL13	Xcslrgh3.1	Xme4em3	Xs10-6	Xtsm86	Xtsm435	Xubp18
QTL21	Xcslrgh3.2	Xme4em5	Xs10-11	Xtsm92	Xtsm460	Xubp19
QTL24	Xeacc-mctg424	Xmsh2	Xs10-16a	Xtsm94	Xtsm461	Xucr2
QTL42	Xeacg-mcgt189	Xmwg60	Xs10-16b	Xtsm103	Xtsm468	Xucr3
QTL65	Xeata-mcgc410	Xmwg68	Xs10-16c	Xtsm104	Xtsm469	Xucr4
QTL79	Xecgc-mcag163	Xmwg506	Xs17d-3a	Xtsm106	Xtsm471	Xucr5
QTL89	Xem1.1	Xmwg837	Xs17d-3b	Xtsm108	Xtsm472	Xucr6
rd	Xem1.2	Xmwg913a	Xs17d-3c	Xtsm109	Xtsm480	Xucr7
Rga	Xem1.5	Xmwg913b	Xs17d-4	Xtsm111	Xtsm492	Xucr
S5	Xem1.6	Xmwg938	Xs17d-7a	Xtsm118	Xtsm497	Xulp18
S6	Xem1.8	Xmwg2062a	Xs17d-7c	Xtsm120	Xtsm502	Xwg184
Sec-1	Xem1.9	XNor1	Xs17d-9	Xtsm121	Xtsm520	Xwg876
Sec-3	Xem1.10	Xopr2		Xtsm123	Xtsm535	Yr1
Sec-4	Xem1.11	Xora3				
	Xem1.15					
centro	msh2 (Xc11)	pAWRC.1	Pgi	S5	Xcdo344	Xcsih69	Xucr1

1RL				
	Alp1	Xbcd98	Xgwm861	Xpsr159
	Alp3	Xbcd115	Xgwm1166	Xpsr162
	Bm2	Xbcd200	Xgwm1223	Xpsr313
	Dnr2	Xbcd207	Xgwm1300	Xpsr325
	fs	Xbcd265	Xhvaba	Xpsr330
	Lap1	Xbcd304	Xhvhava1	Xpsr361
	Lr4	Xbcd304r	Xiag22	Xpsr391
	Lr5	Xbcd310	Xiag23(1)	Xpsr393S
	Lr-g	Xbcd342	Xiag79	Xpsr465
	Mdh1	Xbcd342(u)	Xiag111	Xpsr544

(Continued)

TABLE 5.4A

(Continued) A Selection of Genes and Markers Associated with Long and/or Short Arm of Chromosome 1R

Mdh2a	*Xbcd371*	*Xiag138*	*Xpsr547b*
Pgd3	*Xbcd386*	*Xiag140*	*Xpsr568*
Pr4	*Xbcd442*	*Xiag186*	*Xpsr601*
Pr5	*Xbcd446*	*Xiag220*	*Xpsr626*
pSc119.2	*Xbcd448*	*Xiag249*	*Xpsr653*
pSc200	*Xbcd454*	*Xir29re9*	*Xpsr953*
pSc503	*Xbcd592*	*Xir32*	*Xpsr958*
QTL6-1RL	*Xbcd738*	*XksuD49*	*Xpur*
QTL14	*Xbcd808*	*XksuE8*	*Xrms1010*
QTL19	*Xbcd1261*	*Xlec1*	*Xrems1280*
QTL60	*Xcdo77*	*Xlprm3*	*Xrms1107*
QTL66	*Xcdo98*	*Xlprm15*	*Xscg24*
QTL88	*Xcdo99e*	*Xlprm54*	*Xscg45*
S1	*Xcdo105*	*Xme1em3-149*	*Xscg70*
Sec-3	*Xcdo473*	*Xme2em4-510*	*Xscg78*
Thi	*Xcdo572*	*Xme3em5-500*	*Xscm1*
Xab06	*Xcdo658*	*Xm*	*Xscm4*
Xabc160	*Xcdo1396*	*Xmsu488*	*Xscm39*
Xadh	*Xcslh69.11*	*Xmwg506*	*Xscm39*
Xadpg2(r)	*Xcslh69.12*	*Xmwg912*	*Xscm83*
Xaf07	*Xcslh69.14*	*Xmwg2077*	*Xscm107*
Xapr1.1	*Xeacc-mctg302*	*Xopa17-466*	*Xscm126*
Xapr1.4	*Xeacg-mccg117*	*Xopy11-1400H*	*Xscm136*
Xapr1.7	*Xeata-mcct259*	*Xp12/m52-f-306*	*Xscm165*
Xapr1.11	*Xeata-mcta158*	*Xp12/m53-f-138*	*Xscm171*
Xapr1.12	*Xecgc-mctc119*	*Xp12/m54-f-216*	*Xscm177*
Xapr1.13	*Xecgg-mcct83*	*Xp12/m61-f-194*	*Xscm247*
Xapr1.14	*Xecgg-mctc99*	*Xp15/m53-f-277*	*Xscm274*
Xapr1.15	*Xecgg-mctg239*	*Xp15/m53-f-284*	*Xscm340*
Xapr1.16	*Xem1*	*Xp16/m58-f-167*	*Xscsz732*
Xapr1.17	*Xem1.3*	*Xp16/m58-f-366*	*Xsec3*
Xapr1.18	*Xem1.4*	*Xp17/m57-f-187*	*Xtam2*
Xapr1.19	*Xem1.7*	*Xp17/m60-f-156*	*Xtam22*
Xapr1.20	*Xem1.12*	*Xp17/m60-f-237*	*Xter1*
Xawpc-a4	*Xem1.13*	*Xpgk1*	*Xuah39(a)*
Xawpc-b4e	*Xem1.14*	*Xppdk1*	*Xubp24*
Xawpc-aftp327	*Xem1.16*	*Xpsr12*	*Xwg180*
Xbcd12(W)	*Xem1.17*	*Xpsr19*	*Xwg241*
Xbcd22(W)	*Xem1.23*	*Xpsr78*	*Xwg605*
Xbcd12	*Xem1.26*	*Xpsr95*	*Xwg983*
Xbcd22	*Xgwm752*	*Xpsr158*	*Xwye838*

Source: Additional loci and details can be taken from Schlegel, R., and Korzun, V., Genes, markers and linkage data of rye (*Secale cereale* L.), 9th updated inventory, V. 03.13, 2013. With permission.

Note: The prefix "*X*" designates molecular markers of different origin.

TABLE 5.4B

A Selection of Genes and Markers Associated with Long and/or Short Arm of Chromosome 2R

2RS	asc3	Xapr2.7	Xem2.2	Xmelem5-336	Xpsr112(L)
	Embp.1	Xapr2.8	Xem2.3	Xmelem6-106	Xpsr130
	Embp.2	Xapr2.10	Xem2.4	Xmsh3	Xpsr130-220
	Est6	Xapr2.11	Xem2.5	Xmwg858	Xpsr131
	Gdh1	Xapr2.18	Xem2.7	XnabgmpD17	Xpsr143
	I(as)1	Xapr2.19	Xem2.8	Xopa5-530	Xpsr160
	L2Per2	Xapr2.21	Xem2.9	Xopa8	Xpsr331
	L2Per3a	Xapr2.22	Xem2.10	Xopb12	Xpsr386
	L2Per 3b	Xapr2.23	Xem2.12	Xopc7	Xpsr489(S)
	L2Per4	Xapr2.24	Xem2.13	Xopf01-263	Xpsr540
	Lr6	Xbcd102	Xem2.14	Xopi9	Xpsr549
	Lr8	Xbcd111	Xem2.15	Xopj13-450	Xpsr666
	mo1	Xbcd120	Xem2.16	Xopj13-1670	Xpsr900
	Per1	Xbcd266	Xem2.17	Xopj13-1730	Xpsr932
	Per2	Xbcd880	Xem2.18	Xopj18	Xper2
	Per3	Xbcd292	Xem2.20	Xopm1	Xrsq805
	Per4	Xbcd1086	Xem2.21	Xopo10	Xscg46
	pSc200	Xbzh834	Xem2.22	Xopp6	Xscg60.2
	pSc503	Xcdo59	Xem2.23	Xopq14-1205	Xscm10
	QTL25	Xcdo78	Xem2.24	Xopq14-1270	Xscm31
	QTL39	Xcdo718	Xem2.25	Xopt12-1060	Xscm32
	QTL55	Xcdo795	Xem2.26	Xopt12-1120	Xscm35
	QTL67	Xcdo1328b	Xem2.27	Xopy7	Xscm43
	R21800	Xcmwg660	Xiag33	Xopy20	Xscm73
	Rf2	Xcmwg699	Xiag57	Xp12-m54-f-143	Xscm75
	Sec-2	Xcmwg720	Xiag61a	Xp12-m59-f-131	Xscm118
	Sec-5	Xeacc-mctg216	Xiag167	Xp15-m53-f-134	Xscm153
	Sod1	Xeacg-mctc424	Xiag233	Xp16-m56-f194	Xscm169
	Sod2	Xeacg-mcgt92	Xiag240a	Xp16-m58-f321	Xscm188
	Ss2-S	Xeacg-mctt486	Xiag666	Xp17-m57-f-58	Xscm233
	Xa5530	Xecgc-mccg332	Xir30-1300	Xp17-m57-f-139	Xscm290
	Xad60-850	Xecgc-mcgc362	XksuD17	Xp17-m60-f-504	Xscm298
	Xapr2.2	Xecgc-mctg315	XksuF41	Xpsr107	Xscm357
	Xapr2.3	Xecgg-mcta58	Xlprm10	Xpsr109	Xscm363
	Xapr2.5	Xem2.1	Xmelem1-369	Xpsr109a	Xss2
	Xapr2.6				
centro	pAWRC.1		Xscm43		Xsi1
2RL	al	Xbcd266	Xiag229d	Xopq14	Xrems1130
	asc	Xbcd512	Xiag233	Xopr2-1800	Xrems1132
	beta-Glu1	Xcdo36	Xiag235	Xopr3	Xrems1138
	el	Xcdo78	Xiag240b	Xops9	Xrems1194
	Est3/5	Xcdo147c	Xiag240d	Xops16	Xrems1203

(Continued)

TABLE 5.4B
(Continued) A Selection of Genes and Markers Associated with Long and/or Short Arm of Chromosome 2R

Est6	Xcdo595	Xiag267a	Xopt8	Xrems1208
Est7	Xcmwg609	Xir10-690	Xp12-m52-f-232	Xrems1230
F3h	Xcmwg699	Xir12re2-1350	Xp12-m53-f-129	Xrems1238
H1	Xcmwg720	Xir13-1400	Xp12-m53-f-146	Xrems1251
H2l	Xeacc-mcac199	Xir21re7-900	Xp12-m57-f-335	Xrsq803
Mdh1	Xeacc-mcct454	Xis2.1	Xp15-m53-f-150	Xrsq805.1
msh3 (Xs2)	Xeacg-mcgt167	Xisa1	Xp16-m57-f-176	Xrsq805.2
Pgd4	Xeacg-mcgt288	Xissr1	Xp17-m57-f-99	Xrz698
Pm2	Xeacg-mctt443	XksuF41	Xp17-m57-f-440	Xscg50
PmJZHM	Xeacg-mctt510	Xmelem3-335	Xp17-m60-f-102	Xscg57
pSc34	Xeata-mcgt168	Xmelem4-106	Xp17-m60-f-120a	Xscm41
pSc200	Xeata-mctg270	Xme2em5-348	Xp17-m60-f-173	Xscm43
pScJNK1	Xecgc-mccc113	Xme2em5-365	Xp17-m60-f-31	Xscm61
QTL1	Xem2.6	Xme4em1-213	Xpsr102	Xscm73
QTL36	Xem2.11	Xme4em4-198	Xpsr107	Xscm75
QTL43	Xem2.19	Xme4em5-550	Xpsr109	Xscm186
QTL49	Xem2.28	Xme5em3-261	Xpsr112	Xscm215
QTL61	Xembp(1)	Xmsh3	Xpsr112L	Xscm254
QTL75	Xembp(2)	Xmwg829	Xpsr143	Xscm276
QTL85	Xest6	Xmwg949	Xpsr151	Xscm328
Si1	Xgwm132	Xnsft03p	Xpsr331	Xsi1
Sod1	Xgwm526	Xopd12-600	Xpsr349	Xtc101821
tg	Xgwm791	Xopd12-930	Xpsr388	Xtc108778
Wsp1	Xgwm877	Xopd16	Xpsr390	Xtc116908
Xab06-690	Xgwm911	Xopf08-311	Xpsr489S	Xtc17178
Xapr2.1	Xgwm912	Xopf13	Xpsr540	Xtc31342
Xapr2.4	Xgwm959	Xopf19-221	Xpsr609	Xtc32601
Xapr2.6	Xgwm991	Xopf19-990	Xpsr630	Xtc35485
Xapr2.9	Xgwm1048	Xopo15-1200E	Xpsr666	Xtc77238
Xapr2.12	Xhvpaf	Xopm3	Xpsr687	Xtc89057
Xapr2.13	Xiag57	Xopm6-490	Xpsr900	Xtc89869
Xapr2.14	Xiag61b	Xopm6-1780	Xpsr901	Xtc98482
Xapr2.15	Xiag68	Xopm9	Xpsr932	Xwscm75
Xapr2.16	Xiag69c	Xopm20-850E	Xpsr934	Yr2
Xapr2.17	Xiag167	Xopq10	Xpsr1200	Z
Xapr2.20				

Source: Additional loci and details can be taken from Schlegel, R., and Korzun, V., Genes, markers and linkage data of rye (*Secale cereale* L.), 9th updated inventory, V. 03.13, 2013. With permission.

Note: The prefix "*X*" designates molecular markers of different origin.

TABLE 5.4C
A Selection of Genes and Markers Associated with Long and/or Short Arm of Chromosome 3R

3RS	*Aat1*	*Xapr3.12*	*Xem3.6*	*Xme5em3-478*	*Xpsr902S*
	Aat4	*Xapr3.13*	*Xem3.8*	*Xmwg988*	*Xpsr903*
	Alt2	*Xapr3.14*	*Xem3.10*	*Xopb12-400H*	*Xpsr910*
	Dnr3	*Xapr3.15*	*Xem3.12*	*Xopb17-315*	*Xpsr1060*
	hsp17.3	*Xapr3.16*	*Xem3.16*	*Xopb18-330*	*Xpsr1077*
	I(a1)	*Xapr3.17*	*Xem3.18*	*Xopf01-570*	*Xpsr1196*
	I(dha1)	*Xapr3.18*	*Xem3.22*	*Xopf05-257*	*Xrems1135*
	Mal1	*Xbcd298a*	*Xem3.23*	*Xopo7-950*	*Xrems1254*
	Ndh4	*Xbcd349*	*Xem3.25*	*Xopo18*	*Xrems1323*
	Pm3	*Xbcd402a*	*Xem3.26*	*Xops10*	*Xscb17b*
	QTL15	*Xbcd402b*	*Xem3.28*	*Xopt18*	*Xscb33*
	QTL26	*Xbcd402u*	*Xfbp*	*Xp12-m57-f-120*	*Xscg9u*
	QTL58	*Xbcd828*	*Xgmw779*	*Xp12-m61-f-440a*	*Xscg23*
	QTL59	*Xbcd1532e1*	*Xgmw791*	*Xp12-m61-f-440a*	*Xscg24*
	QTL68	*Xbcd1823a*	*Xgmw1059*	*Xp16-m56-f-65*	*Xscg26.1*
	QTL80	*Xcdo118*	*Xgmw1296*	*Xpsr56*	*Xscg81*
	pScT7	*Xcdo348*	*Xgwm299*	*Xpsr56L*	*Xscg145b*
	pSc200	*Xcdo395*	*Xiag111*	*Xpsr74*	*Xscg186*
	Sec-4	*Xcdo419a*	*Xem3.30*	*Xpsr78*	*Xscg202u*
	Sr2	*Xcdo460*	*Xiag120*	*Xpsr170.2*	*Xscg206e*
	5S-rDNA3	*Xcdo1396sc*	*Xiag180*	*Xpsr305*	*Xscg223*
	Xabc465	*Xcmwg691*	*Xiag241b*	*Xpsr549*	*Xscm84*
	Xafl4-380	*Xeacg-mccg111*	*Xir26-810*	*Xpsr549.1*	*Xscm162*
	Xapr3.1	*Xeacg-mcgt315*	*Xir26re7*	*Xpsr578*	*Xscm283*
	Xapr3.2	*Xecgc-mccg271*	*XksuG53*	*Xpsr598*	*Xttu935*
	Xapr3.5	*Xecgg-mctg208*	*Xlprm36*	*Xpsr689*	*Xttu1935*
	Xapr3.6	*Xecgg-mctg367*	*Xlprm45*	*Xpsr754*	*Xwg222*
	Xapr3.7	*Xem3.2*	*Xmal1*	*Xpsr902*	*Xwg110a*
centro	*pAWRC.1*	*Xpos10*	*Xpsr578*	*Xpsr598*	*Xpsr1060*
3RL	*Aat3*	*Xapr3.9*	*Xgmw156*	*Xp15-m53-f-135*	*Xpsr1077*
	Acc2	*Xapr3.10*	*Xgmw299*	*Xp16-m56-f-309*	*Xpsr1149*
	Cxp1	*Xapr3.11*	*Xiag111a*	*Xp16-m56-f-385*	*Xrems1261*
	En(cg)2	*Xapr3.19*	*Xiag120*	*Xp16-m58-f-179*	*Xrz538*
	Est1	*Xapr3.20*	*Xiag180*	*Xp17-m60-f-81*	*Xsbp*
	Est2	*Xapr3.21*	*Xiag241b*	*Xp17-m60-f-369a*	*Xscg9e1*
	Aat4	*Xbcd131.1*	*XksuG59*	*Xp17-m60-f-273*	*Xscg143u*
	Mal1	*Xbcd240*	*XksuG62*	*Xpsr74*	*Xscg187u*
	Mdh2b(2)	*Xbcd372*	*Xmwg514*	*Xpsr78*	*Xscm5**
	Per2	*Xcmwg691*	*Xmwg932*	*Xpsr116*	*Xscm57*
	Per4	*Xcxp1*	*Xmwg988*	*Xpsr131*	*Xscm87*
	pSc200	*Xem3.3*	*Xmwg2053a*	*Xpsr156*	*Xscm117*
	QTL27	*Xem3.4*	*Xopb04-325*	*Xpsr170*	*Xscm162*

(Continued)

TABLE 5.4C

(Continued) A Selection of Genes and Markers Associated with Long and/or Short Arm of Chromosome 3R

QTL33	Xem3.5	Xopb07-276	Xpsr170.1	Xscm213
QTL37	Xem3.7	Xopb16-175	Xpsr170.2	Xscm239
QTL40	Xem3.9	Xopo10-1000	Xpsr473	Xscm294
QTL44	Xem3.11	Xopx12-900	Xpsr475	Xscm369
QTL51	Xem3.13	Xp12-m52-f-83	Xpsr54	Xscm371
QTL56	Xem3.14	Xp12-m52-f-235	Xpsr570	Xtavp1
QTL76	Xem3.15	Xp12-m53-f-226	Xpsr598	Xtip
Sbp	Xem3.17	Xp12-m53-f-560	Xpsr754	Xubp20
Tpi1	Xem3.19	Xp12-m58-f-136	Xpsr754-218	Xumc60
Xabc160u	Xem3.20	Xp12-m59-f-331	Xpsr804	Xwawl023
Xapr3.1	Xem3.21	Xp12-m61-f-88	Xpsr902	Xwgl10
Xapr3.3	Xem3.24	Xp12-m61-f-101	Xpsr902S	Xwgl10b
Xapr3.4	Xem3.27	Xp12-m61-f-344-	Xpsr1060	Xwia483
Xapr3.8	Xem3.29			

Source: Additional loci and details can be taken from Schlegel, R., and Korzun, V., Genes, markers and linkage data of rye (*Secale cereale* L.), 9th updated inventory, V. 03.13, 2013. With permission.

Note: The prefix "*X*" designates molecular markers of different origin.

TABLE 5.4D

A Selection of Genes and Markers Associated with Long and/or Short Arm of Chromosome 4R

4RS	Adh1	Xbgl485	Xgwm1091	Xpsr152L
	Amp2	Xbgl485.1	Xgwm1266	Xpsr485
	Dia2	Xbgl485.2	Xiag32	Xpsr584
	Ep2	Xcdo358	Xiag68	Xpsr584S
	Ger1	Xcdo836	Xiag73	Xpsr681
	GloB	Xcdo942	Xiag94	Xpsr921
	Nca1	Xcdo1328	Xiag115	Xpsr954
	Ndh1	Xeacc-mcct411	Xiag120	Xrems1160
	O20400	Xeacg-mcgt282	Xiag229c	Xrsq808
	Pgm1	Xeata-mcga241	Xir29re9	Xrz14
	pSc200	Xecgc-mcat515	Xlprm19	Xscb230
	QTL45	Xecgc-mcgc315	Xme1em3-321	Xscb258a
	QTL47	Xem4.7	Xme1em6-227	Xscg143
	QTL69	Xem4.8	Xme2em5-190	Xscm6
	Rf4a	Xem4.9	Xmwg77	Xscm47
	Ssp3	Xem4.11	Xmwg573	Xscm66
	Xabc454u	Xem4.12	Xmwg634	Xscm139
	Xapr4.3	Xem4.13	Xmwg643	Xscm147
	Xapr4.7	Xem4.14	Xmwg2299	Xscm155

TABLE 5.4D

(Continued) A Selection of Genes and Markers Associated with Long and/or Short Arm of Chromosome 4R

	Xapr4.9	Xem4.15	Xnpi253	Xscm251
	Xapr4.11	Xem4.19	Xopa05-385	Xscm327
	Xapr4.13	Xem4.20	Xopr4R11	Xuah5
	Xapr4.14	Xem4.24	Xops19B	Xuah52
	Xapr4.15	Xem4.26	Xp12-m54-f-173-p2	Xuah74
	Xapr4.16	Xem4.28	Xp12-m54-f-176-p2	Xuah82
	Xapr4.17	Xem4.29	Xp12-m61-f-118-p1	Xuah82A
	Xapr4.18	Xem4.30	Xp15-m53-f-48-p2	Xuah82B
	Xbcd93	Xger(1)	Xphp10005	Xuah82C
	Xbcd115	Xger(2)	Xpsr103	Xuah103B
	Xbcd327	Xglo	Xpsr108	Xwg9
	Xbcd38	Xgwm131	Xpsr150	
centro	pAWRC.1	Xnpi253	Xpsr152L Xpsr584S	Xpsr681
4RL	Aat	Xapr4.25	Xhvm4	Xphp10005
	Aat1	Xapr4.26	Xiac66	Xpo1
	Aat4	Xapr4.27	Xiac67	Xpr2/450bp
	Alt3	Xapr4.28	Xiac69	Xpr23/500bp
	an1	Xapr4.29	Xiac70	Xpr319/550bp
	Cxp3	Xb1	Xiac71	Xpr334/700bp
	Eper1	Xb4	Xiac74	Xpr502/750bp
	EPer2	Xb6	Xiac76	Xpr530/800bp
	EPer3	Xbcd93	Xiac81a	Xpr530/850bp
	EstB	Xbcd115	Xiac81b	Xpr670/900bp
	Est10	Xbcd265	Xiac82	Xpr670/950bp
	lg1	Xbcd1230u	Xiac83	Xpr743/750bp
	ms1	Xcdo358	Xiac84	Xpr777/700bp
	np	Xcdo836	Xiac85	Xpr794/800bp
	Nra1	Xcdo942	Xiac501	Xpsr8
	Nra2	Xcdo1328	Xiag32	Xpsr108
	Nra3	Xcmwg652	Xiag68	Xpsr119
	O2H1100	Xcxp3	Xiag73	Xpsr119-1000
	Pc	Xeacc-mcac118	Xiag94	Xpsr131u
	Per1	Xeacc-mcag236	Xiag115	Xpsr150
	Per2	Xeacc-mcag240	Xiag176	Xpsr167
	Per3	Xeacc-mcgt543	Xiag220u	Xpsr392
	Per4	Xeacc-mcta92	Xiag229a	Xpsr485
	Pgd1	Xeacc-mcta318	Xir11re18-500	Xpsr490
	pSc34	Xeacc-mctg213	Xir21re5-630	Xpsr566
	pSc119.2	Xeacg-mccc309	Xir26-1700	Xpsr604
	pSc200	Xeacg-mctc445	Xir41re9-980	Xpsr662
	QTL2	Xeacg-mctg152	Xis4.1	Xpsr833
	QTL10	Xeata-mcgc168	Xis4.2	Xpsr899
	QTL16	Xeata-mcgc171	Xis4.3	Xpsr943
	QTL50	Xeata-mcgc246	XksuD15	Xpsr954

(Continued)

TABLE 5.4D

(Continued) A Selection of Genes and Markers Associated with Long and/or Short Arm of Chromosome 4R

QTL70	Xeata-mcgc400	XksuF48	Xr11
QTL77	Xeata-mcgt198	Xlprm2	Xrems1154
QTL81	Xecgc-mccc318	Xlprm51	Xrms1007
Rc	Xecgc-mcga76	Xme1em6-264	Xrms1024
Rf4b	Xecgc-mcgc255	Xme4em3-415	Xrms1117
Rfc4	Xecgc-mctc139	Xme4em3-421	Xrz14
Rfcla	Xecgg-mcct99	Xme4em4-266	Xs10
Rfg1	Xecgg-mcgc183	Xme4em4-549	Xscb21
Rfp1	Xecgg-mcta106	Xme5em5-280	Xscb52
Rfp2	Xecgg-mcta218	Xmwg59	Xscb53
Sec7	Xem4.1	Xmwg89	Xscg143
Ss1	Xem4.2	Xmwg530	Xscg206
Ssp2	Xem4.3	Xmwg539	Xscad04
w	Xem4.4	Xmwg573	Xscm80
W16500	Xem4.5	Xmwg2053b	Xscm116
Xac03-700	Xem4.6	Xmwg2202	Xscm155
Xak466	Xem4.7	XnabgmpA9U	Xscm219
Xak466.1	Xem4.10	Xnpi53	Xscm245
Xak466.2	Xem4.16	Xopf14-370	Xscp12M56
Xak466.3	Xem4.17	Xopo2-1100H	Xscp14M55
Xapo5	Xem4.18	Xopo7-950	Xscp15m5
Xapr4.1	Xem4.21	Xopr4R11	Xscp15m55
Xapr4.2	Xem4.22	Xops4	Xscp16m58
Xapr4.4	Xem4.23	Xops14-705	Xscs4-765
Xapr4.5	Xem4.25	Xopw16-500	Xscxx04
Xapr4.6	Xem4.27	Xp12-m52-f-184-p2	Xscy03
Xapr4.8	Xem4.31	Xp12-m54-f-101-p2	Xscy09d
Xapr4.10	Xe36m61-191	Xp12-m57-f-124-p2	Xscsz502-900
Xapr4.12	Xe36m61-192	Xp12-m61-f-531-p2	Xtc80147
Xapr4.17	Xgwm130	Xp15-m53-f-60-p2	Xuah5
Xapr4.18	Xgwm156	Xp16-m56-f-218-p2	Xuah52
Xapr4.19	Xgwm313	Xp16-m57-f-130-p2	Xuah74
Xapr4.20	Xgwm676	Xp16-m58-f-91-p2	Xuah82
Xapr4.21	Xgwm720	Xp16-m58-f-150-p2	Xuah103
Xapr4.22	Xgwm941	Xp17-m60-f-84-p2	Xuah103B
Xapr4.23	Xgwm1154	Xphp1000	Xumc105
Xapr4.24	Xgwm1255		

Source: Additional loci and details can be taken from Schlegel, R., and Korzun, V., Genes, markers and linkage data of rye (*Secale cereale* L.), 9th updated inventory, V. 03.13, 2013. With permission.

Note: The prefix "*X*" designates molecular markers of different origin.

TABLE 5.4E
A Selection of Genes and Markers Associated with Long and/or Short Arm of Chromosome 5R

5RS			
alpha-Amy4	Xapr5.15	Xhvm40a	Xpsr312
br1	Xapr5.16	Xhvm40g	Xpsr326
En(gi)	Xapr5.17	Xhvsut2	Xpsr560
Gpd	Xapr5.18	Xiag5	Xpsr574
I(ncw)	Xbcd1117e	Xir32-1400	Xpsr628
pSc200	Xbcd202e	Xir32re1-1600	Xpsr929S
pScT7	Xcdo749	Xme5em3-695	Xpsr945
QTL28	Xcdo122	Xmwg502	Xpsr945a
QTL53	Xcdo419C	Xmwg2225	Xpsr1327
QTL71	Xcdo1081	Xopa01-570	Xrems1167
Skd	Xeacc-mcta150	Xopa20-564	Xrz261
5S-rDNA2	Xeata-mcct106	Xopw13-780	Xscg25
Ti	Xeata-mcct247	Xp12-m52-f-187-p2	Xscg130u
Xabl127	Xecgc-mcag342	Xp12-m59-f-180-p2	Xscg173
Xabl141	Xem5.3	Xp12-m59-f-256-p2	Xscm90
Xapr5.1	Xem5.4	Xp12-m59-f-354-p2	Xscm109
Xapr5.3	Xem5.5	Xp12-m61-f-186-p2	Xscm138(a)
Xapr5.4	Xem5.11	Xp15-m53-f-279-p2	Xscm159
Xapr5.6	Xem5.14	Xp15-m53-f-299-p2	Xscm166
Xapr5.7	Xgwm205	Xp15-m53-f-321-p2	Xscm268
Xapr5.9	Xgwm601	Xp17-m57-f-62-p2	Xuah58B
Xapr5.10	Xgwm1284	Xp17-m60-f-199-p2	Xuah97
Xapr5.12	Xgwm1296	Xp17-m60-f-235-p2	Xubp3
Xapr5.13	Xgbs613	Xp17-m60-f-444-p2	Xwg364A
Xapr5.14	Xh13-650	Xp17-m60-f-517-p2	Xwms6

centro					
pAWRC.1	Xbcd1130	Xmsh6	Xpsr109b	Xpsr360
UbaL	Xksu26	Xpsr100	Xpsr120.1	Xpsr860

5RL			
Aadh1	Tw1	Xgwm131	Xpsr574
Acl1.2	Tys	Xgwm179	Xpsr604u
Aco1(2)	Xab04-1900	Xgwm212	Xpsr628
Aco2	XAcl2	Xgwm271	Xpsr637
alpha-Amy2	Xad16-730	Xgwm335	Xpsr806
alpha-Amy3	Xadpg	Xgwm538	Xpsr860
alpha-Amy7/18/20	Xaf14-600	Xgwm996	Xpsr906
An5	Xalpha-Amy3	Xgwm1011	Xpsr911
beta-Amy1	Xapr5.2	Xgwm1059	Xpsr912
Cat	Xapr5.5	Xgwm1122	Xpsr918
Ce	Xapr5.8	Xgwm1165	Xpsr1194
Chi	Xapr5.11	Xgwm1205	Xpsr1201
Chl	Xapr5.19	Xgwm1226	Xrems1174
ct2	Xapr5.20	Xgwm1266	Xrems1186
dw8	Xapr5.21	Xhvm40e	Xrems1205

(Continued)

TABLE 5.4E

(Continued) A Selection of Genes and Markers Associated with Long and/or Short Arm of Chromosome 5R

Dw1	Xapr5.22	Xhvsut2	Xrems1218
Est1	Xapr5.23	Xiag10	Xrems1237
Est2	Xapr5.24	Xiag10(1)	Xrems1264
Est3	Xapr5.25	Xiag10(2)	Xrems1266
Est4	Xbeta-Amy	Xiag88	Xrms1115
Est5	Xcat	Xiag170	Xscb17a
Est6	Xbcd9	Xiag170(1)	Xscb35
Est7	Xbcd147e	Xiag220	Xscb169
Gai1a	Xbcd221	Xiag628	Xscg1.2
Gai1b	Xbcd298	Xibf-R1	Xscg170
Gai1c	Xbcd351	Xir30-710	Xscg219u
gr1	Xbcd451	Xis5.1	Xscg222
Ha1	Xbcd508	Xksu8	Xscg244
Ibf1	Xbcd603	Xksu26	Xscm12
l(et)2	Xbcd808	Xlbf1	Xscm54
Mdh3	Xbcd1302	Xme1em4-479	Xscm58
msh6	Xcdo419B	Xme1em5-510	Xscm72
Nca2	Xcdo456	Xme2em3-593	Xscm74
Ner1	Xcdo687	Xme4em5-810	Xscm76
Ph	Xcdo749	Xmsh6	Xscm77
Ph2	Xcdo786A	Xmtd862	Xscm85
Pl	Xcdo1417A	Xmwg42	Xscm133
Pm4	Xdw	Xmwg77b	Xscm137
q1	Xeacc-mcgc220	Xmwg502	Xscm138a
QTL1	Xeacc-mcgt-242	Xmwg989	Xscm159
QTL3	Xeacc-mcgt-243	Xmwg2112	Xscm172
QTL4	Xeacg-mccc182	Xmwg2062b	Xscm174
QTL6	Xeacg-mccg287	Xmwg2112	Xscm179
QTL7	Xeacg-mcgt341	Xmwg2225	Xscm193
QTL8	Xeacg-mctt407	Xp12-m59-f-140-p2	Xscm251
QTL4	Xeata-mcga316	Xp16-m56-f-108-p2	Xscm260
QTL9	Xeata-mcct205	Xp16-m56-f-246-p2	Xscm309
QTL17	Xeata-mcga252	Xp16-m57-f-96-p2	Xscm312
QTL18	Xeata-mcgc302	Xp16-m57-f-146-p2	Xscm365
QTL23	Xecgg-mcta441	Xp16-m58-f-206-p2	Xtria
QTL29	Xecgg-mcgt195	Xpsb43	XtriaIII
QTL34	Xem5.1	Xpsc34	Xuba
QTL35	Xem5.2	Xpsc74	Xuah48
QTL38	Xem5.6	Xpsc119.2	Xuah58A
QTL41	Xem5.7	Xpsc200	Xuah97
QTL46	Xem5.8	Xpsr79	Xuah105
QTL48	Xem5.9	Xpsr100	Xwaxc4
QTL52	Xem5.10	Xpsr109b	Xwg114
QTL57	Xem5.12	Xpsr120a(1)	Xwg180

TABLE 5.4E

(Continued) A Selection of Genes and Markers Associated with Long and/or Short Arm of Chromosome 5R

QTL62	Xem5.13	Xpsr120b(2)	Xwg199
QTL72	Xem5.15	Xpsr145	Xwg380
QTL86	Xem5.16	Xpsr164	Xwg644
QTL87	Xem5.17	Xpsr326	Xwm410-135
3Rt	Xem5.18	Xpsr335	Xwms6
Sf5	Xem5.19	Xpsr360	Xwr1026
Sp1 = Ssp3	Xembp	Xpsr370	Xx56785
Ssp5	Xgbs712	Xpsr426	Xz11772
Su(cp)	Xgwm6	Xpsr484	
T	Xgwm126		

Source: Additional loci and details can be taken from Schlegel, R., and Korzun, V., Genes, markers and linkage data of rye (*Secale cereale* L.), 9th updated inventory, V. 03.13, 2013. With permission.

Note: The prefix "*X*" designates molecular markers of different origin.

TABLE 5.4F

A Selection of Genes and Markers Associated with Long and/or Short Arm of Chromosome 6R

6RS					
	alpha-Amy1.1	Xapr6.21	Xbcd276	Xopa20-450	Xpsr1205
	alpha-Amy1.2	Xapr6.22	Xcdo405	Xopb8-1300	Xpsr1327
	Alt1	Xapr6.23	Xcdo836	Xopg7-550	Xrems1259
	Amp1	Xapr6.24	Xcdo1081	Xopr01-600	Xrz87
	Axc	Xapr6.25	Xcinau142	Xopt6	Xscb15
	Co	Xapr6.27	Xcinau143	Xopx12-500	Xscg9
	Est8	Xapr6.28	Xedm129	Xp12-m57-f-348-p2	Xscg33.2
	pSc200	Xapr6.30	Xem6.2	Xp12-m58-f-105-p2	Xscg91
	QTL11	Xapr6.31	Xem6.3	Xp12-m61-f-89-p2	Xscg187
	QTL78	Xapr6.33	Xem6.5	Xp15-m53-f-303-p2	Xscg202
	QTL83	Xapr6.34	Xem6.6	Xp16-m56-f-57-p2	Xscm112
	Rfc2	Xapr6.35	Xem6.10	Xp16-m56-f-400-p2	Xscm135
	Wh	Xapr6.37	Xem6.11	Xp17-m60-f-337-p2	Xscm142
	Xabcd175	Xapr6.38	Xem6.12	Xpr433/800bp	Xscm168
	Xapr6.1	Xapr6.39	Xgal	Xpr777/600bp	Xscm176
	Xapr6.2	Xapr6.40	Xgwm37	Xpsr106(1)	Xscm275
	Xapr6.3	Xapr6.41	Xgwm63	Xpsr106-1-350	Xscm280
	Xapr6.4	Xapr6.42	Xgwm232	Xpsr149L	Xscm304
	Xapr6.5	Xapr6.43	Xgwm959	Xpsr154	Xscm313
	Xapr6.8	Xapr6.44	Xiag23	Xpsr160	Xscr01
	Xapr6.12	Xapr6.45	Xiag94e	Xpsr312	Xscsz841L-450

(Continued)

TABLE 5.4F

(Continued) A Selection of Genes and Markers Associated with Long and/or Short Arm of Chromosome 6R

Xapr6.13	Xapr6.46	Xis6.2	Xpsr312.1(S)	Xscsz980L-650
Xapr6.14	Xapr6.47	Xis6.5	Xpsr312.2	Xuah30
Xapr6.15	Xapr6.48	Xis6.6	Xpsr371-700	Xuah30e
Xapr6.16	Xbcd1	Xlprm13	Xpsr627	Xwg110c
Xapr6.17	Xbcd147	Xmwg65	Xpsr915	

centro	pAWRC.1		Xa20450	
6RL	Aadh2	Xalpha-Amy1-2	Xecgc-mcga207	Xp16-m58-f-460-p2
	Aat2	Xapr6.1	Xecgg-mcta225	Xp17-m57-f-48-p2
	AawC2	Xapr6.3	Xecgg-mctg383	Xp17-m57-f-237-p2
	AawS5/FG3-170	Xapr6.4	Xedm10	Xp17-m60-f-342-p2
	AawS5/FG3-280	Xapr6.6	Xedm93	XpcrS5
	AawS5/FG4-2	Xapr6.7	Xem6.1	XpcrC2
	AawS5/FG4-1	Xapr6.9	Xem6.4	XpcrC4/R1
	Aco1	Xapr6.10	Xem6.5	XpcrR1
	alpha-Amy1	Xapr6.11	Xem6.7	XpcrS5/FG3
	alpha-Amy1-1	Xapr6.16	Xem6.8	XpcrS5/FG4
	alpha-Amy1-1-1	Xapr6.17	Xem6.9	XpcrS5/S6
	alpha-Amy1-1-2	Xapr6.18	Xem6.10	XpcrS14/S15
	alpha-Amy1-2	Xapr6.19	Xep1	Xpsr106(2)
	Ans	Xapr6.20	Xest5	Xpsr142
	Axc	Xapr6.21	Xfo37	Xpsr148
	Axc2	Xapr6.22	Xgbs526	Xpsr149
	cb	Xapr6.23	Xglb3	Xpsr154
	Crn	Xapr6.24	Xglb33	Xpsr160
	Dia1	Xapr6.26	Xgwm114	Xpsr454
	Dia2	Xapr6.29	Xgwm131	Xpsr454L
	Eml-R1b	Xapr6.32	Xgwm247	Xpsr627
	En(alpha-Amy)	Xapr6.36	Xgwm340	Xpsr687
	Ep1	Xapr6.49	Xgwm391	Xpsr915
	Est5	Xapr6.50	Xgwm732	Xpsr965
	Est6	Xapr6.51	Xgwm751	Xpsr1149b
	Est7	Xapr6.52	Xgwm861	Xpsr1203
	Est8	Xapr6.53	Xgwm984	Xpsr1205
	Est9	Xapr6.54	Xgwm1103	Xpsr1327
	Est10	Xapr6.55	Xiag85	Xrems1152
	Est13	Xapr6.56	Xiag111(1)	Xrems1154
	EstA	Xapr6.57	Xiag111(2)	Xrems1247
	Gal	Xapr6.58	Xiag119	Xrems1250
	Glb3	Xaw15	Xiag186b	Xrms1121
	Glb33	Xawbm15	Xiag229	Xrz995
	gd	Xbcd1	Xiag267b	Xsca08
	H2	Xbcd102	Xir10-650	Xscb03
	Ha3	Xbcd131	Xir13-re7-850	Xscg26-2
	I(scx)	Xbcd147	Xis6.1	Xscg143

TABLE 5.4F

(Continued) A Selection of Genes and Markers Associated with Long and/or Short Arm of Chromosome 6R

Lap2	Xbcd265.2	Xis6.2	Xscg146
Lr syn. Prl	Xbcd269	Xis6.3	Xscg184sc
Lra	Xbcd276	Xis6.4	Xscg217
Lrb	Xbcd340	XksuD2	Xscm2
Mn	Xbcd385	XksuD12	Xscm28
Ndh2	Xbco496	XksuD27	Xscm46
Ndh3	Xbcd758	XksuF37	Xscm68
Per1	Xbcd1426	XksuF37a	Xscm78
Per2	Xbco1380	Xme3em5-328	Xscm168
Pgd2	Xcdo244	Xme5em2-178	Xscm176
Pm	Xcdo345d	Xmwg669	Xscm180
Pm5	Xcdo405	Xmwg934	Xscm214
Pm20	Xcdo676	Xmwg2162	Xscm261
pSc34	Xcdo772	Xopa02-304	Xscm270
pSc74	Xcdo780	Xopa08-415	Xscm310
pSc119.1	Xcdo836	Xopa13-230	Xscsz1169L680
pSc119.2	Xcdo1380	Xopb15-365	Xtatam17
QTL2	Xcdo1396	Xopf05-257	Xtatam25
QTL9	Xcinau145	Xops2	Xtatam30
QTL30	Xcinau146	Xopt6	Xtatam36
QTL63	Xcinau346	Xp12-m54-f-79-p2	Xuah61
QTL73	Xcinau450	Xp12-m54-f-204-p2	Xuah103A
Reg	Xcinau660	Xp12-m57-f-106-p2	Xwg180
Rf68	Xcinau895	Xp12-m57-f-230-p2	Xwg933
Rog	Xdoo1.45	Xp12-m57-f-241-p2	Xwia482
Ws1	Xeacc-mcac270	Xp12-m59-f-136-p1	Xwia807
Xaa11-630	Xeacc-mcag459	Xp12-m61-f-205-p2	Xwmc753
Xad14-880	Xeacg-mctt108	Xp15-m53-f-96a-p2	Xwsp
Xalpha-Amy1-1-1	Xeacg-mctg293	Xp16-m56-f-294-p2	Yr3
Xalpha-Amy1-1-2	Xeata-mcta335		

Source: Additional loci and details can be taken from Schlegel, R., and Korzun, V., Genes, markers and linkage data of rye (*Secale cereale* L.), 9th updated inventory, V. 03.13, 2013. With permission.

Note: The prefix "X" designates molecular markers of different origin.

TABLE 5.4G

A Selection of Genes and Markers Associated with Long and/or Short Arm of Chromosome 7R

7RS	Acp2	Xapr7.24	Xhvsut1	Xp17-m57-f-220-p2	Xrsq805
	Acp3	Xapr7.25	Xiag89	Xp17-m60-f-214-p2	Xscarq4-283
	Acp4	Xb1	Xiag300	Xp17-m60-f-223-p2	Xscb85
	Acp5	Xb4	Xiag301	Xpsb37	Xscb101b

(Continued)

TABLE 5.4G

(Continued) A Selection of Genes and Markers Associated with Long and/or Short Arm of Chromosome 7R

	Acp6	*Xb26*	*Xis7.1*	*Xpsb115*	*Xscg27*
	Acp7	*Xbcd207*	*Xis7.3*	*Xpsr39*	*Xscg33*
	Acp8	*Xbcd265(2)*	*XksuD9*	*Xpsr59*	*Xscg145*
	Acp9	*Xbcd808.2*	*XksuF8*	*Xpsr104*	*Xscg206*
	Almt1	*Xbcd1092*	*Xlprm4*	*Xpsr115*	*Xscim810-621*
	Alp2	*Xbnl5.09*	*Xmwg584*	*Xpsr115-219*	*Xscim811-1376*
	Alp4	*Xbzh834*	*Xnpi209*	*Xpsr117*	*Xscim812-626*
	Alp5	*Xcdo57*	*Xnpi427*	*Xpsr163*	*Xscim812-1138*
	Alt3	*Xcdo484*	*Xopc13-1500*	*Xpsr303*	*Xscim819-1434*
	Alt4	*Xcdo545*	*Xopd1-858*	*Xpsr311*	*Xscim823-853*
	Fbp	*Xem7.1*	*Xopf20-483*	*Xpsr567*	*Xscim823-2826*
	mp	*Xem7.2*	*Xopf20-579*	*Xpsr577*	*Xscim881*
	Ndh2	*Xem7.5*	*Xopf20-709*	*Xpsr580*	*Xscm19*
	Per	*Xem7.6*	*Xopj3-857*	*Xpsr580-200*	*Xscm40*
	Per3	*Xem7.7*	*Xopj17-1245*	*Xpsr649*	*Xscm44*
	Per5	*Xem7.12*	*Xopm4-700*	*Xpsr690*	*Xscm49*
	Per6	*Xem7.14*	*Xopn1-667*	*Xpsr914*	*Xscm86*
	pSc200	*Xem7.18*	*Xopn1-708*	*Xpsr944(S)*	*Xscm92*
	QTL74	*Xem7.19*	*Xopn8-1200*	*Xpsr946*	*Xscm102*
	QTL84	*Xem7.20*	*Xopo2-633*	*Xpsr954*	*Xscm138b*
	Sec-5	*Xem7.21*	*Xopo2-984*	*Xpsr1051*	*Xscm150*
	Wsp	*Xem7.22*	*Xopo7-666*	*Xpsr1312*	*Xscm205*
	Xacl3	*Xfbp*	*Xopq4-578(a)*	*Xpsr1318*	*Xscm320*
	Xapr7.1	*Xgwm269*	*Xopq4-725*	*Xpsr1327*	*Xscm322*
	Xapr7.2	*Xgwm565*	*Xp12-m52-f-292-p2*	*Xrems1162*	*Xscm332*
	Xapr7.11	*Xgwm781*	*Xp12-m57-f-252-p2*	*Xrems1187*	*Xuah6a*
	Xapr7.19	*Xgwm1081*	*Xp16-m57-f-182-p2*	*Xrems1188*	*Xubp2*
	Xapr7.22	*Xgwm1166*	*Xp17-m57-f-134-p2*	*Xrms1012*	*Xubp9*
	Xapr7.23	*Xgwm1302*	*Xp17-m57-f-197-p2*	*Xrms1112*	*Xwg110c*
centro	*Acl1.3(S)*	*Xaat2*	*Xnpi209*	*Xpsr566*	*Xscm40*
	Adpg3	*Xcmwg682*	*Xnpi427*	*Xpsr690L*	*Xwaxc2*
	pAWRC.1	*Xmwg2076*	*Xpepco*	*Xpsr1051*	*Xwye1958*
7RL	*Aat1*	*Xapr7.3*	*Xcdo766C*	*XksuD9*	*Xpsr649(1)*
	Aat2	*Xapr7.4*	*Xcmwg682*	*Xlprm6*	*Xpsr690*
	Acp1	*Xapr7.5*	*Xeacc-mcag358*	*Xlprm7*	*Xpsr928*
	Adk1	*Xapr7.6*	*Xeacc-mcct208*	*Xme1em3-281*	*Xpsr944s*
	alpha-Amy2	*Xapr7.7*	*Xeacg-mcgt111*	*Xme2em3-557*	*Xrems1135*
	alpha-Amy2-1	*Xapr7.8*	*Xeata-mcgt246*	*Xme5em2-153*	*Xrems1188*
	alpha-Amy2-2	*Xapr7.9*	*Xecgc-mcag347*	*Xmwg2076*	*Xrems1197*
	alpha-Amy5	*Xapr7.10*	*Xecgg-mctc118*	*Xnashor1*	*Xrems1234*
	an1	*Xapr7.12*	*Xecgg-mctc216*	*Xopa10-370*	*Xrems1235*
	asc1	*Xapr7.13*	*Xem7.3*	*Xopd1-858*	*Xrems1253*
	asc3	*Xapr7.14*	*Xem7.4*	*Xopf01-476*	*Xrems1281*
	ct1	*Xapr7.15*	*Xem7.8*	*Xope5*	*Xrms1018*

TABLE 5.4G
(Continued) A Selection of Genes and Markers Associated with Long and/or Short Arm of Chromosome 7R

Embp1	*Xapr7.17*	*Xem7.9*	*Xp12-m53-f-54-p2*	*Xrsq805* syn. *Embp*
Ep1	*Xapr7.18*	*Xem7.10*	*Xp12-m57-f-138-p2*	*Xscb11(3)*
Gai2a	*Xapr7.20*	*Xem7.11*	*Xp12-m61-f-85-p2*	*Xscb101a*
Gai2b	*Xapr7.21*	*Xem7.13*	*Xp15-m53-f-313-p1*	*Xscg27*
Lr2 syn. *Pr2*	*Xapr7.26*	*Xem7.15*	*Xp15-m53-f-320-p2*	*Xscg72*
Ner2	*Xapr7.27*	*Xem7.16*	*Xp16-m58-f-391-p2*	*Xscg78*
Per10	*Xapr7.28*	*Xem7.17*	*Xp17-m57-f-218-p2*	*Xscg144*
pSc34	*Xapr7.29*	*Xem7.23*	*Xp17-m57-f-270-p2*	*Xscim823-853*
pSc119.2	*Xapr7.30*	*Xembp*	*Xp17-m57-f-321-p2*	*Xscim823-2826*
pSc200	*Xapr7.31*	*Xgwm131*	*Xpepco*	*Xscm102*
QTL20	*Xapr7.32*	*Xgwm1205*	*Xper3*	*Xscm122*
QTL31	*Xapr7.33*	*Xgwm1266*	*Xprx10*	*Xscm140*
QTL32	*Xapr7.34*	*Xhvsut1*	*Xpsr39*	*Xscm181*
QTL54	*Xapr7.35*	*Xiag89*	*Xpsr56*	*Xscm183*
QTL64	*Xapr7.36*	*Xiag111b*	*Xpsr117*	*Xscm208*
wa1	*Xapr7.37*	*Xiag119*	*Xpsr129*	*Xscm272*
Xab04-1500	*Xapr7.38*	*Xiag229c*	*Xpsr150*	*Xscm354*
Xab09-630	*Xbcd110*	*Xir11re4-680*	*Xpsr169*	*Xwg522*
Xac03-980	*Xbcd147a*	*Xir11re6-1000*	*Xpsr303*	*Xwyc1958*
Xalpha-Amy2-1	*Xbcd207*	*Xir13re4-480*	*Xpsr305*	*Xuah6a*
Xalpha-Amy2-2	*Xbcd265.1*	*Xir35re1-1700*	*Xpsr311*	*Xubp2*
Xabc310.1	*Xbzh834*	*Xis7.2*	*Xpsr492S*	*Xubp9*
Xabc310.2	*Xcdo545*	*Xis7.3*	*Xpsr547*	*Xumc114*
Xabc455	*Xcdo783*	*Xis7.4*	*Xpsr566*	*Xwye1958*
Xabg463	*Xcdo786B*	*Xis7.5*		

Source: Additional loci and details can be taken from Schlegel, R., and Korzun, V., Genes, markers and linkage data of rye (*Secale cereale* L.), 9th updated inventory, V. 03.13, 2013. With permission.

Note: The prefix "*X*" designates molecular markers of different origin.

and linkage are not included. These can be taken from Schlegel and Korzun (1997) and from the author's web site, www.rye-gene-map.de, which is annually updated.

Although there has been big progress in rye molecular genetics, the availability of DNA marker is much smaller than that in autogamous barley or wheat. The first DNA-based linkage maps in rye were published by Senft and Wricke (1996) and Saal and Wricke (1999) involving RFLP and RAPD markers (see Table 5.5). Due to the close evolutionary relationship of rye, wheat, barley, rice, or others, some known markers of the latter species [RFLP, RAPD, single-nucleotide polymorphism (SNP), SSR, cleaved amplified polymorphic sequence (CAPS)] could be exploited for rye linkage maps (Masojć et al. 2001). The length of the maps ranges over the years between 340 cM (Loarce et al. 1996) and more than 1500 cM (Mysków et al. 2012). It shows that still better saturated maps are needed in rye.

TABLE 5.5

A Selection of Recent Marker and Mapping Populations in Rye Marker Type

RFLP	RAPD	SSR	AFLP	ISSR-SCIM	SCAR	SAMPL	STS-EST	SNP	DArT	Cross Population
**	**									Dank × Halo
**	**									Inbred × Inbred
**	**									Ex
**		**								P87 × P105
**	**	**	**							Inbred × Inbred
**		**								Kings × Imperial
**	**									DS2 × RXL10
**	**		**							DS2 × RXL10
**		**								BC1 pop. 9953
**										P87 × P105
				**						Ailes × Riodeva
**	**	**	**				**			DS2 × RXL10
	**	**	**	**	**	**	**	**		L318 × L9
	**	**	**				**			5410tl3
	**	**	**	**			**			DS2R × L10
**	**						**			P87 × P105
		**	**				**			L2053 × Altevogt14160
									**	S120 × S76
	**	**					**		**	EM-1 DH × Voima DH
									**	S120 × S76

Note: AFLP, amplified fragment length polymorphism; DArT, Diversity Arrays Technology; ISSR-SCIM, intersingle-sequence repeat-*Secale cereale* intermicrosatellite; RAPD, random amplified polymorphic DNA; RFLP, restriction fragment length polymorphism; SAMPL, selective amplification of microsatellite polymorphic locus; SCAR, sequence-characterized amplified region; SNP, single-nucleotide polymorphism; SSR, single-sequence repeat; STS–EST, sequence-tagged site–expressed sequence tag.

Recent advances in molecular marker technology and the development of high-density molecular marker linkage maps have provided powerful tools for elucidating the genetic basis of quantitatively inherited traits. The first high-resolution map of rye constructed using DArT was created in 2009 (Bolibok-Bragoszewska et al. 2009). The next high-density linkage map of rye, based on DArT and PCR markers, was recently used for detecting QTLs for preharvest sprouting (PHS) and heading earliness (Mysków 2010, Mysków et al. 2012).

5.4 LINKAGES

The association of genes that results from their being on the same chromosome is called linkage. It is detected by the greater association in the inheritance of two or more nonallelic genes than would be expected from independent assortment.

The nearer such genes are to each other on a chromosome, the more closely linked they are, and the less often they are likely to be separated in future generations by crossing over. All genes in one chromosome form one linkage group. Linkage groups within rye were first confirmed by trisomic analysis (Sturm 1978)—a method for mapping gene loci on individual chromosomes by comparing disomic and trisomic segregation patterns of a series of individuals. Meanwhile, numerous genetic and molecular studies revealed the linkage of genes and markers that can be utilized in practical breeding. An example for chromosome 3R is given in Table 5.6. The uneven distribution of coding genes along a chromosome is now compensated by almost even and dense distribution of molecular markers (Philipp et al. 1994).

TABLE 5.6
Linkage Relationships between Genes, Genes and Markers, and Markers as Well as Markers as Example for Rye Chromosome 3R

Chromosome 3R	Value	Variation	Unit (%) or (cM)	References
Aat1–Xapr3.2	1.4		cM	412
Aat1–Xem3.28	5.5		cM	412
Aat3–Centr	20.6	±2.4	cM	16
Aat3–Est1(2)	5.6	±3.1	cM	38
	15.4		%	16, 41
Aat3–Mdh2	5		cM	216
	6.0		%	223
	21.5	±3.4	%	264
Aat3–Mdh2(b)	4.2	±1.5	%	264
	5	±2.0	%	167
	16	±3.0	%	167
	16.6	±1.7	%	162
	21.0		%	16, 41
Aat3(Got4)–Tpi1	7	±2.0	%	167, 216
	18		%	169
Aat4–In	13.9	±1.9	%	351
Aat4–Xapr3.2	1.4		cM	355
	2.2		%	299
Aat4–Xem3.28	5.5		cM	355
Aat4–XksuG53	15.3		cM	261
Aat4–Xopo7-950	0		cM	274
Aat4–Xopo10-1000	3.3		cM	274
Aat4–Xtto1935	6.5		cM	261
Centr–Mdh2	1.9	±1.1	%	217
En(cg)2–Xpsr598	21.9		%	263
Est1–Mdh1	0.0		cM	261
Est1(2)–Mdh2b	5.6		%	16, 41
	7.4	±2.9	%	38

(Continued)

TABLE 5.6
(Continued) Linkage Relationships between Genes, Genes and Markers, and Markers as Well as Markers as Example for Rye Chromosome 3R

Chromosome 3R	Value	Variation	Unit (%) or (cM)	References
Est1–Xpsr156	0.0		cM	261
	7.1		cM	274
I(a1)–Xapr3.1	2.4		cM	355, 412
I(a1)–Xapr3.2	1.5		cM	355, 412
	2.1		cM	299
LPer2, Xem3.15–Xem3.20, Xem3.19, Xem3.9, Xem3.7	2.2		cM	412
LPer2, Xem3.15–Xem3.21	1.6		cM	412
LPer4–Mdh2	30.7	±5.8	%	162
LPer4, Xem3.15–Xem3.20, Xem3.19, Xem3.9, Xem3.7	2.2		cM	355
LPer4, Xem3.15–Xem3.21	1.6		cM	355
Mal1–Xpsr305	26.0		cM	186
Mal1–Xpsr598	5.7		cM	186
Mal1–Xpsr1060	5		cM	206
Mal1–Xpsr1196	16		cM	206
Mdh2–XiagA03	21		cM	216
Mdh2b(2)–Tpi1	5		%	169
	9	±1.6	%	167
Mdh2b(2)–XiagA03	22		%	169
QTL68–Xapr3.14	0		cM	430
QTL76–Xapr3.3	2		cM	430
QTL80–Xapr3.14	5		cM	430
Tpi1–Xiag20	23		cM	216
Xabc160–Xbcd131.1	0.0		cM	300
Xabc160–Xscg143	2.8		cM	300
Xabc465–Xbcd349	1.4		cM	300
Xabc465–Xbcd402	2.5		cM	300
Xapr3.1–Xapr3.6	14.1		cM	299
Xapr3.1–Xapr3.7	8.8		cM	355, 412
	14.1		cM	299
Xapr3.3–Xem3.17	1.8		cM	355, 412
Xapr3.3–Xpsr473	5.2		cM	299
Xapr3.3–Xubp20	5.5		cM	355, 412
	6.4		cM	299
Xapr3.4–Xem3.29	20.9		cM	299, 412
Xapr3.4–Xwia483	25.9		cM	299
Xapr3.5–Xpsr305	20.5		cM	299
	24.3		cM	355, 412

TABLE 5.6
(Continued) Linkage Relationships between Genes, Genes and Markers, and Markers as Well as Markers as Example for Rye Chromosome 3R

Chromosome 3R	Value Variation	Unit (%) or (cM)	References
Xapr3.6–Xapr3.7	0.0	cM	299
	4.3	cM	355, 412
Xapr3.6–Xem3.2	4.5	cM	355, 412
Xapr3.6–Xpsr549	8.7	cM	299
Xapr3.6–Xpsr1196	8.7	cM	299
Xapr3.6–XksuG53	8.7	cM	299
Xapr3.7–Xpsr549	8.7	cM	299
Xapr3.7–Xpsr1196	8.7	cM	299
Xapr3.10, Xapr3.8, Xwia483–Xem3.3, Xem3.14	14.4	cM	355, 412
Xapr3.10, Xapr3.8, Xwia483–Xem3.5	2.7	cM	355, 412
Xapr3.11–Xapr3.21	5.8	cM	412
Xapr3.11–Xpsr754-218	8.0	cM	412
Xapr3.12–Xem3.13	3.2	cM	412
Xapr3.12–Xem3.17	13.8	cM	412
Xapr3.13–Xem3.18	10.5	cM	412
Xapr3.14–Xem3.16, Xem3.15	16.2	cM	412
Xapr3.18–Xpsr598-206	15.5	cM	412
Xapr3.19–Xapr3.20	14.8	cM	412
Xapr3.20–Xapr3.21	9.0	cM	412
Xem3.16, Xem3.15–Xapr3.17	10.8	cM	412
Xbcd131.1, Xabc160u–Xbcd372	1.3	cM	437
Xbcd131.1, Xabc160u–Xpsr954	1.2	cM	437
Xbcd131.1–Xrz538	7.3	cM	300
Xbcd240–Xrz538	7.5	cM	300
Xbcd240–Xscg187	3.5	cM	300
Xbcd349–Xcdo1396	1.3	cM	300, 437
Xbcd349–Xiag241b, Xabc465	1.4	cM	437
Xbcd372–Xpsr170	6.6	cM	266, 308
Xbcd372–Xrz538	6.0	cM	437
Xbcd372–Xwaw1023	4.0	cM	308
	4.3	cM	266
Xbcd402u–Xiag241b, Xabc465	3.2	cM	437
Xbcd402–Xscg223	8.6	cM	300, 437
Xbcd828–Xbcd1532	8.0	cM	300
Xbcd828–Xscg9	3.7	cM	300
Xbcd1532–Xscg24	0.0	cM	300
Xbcd1823–Xcdo1396	3.0	cM	300, 437
Xbcd1823–Xpsr305	1.3	cM	437
Xbcd1823–Xscg26.1	2.4	cM	300

(Continued)

TABLE 5.6

(Continued) Linkage Relationships between Genes, Genes and Markers, and Markers as Well as Markers as Example for Rye Chromosome 3R

Chromosome 3R	Value	Variation	Unit (%) or (cM)	References
Xcdo118–Xscg81	1.0		cM	300, 437
Xcdo118–Xscg186	3.1		cM	300, 437
Xcdo348, Xpsr170.2–Xscg206e	0.7		cM	437
Xcdo348–Xcdo419a, Xbcd298a	12.8		cM	267, 300
Xcdo395–Xcdo460	9.9		cM	267
Xcdo395–Xopo18	7.6		cM	267
Xcdo419a, Xbcd298a–Xcdo460	19.2		cM	267
Xcdo419a, Xbcd828–Xscg9u	3.9		cM	437
Xcdo419a, Xbcd828–Xiag180, Xiag120	3.1		cM	437
Xcdo460, Xpsr1077, Xscb17b, Xcdo395–Xpsr78	2.2		cM	437
Xcdo460, Xpsr1077, Xscb17b, Xcdo395–Xpsr754, Xpsr74	0.5		cM	437
Xcmwg691–XksuG53	24.3		cM	308
Xcmwg691–Xpsr156	3.4		cM	437
Xcmwg691–Xpsr475, Xbcd240	1.0		cM	437
Xcmwg691a–Xpsr1196	7.1		cM	308
Xxxp1–Xpsr170(2)	34.5		cM	186
Xcxp1–Xtlp	20		cM	206
Xem3.1, Xpsr56L, Xpsr116–Xem3.27, Est2, Xpsr156, Mal1	1.2		cM	355
Xem3.1, Xpsr56L, Xpsr116–Xem3.27, Est1, Xpsr156, Mal1	1.2		cM	412
Xem3.1, Xpsr56L, Xpsr116–Xttu1935, Xpsr598, Xpsr1060, Xpsr902S	1.3		cM	355, 412
Xem3.2–Xem3.10	1.6		cM	355, 412
Xem3.3, Xem3.14–Xwg110	2.0		cM	355, 412
Xem3.4, Xpsr78–Xpsr170.1, Xpsr170.2, Xapr3.9	2.4		cM	355, 412
Xem3.4, Xpsr78–Xpsr1077, Xpsr804, Xpsr74, Xpsr754	1.7		cM	355, 412
Xem3.5–Xem3.29	6.8		cM	355, 412
Xem3.6, Xem3.16, Xem3.18, Xpsr1196–Xem3.12, Xpsr549.1, XksuG53	2.0		cM	355
Xem3.6, Xem3.16, Xem3.18, Xpsr1196–Xem3.22	1.3		cM	355
Xem3.8–Xem3.22	3.3		cM	355, 412
Xem3.8–Xem3.30	2.9		cM	355, 412
Xem3.10–Xem3.12, Xpsr549.1, XksuG53	1.6		cM	355
Xem3.10–Xpsr549, XksuG53, Xem3.18	1.6		cM	412
Xem3.13, Xem3.11–Xem3.21	1.2		cM	355, 412
Xem3.13, Xem3.11–Xpsr1077, Xpsr804, Xpsr74, Xpsr754	2.3		cM	355, 412

TABLE 5.6

(Continued) Linkage Relationships between Genes, Genes and Markers, and Markers as Well as Markers as Example for Rye Chromosome 3R

Chromosome 3R	Value Variation	Unit (%) or (cM)	References
Xem3.13, Xem3.11–Xem3.21	1.2	cM	355, 412
Xem3.16, Xpsr1196, Xem3.6, Xem3.18–Xpsr549, XksuG53, Xem3.12	2.0	cM	412
Xem3.17–Xpsr475	2.0	cM	355, 412
Xem3.20, Xem3.19, Xem3.9, Xem3.7–Xpsr570, Xpsr1149, Xpsr543	2.8	cM	355, 412
Xem3.24–Xpsr170.1, Xpsr170.2, Xapr3.9	3.4	cM	355, 412
Xem3.24–Xubp20	2.3	cM	355, 412
Xem3.25, Xem3.26, Xem3.23–Xem3.28	1.7	cM	355, 412
Xem3.25, Xem3.26, Xem3.23–Xttu1935, Xpsr598, Xpsr1060, Xpsr902S	3.1	cM	355, 412
Xem3.27, Est1, Xpsr156, Mal1–Xpsr570, Xpsr1149, Xpsr543	1.1	cM	412
Xem3.27, Est2, Xpsr156, Mal1–Xpsr570, Xpsr1149, Xpsr543	1.1	cM	355
Xem3.30–Xpsr305	4.3	cM	355, 412
Xfbp–Xscg23	0.0	cM	300
Xfbp–Xscg81	5.0	cM	300
Xgwm299–Xpsr475, Xbcd240	1.6	cM	437
Xgwm299–Xscb17b	2.7	cM	266
Xgwm299–Xscg187u	1.6	cM	437
Xgwm299–Xwg110a	2.1	cM	266
Xgmw1296-3R–XksuG59	2.1	cM	308
Xgmw1296-3R–Xwg110a	7.6	cM	308
Xgwm156-3R–Xmwg514	11.0	cM	308
Xgwm779-3R–Xpsr1196	7.3	cM	308
Xgwm779-3R–Xrems1254-3R	16.2	cM	308
Xgwm1059-3R–Xpsr899b	21.3	cM	308
Xiag111–Xscg145	6.2	cM	300, 437
Xiag111–Xscg223	10.5	cM	300, 437
Xiag111a–Xiag120	0	cM	274
Xiag111a–Xmwg932	33.7	cM	274
Xiag180, Xiag120–XksuG53, Xpsr549	1.6	cM	437
Xiag120–Xiag180	5.3	cM	274
Xiag241b–Xpsr578	1.5	cM	274
XksuG53–Xpsr1196, Xopt18	1.7	cM	437
	6.7	cM	308
XksuG59–Xpsr170	12.0	cM	308
XksuG62–Xmwg932	10.2	cM	308

(Continued)

TABLE 5.6

(Continued) Linkage Relationships between Genes, Genes and Markers, and Markers as Well as Markers as Example for Rye Chromosome 3R

Chromosome 3R	Value Variation	Unit (%) or (cM)	References
XksuG62–Xmwg2053a	16.0	cM	308
Xlprm36–Xscm162	1.0	cM	431, 439
Xlprm45–Xp12-m57-f-120-p2	1.5	cM	431
Xlprm45–Xp16-m56-f-65-p2	0.6	cM	431
Xmal1–Xcdo348, Xpsr170.2	0.9	cM	437
Xmal1–Xscg202u, Xwg222	0.8	cM	437
Xmwg514–Xmwg2053a	3.2	cM	308
Xmwg932–Xmwg2053a	25.2	cM	308
Xmwg932–Xpsr687c	9.1	cM	308
	13.9	cM	308
Xmwg932–Xscm5	5.8	cM	308, 439
Xmwg932–Xwg110a	10.3	cM	308
	11.4	cM	308
	14.4	cM	308
Xmwg988, Xpsr116–Xpsr578	1.6	cM	437
Xmwg988, Xpsr116–Xpsr1149, Xpsr570, Xscg9e1	1.1	cM	437
Xmwg2053a–Xpsr687c	4.1	cM	308
	5.0	cM	308
Xopb12-400H–Xpsr689	6.2	cM	274
Xopb12-400H–Xpsr902	4.5	cM	274
Xopo7-950–Xiag241b	0.0	cM	274
Xopo10-1000–Xiag180	16.2	cM	274
Xopo18–Xopt18	14.7	cM	267
Xopo18, Xbcd1532e1, Xpsr910, Xbcd1532e1, Xscg24–Xpsr903	1.7	cM	437
Xopo18, Xbcd1532e1, Xpsr910, Xbcd1532e1, Xscg24–Xpsr1196, Xopt18	0.5	cM	437
Xops10–Xpsr156	12.8	cM	267
Xops10–Xwg222	4.2	cM	267
Xopt18–Xpsr919	6.2	cM	267
Xopx12-900–Xmwg932	13.7	cM	274
Xp12-m53-f-226-p2, Xp12-m52-f-235-p2–Xp17-m60-f-81-n–p2	2.1	cM	431
Xp12-m53-f-226-p2, Xp12-m52-f-235-p2–Xp17-m60-f-273-p2	3.5	cM	431
Xp12-m58-f-136-p2–Xscm213	0.3	cM	431, 439
Xp12-m53-f-560-p2–Xp16-m56-f-385-p2, Xp12-m61-f-101-p2, Xp16-m58-f-179-p2	0.1	cM	431
Xp12-m53-f-560-p2–Xscm369, Xp12-m52-f-83-p2	2.9	cM	431, 439

TABLE 5.6
(Continued) Linkage Relationships between Genes, Genes and Markers, and Markers as Well as Markers as Example for Rye Chromosome 3R

Chromosome 3R	Value	Variation	Unit (%) or (cM)	References
Xp12-m58-f-136-p2–Xp16-m56-f-385-p2, Xp12-m61-f-101-p2, Xp16-m58-f-179-p2	0.1		cM	431
Xp12-m59-f-351-p2–Xscm369, Xp12-m52-f-83-p2	3.6		cM	431, 439
Xp12-m59-f-331-p2–Xp17-m60-f-369a-p2	0.1		cM	431
Xp12-m59-f-331-p2–Xmwg283	0.1		cM	431
Xp12-m59-f-351-p2–Xmwg283	1.4		cM	431
Xp12-m61-f-344-p2–Xp17-m60-f-274-p2	1.3		cM	431
Xp12-m61-f-88-p2–Xp12-m61-f-344-p2	8.9		cM	431
Xp12-m61-f-88-p2–Xscm117	9.5		cM	431, 439
Xp12-m61-f-440a-p2–Xp16-m56-f-65-p2	0.8		cM	431
Xp12-m61-f-440a-p2–Xmwg84	1.6		cM	431
Xp15-m53-f-135-p2–Xscm213	1.5		cM	431, 439
Xp15-m53-f-135-p2–Xscm371	1.6		cM	431, 439
Xp16-m56-f-309-p2, Xscm239–Xp17-m60-f-369a-p2	0.9		cM	431, 439
Xp16-m56-f-309-p2, Xscm239–Xscm5	0.5		cM	431, 439
Xp17-m60-f-81-n-p2–Xscm294	4.5		cM	431, 439
Xpos10, Xpsr598, Xpsr1060–Xpsr578	1.1		cM	437
Xpos10, Xpsr598, Xpsr1060–Xpsr902	0.6		cM	437
Xpsr56L, Xttu1935, hsp17.3–Xpsr78	1.9		cM	437
Xpsr5645–Xpsr156	1.7		cM	186
Xpsr56L, Xttu1935, hsp17.3–Xpsr902	0.6		cM	437
Xpsr5645–Xpsr902	1.4		cM	186
Xpsr74–Xpsr543	3.5		cM	186
Xpsr74–XSbp	1.3		cM	186
Xpsr78–Xpsr170(1)	1.4		cM	186
Xpsr78–XSbp	1.7		cM	186
Xpsr131-350–Xpsr754-218	9.4		cM	412
Xpsr131-350–Xscm162	16.5		cM	412, 439
Xpsr156–Xpsr570	1.6		cM	186
Xpsr156–Xrz538	2.9		cM	437
Xpsr170–Xpsr1077	11.8		cM	293
Xpsr170(1)–Xpsr170(2)	0.6		cM	186
Xpsr170(2)–XTlp	14		cM	206
Xpsr305–Xpsr549	20		cM	206
Xpsr305–Xpsr1196	11		cM	206
Xpsr305–Xscg26.1	1.0		cM	437
Xpsr475–Xwg110	6.6		cM	355, 412
	6.7		cM	299
Xpsr543–Xpsr570	2.3		cM	186

(Continued)

TABLE 5.6

(Continued) Linkage Relationships between Genes, Genes and Markers, and Markers as Well as Markers as Example for Rye Chromosome 3R

Chromosome 3R	Value	Variation	Unit (%) or (cM)	References
Xpsr578–Xpsr902	1.5		cM	274
Xpsr598–Xpsr902	0.9		cM	186
Xpsr598–Xpsr1060	1		cM	206
Xpsr598-206–Xscm162	3.9		cM	412, 439
Xpsr754, Xpsr74–Xscg202u, Xwg222	2.0		cM	437
Xpsr899b–Xpsr902	14.2		cM	308
	22.9		cM	308
Xpsr899b–Xpsr1196	11.3		cM	308
Xpsr902–Xpsr1077	11.1		cM	308
	13.9		cM	308
Xpsr902–Xrems1323-3R	5.3		cM	308
Xpsr903–Xscg206e	0.9		cM	437
Xpsr919–Xwg222	2.5		cM	267
Xpsr954–Xscg143u	1.6		cM	437
Xpsr1077–Xrems1261-3R	12.2		cM	308
Xpsr1077–Xwg110a	18.5		cM	308
Xpsr1149, Xpsr570, Xscg9e1–Xumc60, Xpsr804, Sbp	1.4		cM	437
Xpsr1196–Xrems1135	6.9		cM	308
Xpsr1196–Xscb17b	11.1		cM	308
Xpsr1196–Xwg110a	15.2		cM	276
Xrems1135-3R–Xrems1323-3R	16.6		cM	308
Xrems1261-3R–Xwg110a	20.2		cM	308
Xscb17b–Xwg110a	5.4		cM	308
Xscb33–Xwg110a	6.4		cM	276
Xscb33–Xwg110b	23.9		cM	276
Xscg9u–Xscg26.1	0.8		cM	437
Xscg81–Xscg23, Xfbp	5.0		cM	437
Xscg143u–Xwia483, Cxp1	1.1		cM	437
Xscg145b–Xscg186	6.8		cM	437
Xscg187u–Xumc60, Xpsr804, Sbp	0.8		cM	437
Xscm5–Xscm117	5.0		cM	431, 439
Xscm5–Xwg110b	5.8		cM	308, 439
Xscm87–Xscm162	1.0		cM	431, 439
Xscm87–Xscm371	1.5		cM	431, 439
Xtavp1–Xscm57	11.0		cM	431, 439
Xtavp1–Xscm294	5.2		cM	431, 439
Xwaw1023–Xwg110b	3.3		cM	308

Source: Schlegel, R., and Korzun, V., Genes, markers and linkage data of rye (*Secale cereale* L.), 9th updated inventory, V. 03.13, 2013. With permission.

Note: The reference citations found in the table are originally taken from Schlegel and Korzun (2013).

Nevertheless, the precision of linkage between genes and markers varies often considerably. For example, the linkage values between the genes for isoenzymes *Aat3–Mdh2(b)* range between 4.2 and 21, depending on the author, that is, the method, the precision of the experiment, and the size of segregating populations. However, for an efficient selection linkage values of <1 cM are needed.

5.5 QTL MAPPING

Quantitative genetics is a branch of genetics that deals with the inheritance of QTLs. Sometimes it is also called biometrical or statistical genetics. The inheritance depends upon the cumulative action of many genes, each of which produces only a small effect. The character shows continuous variation (i.e., graduation from one extreme to the other). Thus, QTLs are characteristics in which variation is continuous so that classification into discrete categories is arbitrary.

In rye, most of agronomically important characters are polygenic or quantitative. In contrast to qualitative inherited genes, the search and description of QTLs are more difficult. First, indirect notes on a QTL of rye date from Lukyanenko (1973) when he described the German wheat variety Neuzucht (the donor of 1RS.1BL wheat–rye translocation in the wheat varieties Avrora and Kavkaz; see Section 4.8.3.4) with significant increase of spikelet fertility. Schlegel and Meinel (1994) confirmed this. The association of increased spikelet fertility in wheat with the presence of the short arm of rye chromosome 1R was ascribed to a QTL on 1RS, designated *Spf1* (spikelet fertility). The *Spf1* locus accounted for ~20% of the variance of this character.

Molecular marker technologies and the establishing of saturated genetic maps in rye made the prerequisite for more extensive QTL studies. Using the RFLP marker information of 275 F_2 plants, QTLs determining morphological and yield characters were then analyzed (Börner et al. 2000). The first map constructed contained 113 markers including the major dwarfing gene *Ddw1* with an average distance of ~10 cM between adjacent markers. Of the 21 QTLs detected, 10 were found to map on chromosome 5RL in the region of *Ddw1*. Besides the expected effects on plant height and peduncle length that are most probably due to the presence of the major dwarfing gene, additional effects on yield characters and flowering time were discovered in that region. Pleiotropic effects of *Ddw1* might cause it. An additional gene cluster consisting of four QTLs controlling flowering time and yield components was discovered in the centromere region of chromosome 2R. Further loci were distributed on chromosomes 1R (one), 4R (one), 6R (three), and 7R (one).

Meanwhile, ~100 QTLs were detected on almost all chromosome arms of rye (see Table 5.7). Tenhola-Roininen et al. (2011) described a QTL on chromosome 5R affecting sprouting of rye. By using a doubled haploid mapping population from the cross progeny of Amilo and Voima genotypes segregating for PHS, a linkage map of 289 loci was constructed. AFLP, microsatellite, RAPD, retrotransposon-microsatellite amplified polymorphism (REMAP), interretrotransposon amplified polymorphism (IRAP), inter-SSR (ISSR), and sequence-related amplified polymorphism (SRAP) markers extended the genetic map to 732 cM, that is, one locus in every 2.5 cM. Distorted segregation of markers was detected on all chromosomes. One major QTL affecting α-amylase activity was found, which explained 16.1% of phenotypic variation. The QTL was localized on the long arm of chromosome 5R. Microsatellites SCM74, RMS1115, and SCM77, nearest to the QTL, can now be used for marker-assisted selection to decrease sprouting damage.

Myskóv et al. (2012) presented QTLs for α-amylase activity on the genetic map of a rye recombinant inbred line population (see Table 5.7)—S120 into S76. They compared them for analysis of PHS and heading earliness. Fourteen QTLs for α-amylase activity on all seven chromosomes were identified. The detected

TABLE 5.7
Compilation of QTLs Described in Rye (*Secale cereale*)

QTL	Chromosome	Description	References
QTL1	2RL, 5RL, 7R	Flowering time, days after sowing	Börner et al. (2000)
QTL2	2R	Number of spikes per plant	Börner et al. (2000)
QTL3	5RL2.3	Plant height of main tillers	Börner et al. (2000)
QTL4	5RL2.3	Peduncle length of main tillers	Börner et al. (2000)
QTL5	6R	Number of florets per spike of main tillers	Börner et al. (2000)
QTL6	2R	Number of grains per spike of main tillers	Börner et al. (2000)
QTL7	2R, 5RL	Spike yield of main tillers	Börner et al. (2000)
QTL8	2R	Thousand-grain weight	Börner et al. (2000)
QTL9	5RL, 6RL	Straw yield	Börner et al. (2000)
QTL10	4RL	Rf_{clb} restorer CMS	Stojalowski et al. (2004)
QTL11	6RS	Rf_c2 restorer CMS	Stojalowski et al. (2004)
QTL12	2R, 4R	Drought tolerance	Mohammadi et al. (2003)
QTL13	1RS	Number of seeds per spikelet (in wheat)	Schlegel and Meinel (1994)
QTL14	1RL	Chlorophyll content in leaves	Milczarski and Masojć (2002, 2007)
QTL15	3RS	Chlorophyll content in leaves	Milczarski and Masojć (2002, 2007)
QTL16	4RL	Chlorophyll content in leaves	Milczarski and Masojć (2002, 2007)
QTL17	5RL	Chlorophyll content in leaves	Milczarski and Masojć (2002, 2007)
QTL18	5RL	Sensitivity of seedlings to gibberellic acid	Milczarski and Masojć (2002, 2007)
QTL19	1RL	Sensitivity of seedlings to gibberellic acid	Milczarski and Masojć (2002, 2007)
QTL20	7RL	Sensitivity of seedlings to gibberellic acid	Milczarski and Masojć (2002, 2007)
QTL21	1RS	Sensitivity of seedlings to abscisic acids	Milczarski and Masojć (2002, 2007)
QTL22	2RS	Sensitivity of seedlings to abscisic acids	Milczarski and Masojć (2002, 2007)
QTL23	5RL	Sensitivity of seedlings to abscisic acids	Milczarski and Masojć (2002, 2007)

TABLE 5.7
(Continued) Compilation of QTLs Described in Rye (*Secale cereale*)

QTL	Chromosome	Description	References
QTL24	*1RS*	Activity of alpha-amylase	Milczarski and Masojć (2007); Masojć and Milczarski (2005, 2009)
QTL65	*1RS*	Activity of alpha-amylase; marker intervals *Xiag95–Xapr1.5*, *Xubp19–Xem1.27*	Masojć and Milczarski (2009)
QTL88	*1RL*	Activity of alpha-amylase; marker interval *Xapr1.1–Xbcd442*	Masojć and Milczarski (2009)
QTL66	*1RL*	Activity of alpha-amylase; marker interval *Xapr1.12–Xapr1.1*	Masojć and Milczarski (2009)
QTL25	*2RS*	Activity of alpha-amylase	Milczarski and Masojć (2007); Masojć and Milczarski (2005)
QTL67	*2RS*	Activity of alpha-amylase, marker intervals *Xapr2.24–Xapr2.22*, *Xpsr900–Xem2.10*	Masojć and Milczarski (2009)
QTL75	*2RL*	Activity of alpha-amylase, marker interval *Xapr2.12–Xpsr901*; coincides with preharvest sprouting locus *Amy1*; marker interval *Xpr181-1200–Xis2.1*	Masojć and Milczarski (2009); Mysków et al. (2010)
QTL26	*3RS*	Activity of alpha-amylase	Milczarski and Masojć (2007); Masojć and Milczarski (2005)
QTL68	*3RS*	Activity of alpha-amylase, marker interval *Xapr3.14–Xapr3.16*	Masojć and Milczarski (2009)
QTL76	*3RL*	Activity of alpha-amylase; marker interval *Xubp20–Xapr3.3* coincides with preharvest sprouting Phs1; marker interval *Xpr665-420–Xis09-500*	Masojć and Milczarski (2009); Mysków et al. (2010)
QTL27	*3RL*	Activity of alpha-amylase	Milczarski and Masojć (2007); Masojć and Milczarski (2005)
QTL69	*4RS*	Activity of alpha-amylase, marker interval *Xapr4.15–Xapr4.16*	Masojć and Milczarski (2009)
QTL70	*4RL*	Activity of alpha-amylase; marker interval *Xapr4.25–Xapr4.26*	Masojć and Milczarski (2009)
QTL77	*4RL*	Activity of alpha-amylase; marker interval *Xpsr167–Xpsr899*	Masojć and Milczarski (2009)
QTL28	*5RS*	Activity of alpha-amylase	Milczarski and Masojć (2007); Masojć and Milczarski (2005)

(Continued)

TABLE 5.7

(Continued) Compilation of QTLs Described in Rye (*Secale cereale*)

QTL	Chromosome	Description	References
QTL71	5RS	Activity of alpha-amylase; marker intervals *Xapr5.13–Xapr5.14*, *Xpsr1327–Xubp3*	Masojć and Milczarski (2009)
QTL87	5RL	Activity of alpha-amylase, marker interval *Xapr5.2–alpha-Amy3*, *Xapr5.2–alpha-Amy3*; coincides with preharvest sprouting Phs2; marker interval *Xpr215-480–Xis43-810*	Masojć and Milczarski (2009); Mysków et al. (2010)
QTL29	5RL	Activity of alpha-amylase	Milczarski and Masojć (2007); Masojć and Milczarski (2005)
QTL72	5RL	Activity of alpha-amylase; marker interval *Xapr5.24–Xapr5.2*; 5RL = *Xrms1115*, *Xscm74*, *Xscm77*	Masojć and Milczarski (2009); Tenhola-Roininen et al. (2011)
QTL78	6RS	Activity of alpha-amylase; marker interval *Xpsr106-1–Xem6.11*; coincides with preharvest sprouting Phs3; marker interval *Xapr6.43–Xapr6.41*	Masojć and Milczarski (2009); Mysków et al. (2010)
QTL30	6RL	Activity of alpha-amylase; marker interval *Xscm46–Xapr6.7*	Masojć and Milczarski (2005, 2009); Milczarski and Masojć (2007)
QTL73	6RL	Activity of alpha-amylase; marker interval *Est5–Xpsr454*	Masojć and Milczarski (2009)
QTL74	7RS	Activity of alpha-amylase; marker interval *Xopq4a–Xapr7.25*	Masojć and Milczarski (2009)
QTL31	7RL	Activity of alpha-amylase	Milczarski and Masojć (2007); Masojć and Milczarski (2005)
QTL32	7RL	Activity of alpha-amylase	Milczarski and Masojć (2007); Masojć and Milczarski (2005)
QTL33	3RL	Plant height	Milczarski and Masojć (2003, 2007)
QTL34	5RL	Plant height	Milczarski and Masojć (2003, 2007)
QTL35	5RL	Spike length	Milczarski and Masojć (2003, 2007)
QTL36	2RL	Thousand-grain weight	Milczarski and Masojć (2003, 2007)
QTL37	3RL	Thousand-grain weight	Milczarski and Masojć (2003, 2007)
QTL38	5RL	Thousand-grain weight	Milczarski and Masojć (2003, 2007)
QTL39	2RS	Kernel length	Milczarski and Masojć (2003, 2007)
QTL40	3RL	Kernel length	Milczarski and Masojć (2003, 2007)
QTL41	5RL	Kernel length	Milczarski and Masojć (2003, 2007)
QTL42	1RS	Kernel thickness	Milczarski and Masojć (2003, 2007)
QTL43	2RL	Kernel thickness	Milczarski and Masojć (2003, 2007)

TABLE 5.7
(Continued) Compilation of QTLs Described in Rye (*Secale cereale*)

QTL	Chromosome	Description	References
QTL44	*3RL*	Kernel thickness	Milczarski and Masojć (2003, 2007)
QTL45	*4RS*	Kernel thickness	Milczarski and Masojć (2003, 2007)
QTL46	*5RL*	Kernel thickness	Milczarski and Masojć (2003, 2007)
QTL47	*4RS*	Heading time	Milczarski and Masojć (1999, 2007)
QTL48	*5RL*	Heading time	Milczarski and Masojć (1999, 2007)
QTL49	*2RL*	Length of flag leaf	Milczarski and Masojć (2007)
QTL50	*4RL*	Length of flag leaf	Milczarski and Masojć (2007)
QTL51	*3RL*	Length of flag leaf	Milczarski and Masojć (2007)
QTL52	*5RL*	Length of flag leaf	Milczarski and Masojć (2007)
QTL53	*5RS*	Thickness of second internode	Milczarski and Masojć (2007)
QTL54	*7RL*	Thickness of second internode	Milczarski and Masojć (2007)
QTL55	*2RS*	Number of kernels per spike	Milczarski and Masojć (2007)
QTL56	*3RL*	Number of kernels per spike	Milczarski and Masojć (2007)
QTL57	*5RL*	Number of kernels per spike	Milczarski and Masojć (2007)
QTL58	*3RS*	Number of kernels per spike	Milczarski and Masojć (2007); Masojć and Milczarski (2004)
QTL59	*3RS*	Number of kernels per spike	Milczarski and Masojć (2007); Masojć and Milczarski (2004)
QTL79	*1RS*	Preharvest sprouting; marker interval *Xapr1.5–Xapr1.22, Xem1.20–Xem1.27*	Masojć and Milczarski (2009)
QTL60	*1RL*	Preharvest sprouting; additive gene action; marker interval *Xapr1.18–Xapr1.17*	Masojć and Milczarski (2009); Masojć et al. (2007)
QTL61	*2RL*	Preharvest sprouting; Ot1-3, dominant gene action; marker interval *Xapr2.15–Xapr2.20*	Masojć and Milczarski (2009); Mysków et al. (2010); Masojć et al. (2007)
QTL85	*2RL*	Preharvest sprouting; marker interval *Xapr2.12–Xpsr901*; coincides with preharvest sprouting locus *Amy1*; marker interval *Xpr181-1200–Xis2.1*	Masojć and Milczarski (2009); Mysków et al. (2010)
QTL80	*3RS*	Preharvest sprouting; marker interval *Xapr3.14–Xapr3.16*; coincides with preharvest sprouting Phs1; marker interval *Xpr665-420–Xis09-500*	Masojć and Milczarski (2009); Mysków et al. (2010)
QTL81	*4RL*	Preharvest sprouting; marker interval *Xapr4.25–Xapr4.26*	Masojć and Milczarski (2009)
QTL82	*5RS*	Preharvest sprouting; marker interval *Xapr5.13–Xapr5.14*	Masojć and Milczarski (2009)

(Continued)

TABLE 5.7

(Continued) Compilation of QTLs Described in Rye (*Secale cereale*)

QTL	Chromosome	Description	References
QTL86	5RL	Preharvest sprouting, marker interval *Xmtd862–Dw1*	Masojć and Milczarski (2009)
QTL62	5RL	Preharvest sprouting; dominant gene action; marker intervals *alpha-Amy3–Xis5.1, Xapr5.2–alpha-Amy3*; coincides with preharvest sprouting Phs2; marker interval *Xpr215-480–Xis43-810*	Masojć and Milczarski (2009); Mysków et al. (2010); Masojć et al. (2007)
QTL83	6RS	Preharvest sprouting; marker interval *Xpsr106-1–Xscm112*; coincides with preharvest sprouting Phs3; marker interval *Xapr6.43–Xapr6.41*	Masojć and Milczarski (2009); Mysków et al. (2010)
QTL63	6RL	Preharvest sprouting; Ot1-3, dominant gene action; marker interval *Xscm78–Xapr6.7, Xpsr454–XksuD2*	Masojć and Milczarski (2009); Masojć et al. (2007)
QTL84	7RS	Preharvest sprouting; marker interval *Xopq4a–Xapr7.25*	Masojć and Milczarski (2009)
QTL64	7RL	Preharvest sprouting; Ot1-3, recessive gene action; marker interval *Xapr7.30–Xapr7.31*	Masojć and Milczarski (2009); Masojć et al. (2007)
QTL89	1RS	Rooting ability; the terminal 15% of the rye 1RS arm carries gene(s) for greater rooting ability in wheat	Sharma et al. (2009)

Source: Schlegel, R., and Korzun, V., Genes, markers and linkage data of rye (*Secale cereale* L.), 9th updated inventory, V. 03.13, 2013. With permission.

QTLs were responsible for 6.09%–23.32% of α-amylase activity variation. The lowest limit of detection (LOD) value (~2.2) was achieved by locus QAa4R-M3 and the highest (~7.8) by locus QAa7R-M1. There were six overlapping QTLs for α-amylase activity and PHS (1R, 3R, 4R, 6R, 7R) and the same number for PHS and heading earliness (1R, 2R, 6R, 7R). One interval partially common to all three QTLs was mapped on the long arm of chromosome 1RL. Among the statistically significant markers selected in the Kruskal–Wallis test, there were 55 common ones for PHS and heading earliness (1R, 2R, 6R), 30 markers coinciding between α-amylase activity and PHS (5R, 7R), and 1 marker for α-amylase activity and heading earliness (6R).

A compilation of the available QTLs and their chromosomal distribution is given in Figure 5.4. Generally, QTL mapping can successfully be applied in selection of segregating populations. However, QTL underlying grain yield and several quality

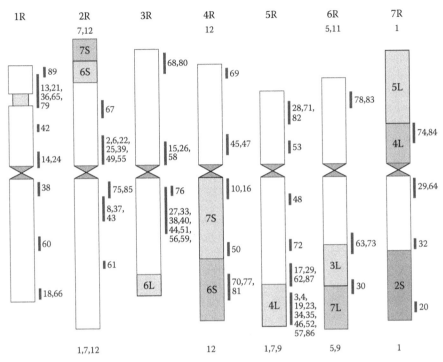

FIGURE 5.4 The estimated distribution of QTLs so far mapped on rye chromosomes compiled from different references; QTLs not associated with a certain chromosomal region but with the chromosome or arm are given as numbers at top or below the chromosomal idiogram; QTLs including references are listed in Table 5.7.

traits have small selective effects. In contrast, 1000-kernel weight, test weight, falling number, and starch content are affected by several major QTLs. These QTLs explain a large proportion of the genotypic variance. They can be exploited in marker-assisted selection programs and are candidates for further genetic dissection.

5.6 PHYSICAL MAPPING

Physical maps include locations of identifiable landmarks on DNA or chromosomes (e.g., restriction enzyme cutting sites, genes, molecular markers) regardless of inheritance. The distance is measured in base pairs or absolute/relative distances. Sybenga et al. at the University of Wageningen (the Netherlands) first studied physical mapping using reciprocal translocations. A translocation tester set of rye was established involving all seven chromosomes by interchanged chromosomes as follows: 1RS–4RL, 1RS–5RL, 1RS–6RS, 2RS–5RS, 2RL–6RL, 3RS–5RL, 4RL–5RL, and 5RL–7RS. They originated from irradiated pollen grains of Petkus rye crossed to several inbred rye lines (de Vries and Sybenga 1984). With the tests, chromosomal localization of 17 monogenically inherited morphological markers was possible.

Later, Gustafson et al. (1990) used a 900 bp sequence from a complementary DNA (cDNA) clone of the rye endosperm storage protein gene *Sec-1* for labeling with biotin and hybridized to Blanco rye chromosomes. Hybridization was seen on the satellite part of the 1RS, where the *Sec-1* gene has been genetically mapped, in approximately 8.5% of the cells analyzed. The clone cross-hybridized at a lower frequency to rye chromosome arms 1RL (4.4%) and 2RS (2.6%), where two additional rye storage protein loci, *Sec-3* and *Sec-2*, respectively, were mapped. A fourth hybridization site was observed on the short arm of chromosome 6R at a frequency of 3.0%. The cross-hybridization is attributed to a combination of residual sequence homology between the protein loci and the low-stringency conditions used in *in situ* hybridization. *In situ* hybridization mapping, in combination with chromosome walking by using molecular techniques, was suggested as a general approach to physical mapping of chromosomes.

By combining cytogenetic and molecular techniques, it was feasible to map the morphological, biochemical, and physiological characteristics on a defined terminal segment of chromosome 5RL (see Figure 5.5). A terminal segment of rye chromosome arm 5RL replaced a faint terminal segment of wheat (cf. Schlegel et al. 1993). On the segment 5RL2.3, the genes could be mapped (see Section 3.4.2).

Among others, Mohan et al. (2007) physically mapped 182 rye expressed sequence tag EST–SSR markers on 21 wheat chromosomes. The mapping involved two approaches: the wet-lab approach involving use of deletion stocks and the

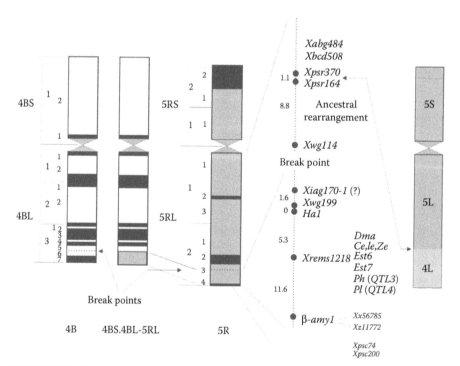

FIGURE 5.5 Karyogram of the wheat and rye chromosomes and/or segments involved in the 4B/5R translocations of Cornell selection and Viking-trans (? = uncertain marker).

in silico approach involving matching with ESTs that were previously mapped. The distribution of loci on the three subgenomes of wheat, the 7 homeologous groups, and the 21 individual chromosomes was nonrandom. Long arms had disproportionately (relative to DNA content) higher number of loci, with more loci mapped to the distal regions of chromosome arms. A fairly high proportion of EST–SSR markers had multiple loci, which were largely (81%) homeoloci. Rye EST–SSRs showed a high level of transferability (~77%) to the wheat genome. Putative functions were assigned to 216 SSR-containing ESTs through homology searches against the protein database.

The best investigated is the short arm of chromosome 1R. It contains agronomically useful genes, particularly for wheat improvement (see Section 4.12.4). Besides numerous other experiments, the latest physical dissection of chromosomes into segments was applied in mapping 1R-specific DNA markers and assembling DNA clones into contig maps. By utilizing the so-called gametocidal system, the rearranged 1R chromosomes of Imperial rye were produced (Zhou et al. 1999, Tsuchida et al. 2008). This method was accomplished by rapid EST isolation from chromosome 1R (Zhou et al. 2008).

More than 55 rearranged chromosomes were studied in the genetic background of wheat. Fifty-two of them had single break points and three had double break points. The 58 break points were distributed in the short arm excluding the satellite (12 break points), in the satellite (4 break points), in the long arm (28 break points), and in the centromere (14 break points). Out of the 55 lines, 9 were homozygous for the rearranged 1R chromosomes, and the remaining lines were hemizygous. Twenty-six PCR-based EST markers were developed which were specific to the 1R chromosome, and nine of them amplified 1R arm-specific PCR products without restriction enzyme digestion. Using the 9 EST markers and 2 previously reported 1R-specific markers, 55 1R dissection lines could be characterized.

5.7   COMPARATIVE MAPPING

Comparative genetics is the science that exploits the results of comparative genetic mapping—two terms that were unknown to plant geneticists 30 years ago. Although the discovery that genes in related species—and as it turns out now, quite distantly related species—tend to be ordered collinearly on chromosomes is quite new, the concept of conservation has a long history in plant genetics.

In the 1920s, N. J. Vavilov had already observed that "similar variations" were to be found in different species. More recently, DNA studies have shown that genes of similar function in different species have remarkably conserved sequences. Over the years, it became evident that different plant characters are coded by homeologous alleles found on homeologous genomes and/or chromosomes. From Table 5.8, it can be seen that most of the alleles correspond to homeologous chromosomes. However, in some case the homeologous loci do not match the corresponding chromosomes. It can be assumed that interchromosomal and intragenomic rearrangements are the reason. This was later confirmed by Devos and Gale (1997) (see Figure 4.6).

The discovery of collinearity by comparative mapping is, however, a function of the new molecular markers employed in plants for the first time in the mid-1980s. Nevertheless, with immense efforts in genetic mapping of morphological mutants and, later, protein loci in plants, it is surprising that gene collinearity

TABLE 5.8

Localization of Different Morphological and Biochemical Characters (Genes) across Homoeologous Triticeae Genomes and/or Chromosomes

Homeologous Genomes, Chromosomes, and Gene Loci in Triticeae

Trait	Rye		Wheat			Barley
	R^{cer}	R^{mon}	A	B	D	H^{vul}
Peroxidase	1R^{cer}S, *Per-R1*			1BS, *Per-B1*	1DS, *Per-D1*	1H^{vul}, *Per-H1*
Glucose phosphate isomerase	1R^{cer}S, *Gpi-R1*		1AS, *Gpi-A1*	1BS, *Gpi-B1*	1DS, *Gpi-D1*	1H^{vul}, *Gpi-H1*
Secalin	1R^{cer}S, *Sec-1*	1R^{mon}		1BS, *Gli-B1*	1DS, *Gli-D1*	1H^{vul}S, *Hor1, Hor2*
NOR	1R^{cer}S, *Nor-R1*			1BS, *Nor-B1*		
5S DNA	1R^{cer}S, *5S-R1*			1BS, *5S-B1*	1DS, *5S-D1*	
Malate dehydrogenase	1R^{cer}L, *Mdh-R1*		1AL, *Mdh-A1*	1BL, *Mdh-B1*	1DL, *Mdh-D1*	1H^{vul}L, *Mdh-H1*
Secalin	1R^{cer}L, *Sec-3*	1R^{mon}L	1AL, *Glu-A1*	1BL, *Glu-B1*	1DL, *Glu-D1*	1H^{vul}L, *Hor3*
Leaf peroxidase	2R^{cer}S, *Per-R2*		2AS, *Per-A2*	2BS, *Per-B2*	2DS, *Per-D2*	2H^{vul}, *Per-H2*
Secalin	2R^{cer}S, *Sec-2*	6R^{mon}				
Superoxide dismutase	2R^{cer}, *Sod-R1*		2AL, *Sod-A1*	2BL, *Sod-B1*	2DL, *Sod-D1*	
Esterase	3R^{cer}, *Est-R1*		3AS, *Est-A1*	3BS, *Est-B1*	3DS, *Est-D1*	
Triosephosphate isomerase	3R^{cer}, *Tpi-R1*		3AS, *Tpi-A1*	3BS, *Tpi-B1*	3DS, *Tpi-A1*	3H^{vul}, *Tpi-H1*
Glutamic oxaloacetic transaminase	3R^{cer}, *Got-R3*		3AL, *Got-A3*	3BL, *Got-B3*	3DL, *Got-D3*	3H^{vul}, *Got-H3*
Phosphoglucomutase	4R^{cer}S, *Pgm-R1*		4AS, *Pgm-A1*	4BL, *Pgm-B1*	4DS, *Pgm-D1*	4H^{vul}, *Pgm-H1*
Alcohol dehydrogenase	4R^{cer}S, *Adh-R1*		4AS, *Adh-A1*	4BL, *Adh-B1*	4DS, *Adh-D1*	4H^{vul}, *Adh-H1*
6-Phospho-gluconate dehydrogenase	4R^{cer}L, *Pgd-R1*			7B, *Pgd-B1*		
Red coleoptile	4R^{cer}L, *an1*	4R^{mon}L		7BS, *Rc2*	7DS, *Rc3*	

Trait						
Purple culm	4R^{cer}L, Pc	4R^{mon}L		7BS, Pc		
Shikimate dehydrogenase	5R^{cer}S, Shdh-R1		5AS, Skdh-A1	5BS, Skdh-B1	5DS, Skdh-D1	
Triosephosphate isomerase	5R^{cer}, Tpi-R2		5AL, Tpi-A2	5BL, Tpi-B2	5DL, Tpi-D2	5H^{vul}, Tpi-H2
β-Amylase	5R^{cer}L, β-Amy-R1		4AL, β-Amy-A1		4DL, β-Amy-D1	4H^{vul}, β-Amy-H1
Corroded	6R^{cer}S, Co			6BS, Co		
α-Amylase	6R^{cer}L, α-Amy-R1	6R^{mon}L, α-Amy-R^{mon}1	6AL, α-Amy-A1	6BL, α-Amy-B1	6DL, α-Amy-D1	6H^{vul}, α-Amy-H1
6-Phosphogluconate dehydrogenase	6R^{cer}L, Pgd-R2		6A, Pgd-A2	6BL, Pgd-B2		
Glutamic oxaloacetic transaminase	6R^{cer}, Got-R2		6AL, Got-A2	6BL, Got-B2	6DL, Got-D2	6H^{vul}, Got-H2
Aconitase	6R^{cer}, Aco-R1					6H^{vul}, Aco-H1
Esterase	6R^{cer}L, Est-R5	6R^{mon}L, Est-R^{mon}5				3H^{vul}, Est-H5
Red grain	6R^{cer}L, Reg		3AS, R2	3B, R3	3DS, R1	
Round grain	6R^{cer}L, Rog				3D, S1	
Acid phosphatase	7R^{cer}S, Acph-R1		4AL, Acph-A4,8	4BS, Acph-A3,2	4DL, Acph-A5,6	4H^{vul}, Acph-H1
Endopeptidase	7R^{cer}L, Ep-R1		7AL, Ep-A1	7BL, Ep-B1	7DL, Ep-D1	4H^{vul}, Ep-H^{chi}1

Source: Zeller, F. J., and Cermeno, M. C., *Chromosome Engineering in Plants: Genetics, Breeding, Evolution*, vol. 2. P. K. Gupta, and T. Tsuchiya (eds.). Developments in Plant Genetics and Breeding. Elsevier Science Publication, Amsterdam, the Netherlands, pp. 313–333, 1991. With permission. Schlegel, R., and Korzun, V., Genes, markers and linkage data of rye (*Secale cereale* L.), 9th updated inventory, V. 03.13, 2013. With permission.

Note: H^{vul}, *Hordeum vulgare*; R^{cer}, *Secale cereale*; R^{mon}, *S. montanum*; A, *Triticum aestivum*; B, *T. aestivum*; D, *T. aestivum*.

remained hidden until the publication of reports in 1988 of the convergence of the maps of the three genomes of hexaploid bread wheat (Chao et al. 1988) and a similar convergence between the maps of tomato and potato. In wheat, evidence for homeology between the three genomes had already been provided by cytogeneticists, particularly by the late E. Sears at Columbia, Missouri, who assembled the first set of aneuploid genetic stocks in wheat. The development of aneuploid stocks in wheat led to the discovery that an extra dose of a particular chromosome could compensate for the absence of another. This compensating ability of chromosomes of different ancestral origin defined their relationship and resulted in the classification of the 21 wheat chromosomes into seven homeologous groups. Similar compensating experiments determined the homeologous relationships between wheat chromosomes and those of other Triticeae species such as rye, barley, and several *Aegilops* species. The ability of chromosomes to substitute for one another suggested that they carried similar genes.

Comparative genetic studies have demonstrated that gene content and orders are highly conserved at both the map and the megabase level between different species within the grass family. Integration of the genetic maps of rice, foxtail millet, sugarcane, sorghum, maize, Triticeae cereals, and oats into a single synthesis reveals that some chromosome arrangements characterize taxonomic groups, while others have arisen during or after speciation (Devos and Gale 1997).

In order to exploit other cereal crops for rye genetics and breeding, Hackauf et al. (2009a) made an inventory of EST-derived DNA markers with known genomic positions in rye, which are related to those in rice. The rice genome has proven a valuable resource for comparative approaches to address individual genomic regions in Triticeae species at the molecular level. As a first inventory set, 92 EST–SSR markers were mapped which had been drawn from a nonredundant rye EST collection representing 5423 unigenes and 2.2 Mbp of DNA. Using a backcross 1 (BC_1) mapping population that involved an exotic rye accession as donor parent, these EST–SSR markers were arranged in a linkage map together with 25 genomic SSR markers as well as 131 AFLP and 4 sequence-tagged site (STS) markers. This map comprises seven linkage groups corresponding to the seven rye chromosomes and covers 724 cM of the rye genome. For comparative studies, additional inventory sets of EST-based markers were included which originated from the rye-mapping data published by other authors. Altogether 502 EST-based markers with known chromosomal localizations in rye were used for Basic Local Alignment Search Tool (BLAST; BLASTN, Nucleotide–nucleotide BLAST) search and 334 of them could be mapped *in silico* in the rice genome. Additionally, 14 markers were included which lacked sequence information but had been genetically mapped in rice. Based on the 348 markers, each of the seven rye chromosomes could be aligned with distinct portions of the rice genome, providing improved insight into the status of the rye–rice genome relationships. Furthermore, the aligned markers provide genomic anchor points between rye and rice, enabling the identification of conserved orthologous set markers for rye.

Another example is given by Khlestkina et al. (2011) studying the *F3h* gene encoding flavanone 3-hydroxylase. The latter is one of the key enzymes of the flavonoid biosynthesis pathway, which is involved in plant defense response. In a recent study, the *F3h* genes were for the first time genetically mapped in rye, wheat, and barley, using microsatellite and RFLP markers. The three wheat *F3h* homeologous copies *F3h-A1*, *F3h-B1*, and *F3h-D1*, and rye *F3h-R1* were mapped close to the microsatellite loci *Xgwm0877* and *Xgwm1067* on chromosomes 2AL, 2BL, 2DL, and 2RL, respectively. Wheat *F3h-G1* and barley *F3h-H1* were also associated with the homeologous *F3h-1* position on chromosomes 2GL and 2HL, respectively. The non-homeologous *F3h* gene (*F3h-B2*) was mapped on wheat chromosome 2BL ~40 cM distal to the *F3h-1* map position. Figure 5.6 shows the relative map positions given for wheat, barley, and rye, demonstrating the conserved linkage over a long period of grass evolution.

5.8 GENE REGULATION

The dynamic behavior of a genome can be regarded in terms of genetic and epigenetic changes. The first concerns modifications of the integrity of DNA sequences themselves, while epigenetic modulation is associated with the regulation and transmission of gene expression patterns. The nucleolar dominance, the first epigenetic process initially disclosed by the Russian botanist Navashin (1934), revealed marked changes in chromatin organization in situations of genome interactions. In both somatic cells of triticale and wheat lines with introgressed rye chromatin, the preferential inactivation of ribosomal genes of rye origin occurs, such imprinting being reprogrammed after meiosis and reestablished in the initial stages of embryogenesis. In contrast, there is an enhancement of wheat rDNA transcriptional activity when 1R chromosome is introgressed in wheat, demonstrating that the interspecific interactions induce transcriptional changes on both parental species.

Recent high-throughput cDNA sequence comparisons between the diploid rye and the hexaploid triticale detected suppression of expression of ~0.5% of rye genes surveyed in the triticale background (Khalil et al. 2013). The expression of 23,503 rye cDNA contig assemblies was analyzed in 454 cDNA libraries obtained from anthers, roots, and stems from both triticale and rye as well as in five 454 cDNA data sets created from ovary, pollen, seed, seedling shoot, and stigma from triticale. Among these, 112 rye cDNA contigs were found to be totally suppressed in all triticale tissues, although their expressions were relatively high in rye tissues. The average DNA sequence identity between rye genes that were suppressed in triticale and their most similar contig in the wheat A and B genomes was only 82%. This degree of similarity is significantly lower than the global average similarity between rye genes and their wheat homeologs, which was found to be 95% identical in a test set of 330 triticale genes.

Ribosomal gene silencing is disrupted through induced DNA hypomethylation that also affects the topology of chromosomes and the relative disposition of parental genomes in polyploids. In rye genomes, there is a typical zebra-like disposition, which is modified to intermingled domains after induced DNA hypomethylation. Analogous changes on chromosome arms disposition confirmed the association between gene silencing, chromatin topology, and epigenetic marks. Ribosomal chromatin topology is also drastically modified when the structure, for example, of chromosome 1R (rDNA loci) is split in its individualized arms (1RS + 1RL). It changes the nuclear positioning of *Nor* loci and greatly reduces the characteristic heteromorphism between homologous.

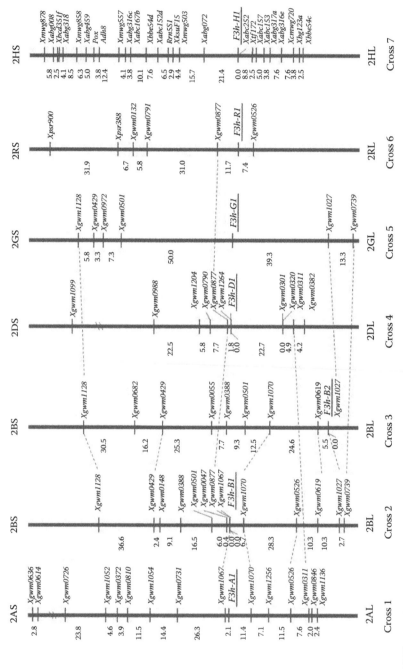

FIGURE 5.6 Genetic mapping of the flavanone 3-hydroxylase (*F3h*) genes in wheat (A, B, D—*Triticum aestivum* and G—*T. timopheevii*), rye (R—*Secale cereale*), and barley (H—*Hordeum vulgare*); chromosome arm designations are shown at the top and bottom of each chromosome; the F3h and DNA marker loci are given to the right, while genetic distances (cM) are shown to the left of each chromosome. (Courtesy of E. Khlestkina.)

5.9 DNA AND GENE TRANSFER

Compared to other cereal and crop plants, rye was less accessible for DNA and/or gene transfer experiments. In addition, rye is known as one of the most recalcitrant species in tissue culture. Neither the tissue culture ability nor the regeneration ability is as good as in wheat, barley, rice, or other grasses. Merely a genotype of *S. vavilovii* (Grosse et al. 1996) showed a higher *in vitro* utilization compared to the average of the *S. cereale* genotypes tested (Flehinghaus et al. 1991a). This strongly hindered the application of the cocultivation approach using *Agrobacterium tumefaciens*.

The first breakthrough was made by Pena et al. (1987). Work on the development of the male germ line of rye showed that 14 days before the first meiotic metaphase the archesporial cells are highly sensitive to caffeine and colchicine injected into the developing floral tillers. At this stage, the authors considered that the archesporial cells might also be permeable to other molecules such as DNA. Therefore, they injected DNA carrying a dominant selectable marker gene into rye plants. It was the plasmid pLGVneo1103 including the aminoglycoside phosphotransferase II reporter gene (APH 3′) under the control of the nopaline synthetase promoter. The plants were injected 2 weeks before meiosis. Because of the high degree of self-incompatibility of the JNK rye (see Table 7.6), seeds were produced by pairwise crossing of infected tillers.

From 3023 seeds screened for kanamycin resistance and derived from 98 plants, 7 seedlings remained green after 10 days of growth on kanamycin-containing medium. These apparently kanamycin-resistant plantlets were assayed for the presence of APH(3′)-II enzymatic activity. Two of the seven resistant seedlings, each resulting from an independent injection experiment, showed APH(3′)-II activity. This was the first unequivocal report on a gene transfer experiment in rye.

Further, Castillo et al. (1994) found a rye inbred line (L22) showing good regeneration response and reasonable transformation. The cocultivation of *Agrobacterium* and rye immature embryos in liquid medium facilitated washing of the cultures to avoid *Agrobacterium* overgrowth and allowed a high throughput. More than 40 independent transgenic plants were regenerated with one to four Southern-positive, independent events from 100 inoculated immature embryos. *Agrobacterium* strain AGL0 supported stable integration of a constitutive *nptII* selectable marker expression cassette into the genome, as indicated by regeneration of plantlets on paromomycin-containing culture medium, Southern blot analysis, Western blot analysis, and the analysis of T-DNA::plant DNA boundary sequences. Transgenic plants were phenotypically normal and fully fertile.

Popelka and Altpeter (2003) described a more efficient transformation system. A total of 45 transgenic plants were regenerated with a transformation efficiency of 1%–4% of the inoculated explants. The cocultivation of *Agrobacterium* strain AGL0, harboring plasmid pJFnptII and rye immature embryos in liquid medium, allowed a high throughput and facilitated washing of the cultures to avoid *Agrobacterium* overgrowth. The selection with paromomycin exclusively during the regeneration allowed the efficient recovery of transgenic events without interfering with somatic embryogenesis. Southern blot analysis confirmed the independent nature of the analyzed plants and indicated single-copy inserts in more than 50% of them. Segregation analysis confirmed single-locus integration and stable transgene expression in most of the lines, while one line with multiple locus integration was also observed.

Even after biolistic gene transfer to callus tissue derived from immature embryos, an efficient and reproducible production of stably expressing transgenic rye plants was feasible (Popelka et al. 2003). The *bar* gene was used as a selectable marker and selection was performed by spraying the regenerated shoots with 0.05% Basta® herbicide solution without any previous selection of tissue cultures. Based on Southern blot analysis, a total of 21 transgenic rye plants with independent transgene integration patterns were produced. A low transgene copy number was observed in most transgenic plants and 40% of the plants had a single-copy transgene insert. All transgenic rye lines with single-copy transgene inserts showed stable transgene expression in sexual progenies, but indications of transcriptional and posttranscriptional gene silencing were also observed in few transgenic lines with multicopy transgene inserts. The identification of 17 transgenic rye plants without using any selectable marker gene by PCR amplification of transgene sequences was also demonstrated. Instant generation of selectable marker-free transgenic rye avoided a negative impact of selective agents on the transgenic tissue cultures.

5.9.1 GLUTENIN

Transformation experiments with genes of breeding interest were initiated during the following period (Herzfeld 2002). The HMW glutenin subunit (HMW-GS) genes *Ax1*, *Dx5*, and *Dy10* from wheat were introduced into rye by biolistic gene transfer and their stable expression in the endosperm of the primary transformants of rye and in their segregating progeny were demonstrated. The unique bread-making characteristic of wheat flour is closely related to the elasticity and extensibility of the gluten proteins stored in the starchy endosperm, particularly the HMW-GS. Rye flour has poor bread-making quality (see Section 9.3), despite the extensive sequence and structure similarities of wheat and rye HMW-GS (see Table 2.9).

Altpeter et al. (2004) demonstrated that genetic engineering significantly alters the polymerization and composition of storage proteins. When they extracted flour proteins by means of a modified Osborne fractionation from wild type (L22) as well as transgenic rye expressing 1Dy10 (L26) or 1Dx5 and 1Dy10 (L8), the amount of transgenic HMW-GS in homozygous rye seeds represented 5.1% (L26) or 16.3% (L8) of the total extracted protein and 17% (L26) or 29% (L8) of the extracted glutelin fraction. The amount of polymerized glutelins was significantly increased in transgenic rye (L26) and more than tripled in transgenic rye (L8) compared to wild type (L22). Gel permeation high-performance liquid chromatography (HPLC) of the nonpolymerized fractions revealed that the transgenic rye flour contained a significantly lower proportion of alcohol-soluble oligomeric proteins compared with the nontransgenic flour.

The quantitative data indicate that the expression of wheat HMW-GS in rye leads to a high degree of polymerization of transgenic and native storage proteins, probably by the formation of intermolecular disulfide bonds. Even γ-40k secalins, which occur in nontransgenic rye as monomers, are incorporated into these polymeric structures. The combination 1Dx5 + 1Dy10 showed stronger effects than 1Dy10 alone (see Table 2.9).

6 Cytoplasm, Cytoplasmic Male Sterility, and Restorer

6.1 CYTOPLASM

Cytoplasmic effects in reciprocal crosses of rye were early described (Nürnberg-Krüger 1951); that is, reciprocal extrachromosomal inheritance is common in rye (Fröst et al. 1970). Breeding characteristics, such as straw length, tillering, and spike fertility, are influenced by it. Individuals from crosses of the same cytoplasm and individuals from crosses of different cytoplasms show almost the same range of variability. Thus, the negative effects of homozygosity in cytoplasms do not seem to be relevant in rye as expected by Heribert-Nilsson (1916a). Environmental conditions can mask cytoplasmic effects. For special crosses and breeding approaches, reciprocal combinations and comparisons are recommended.

Studies using classic genetics as well as restriction fragment length polymorphism (RFLP) analysis have demonstrated that rye, unlike most flowering plants, has biparental inheritance of both plastids and mitochondria (Fröst et al. 1970). Yet later, in in-depth ultrastructural studies no plastids in rye sperm cells were found and DNA-specific staining revealed no cytoplasmic DNA existed in the male gametes (Corriveau and Coleman 1988), that is, it was a purely maternal inheritance of chloroplasts, while more recent experiments demonstrated unambiguous examples of plastids in all cases. The number of plastids per sperm cell varies from 2 to 12. The sperm pair may vary with regard to plastid number; however, these differences are not consistent among the sperm pairs examined (Mogensen and Rusche 2000).

Schnell (1959) provided a first report on rye with cytoplasmic–genic pollen sterility (see Table 6.1). He found numerous offspring showing pollen sterility in selfed lines of European rye varieties. However, the inheritance of the characteristic was not clear (Kobyljanski 1969). More extensive crossing experiments, however, demonstrated the presence of cytoplasmic–genic inheritance of male sterile plants in rye (Geiger and Schnell 1970). Systematic screenings and genetic studies carried out at the University of Hohenheim (Germany) since 1966 in 46 European, 1 Argentinean, and 96 Iranian accessions yielded a cytoplasmic male sterile sample from Argentina, further on designated as the line with P-cytoplasm (see Table 6.1). It became important as the main constituent for the German hybrid-breeding program and behaved analogous to the Texas cytoplasm in maize.

The P-cytoplasm of Argentinean rye is environmentally highly stable (see Table 6.1). It can be identified as mitochondrial sequence-characterized

TABLE 6.1
A Compiled List of Six Sources of Cytoplasmic Male Sterility in Rye (*Secale* sp.)

Type	Description	References
P	Pampa cytoplasm	Dohmen et al. (1994)
V	Cytoplasm of vavilovii type	Melz et al. (2001)
C	Cytoplasm of wild rye, *S. montanum*	Łapiński (1972)
R	Cytoplasm of rye	Kobyljanski and Katerova (1973)
S	Mutated cytoplasm of Kärtner rye	Warzecha and Salak-Warzecha (2003)
G[a]	Mutated cytoplasm of Schlägler Alt, Norddeutscher Champagner	Melz et al. (2001, 2003)

[a] The major recessive gene controlling male sterility is allelic with the male sterility genes of the C and R types; the P type is completely different and is controlled by dominant gene(s) and mitochondrial DNA showing different restriction fragment patterns. The major gene *ms1*(*Rfg1*) is located on chromosome 4RL; the sites of the minor genes *ms2* and *ms3* were found on 3R and 6R, respectively. The hybrids produced by crossing male steriles of G type with inbreds show normal pollination and, therefore, are tolerant to ergot infection, similar to normal rye populations.

amplified region (SCAR) polymorphism (Stojalowski et al. 2006). The authors tested a series of 25 inbred lines with respect to three SCAR polymorphisms. The analysis revealed a close association between marker-determined mitotypes and plasmotypes represented by the inbreds. The mitochondrial markers also confirmed normal (N-) cytoplasmic character of three wild rye species: *Secale montanum*, *S. vavilovii*, and *S. kuprijanovii*. For 186 plants from open-pollinated cultivars of Turkish and South American origin, cytoplasm identification was performed using crossing with double nonrestoring tester (see Table 6.1). In 77 plants, the N-cytoplasm was detected, and for 63 of these polymerase chain reaction (PCR) analysis was performed, producing results that were consistent with the genetic data based on testcrossing and phenotype assessment. The mitochondrial markers also confirmed the presence of sterility-inducing cytoplasm in the remaining 109 plants. Moreover, the markers allowed the differentiation between P(P)- and vavilovii (V)-cytoplasmic individuals and demonstrated widespread occurrence of a sterility-inducing cytoplasm, especially in South American populations (see Table 6.1).

It is easy to find nonrestorer plants for reliable sterility maintenance, whereas fully effective fertility restorers occur rather seldom. On the other hand, there are several other cytoplasms reacting in a similar, but not identical manner as the P-cytoplasm, when tested for male sterility in crosses with specific pollinators. Adolf and Neumann (1981) and Steinborn et al. (1993) detected a second type of cytoplasmic male sterility (CMS) cytoplasm. Compared to the P-cytoplasm, it is classified as vavilovii type (V-cytoplasm). The latter is difficult to maintain, while it was easy to find restorers for it and vice versa for the P-cytoplasm. Melz et al. (2001) described an additional source—the G-cytoplasm found in the old landraces Schlägler Alt and Norddeutscher Champagner. Recently, six different sources are recognized (see Table 6.1). The G-cytoplasm contributes to broaden genetic diversity in European rye breeding. Two successful hybrid varieties were derived from that type, Hellvus (2007) and Helltop (2009), with a multiplication acreage of 600 ha altogether in 2012 (Melz, pers. comm.).

Monogenic inheritance was observed in linkage tests. Using primary trisomics of rye, cv. Esto, the nuclear gene *ms1* was found to be located on chromosome 4R. Modifying genes, probably masked in N-cytoplasm but expressed in male sterility-inducing cytoplasm together with gene *ms1*, were located on chromosomes 3R (*ms2*) and 6R (*ms3*). Mono-, di-, and trigenic inheritance types were found in backcross progenies of trisomics.

Molecular studies of mitochondrial DNA demonstrate a close relationship of the gene pool of the German variety Pluto. Restorer genes were detected not only in exotic accessions but also in several European varieties. The frequency of gametes with restorer genes ranged between 10% and 20% across all accessions. During the last decade, these genes were genetically mapped and described in detail (Glass et al. 1995, Dreyer et al. 1996, Geiger and Miedaner 1996, Miedaner et al. 2000).

Systems of chromosome-induced male sterility have not attained any practical meaning. Although there are *ms* genes, suitable markers, and balanced tertiary trisomics (BTTs) with very low male transmission (as in the BTT system of barley), these prerequisites could not be combined to an efficient breeding program (Sybenga 1985).

6.1.1 Chloroplast

The chloroplast genomes of 44 accessions of rye were surveyed for restriction site polymorphisms (Petersen and Doebley 1993). The accessions were chosen to represent the geographic as well as taxonomic range of the genus. Using 12 restriction enzymes, a total of 348 sites were detected. Twenty-nine mutation sites were phylogenetically informative and used in a cladistic analysis. Further, a 0.1 kb insertion separated *Secale* from the out-group species. Only the annual species *S. silvestre* was distinct from the rest of the taxa. The chloroplast genomes of cultivated rye are mixed with both wild annual and wild perennial accessions. Sequence divergence (p) among taxa of rye was low, varying from 0 to 0.005, again suggesting a rather recent origin of the genus.

The chloroplast genes described are *cpHSP82* (Schmitz et al. 1996); *psbA* and *psbE* (Kolosov et al. 1989); *psbB* and *psbH* (Bukharov et al. 1988); *psbC*, *psbD*, and *psbF* (Bukharov et al. 1989); *psbI* and *trnS* (Kolosov et al. 1991); and *rpoC2* (Ermishev et al. 2002).

The latest results of chloroplast (cp) and mitochondrial (mt) DNA analyses confirm that the level of organellar polymorphism is low among the cultivated rye genotypes (Isik et al. 2007).

6.1.2 Chlorophyll Mutants

In an inbred line (var. Stalrag), a chlorophyll mutant arose spontaneously which displayed both extrachromosomal maternal and paternal inheritance in crosses. Other chlorophyll mutants are summarized by Kubicka et al. (2007), even lethal ones. Striata plants gave rise to albino, striata, and normal offspring when selfed.

When the striata mutant was used as the maternal or paternal parent and crossed with normal chlorophyllous plants, the progenies comprised striata and normal individuals. Upon selfing, the F_1 and F_2 striata plants gave rise to variable non-Mendelian ratios of albino, striata, and normal offspring—the ratios being proportionate to the amount of normal plastid-bearing to defective plastid-bearing tissue in the germ lines (anthers and ovaries). Most of the chlorophyll mutations have been obtained after self-pollination within or between varieties. They are determined by monogenic, recessive alleles.

6.1.3 MITOCHONDRIAL GENES

Mitochondrial genes are known as follows: *atpA* and *atp9* (Hessberg and Tudzynski 1996), *cob* (Dohmen et al. 1994), *nrG* (Steinborn et al. 1993), *nrP* (Börner and Melz 1988), *pol-r* (Dohmen and Tudzynski 1994), *rps2* (Kubo et al. 2004), *rps19* (Fallahi et al. 2005), and *rrn18* (Coulthart et al. 1994).

A 1397-bp fragment corresponding to the *rpoC2* chloroplast RNA polymerase gene was obtained by direct rye DNA amplification (Ermishev et al. 2002). Two rye species, *S. montanum* and *S. cereale*, did not substantially differ in the structure of this DNA fragment. Their nucleotide sequences were 99% identical. The extent of the homology of various stretches of the *rpoC2* rye gene with the corresponding sequences in maize and rice was 81%–95%, whereas the deduced amino acid sequences of *rpoC2* in rye, wheat, maize, and rice were nearly identical (96%–97% of homology). The fragment of the *rpoC2* gene differed from the corresponding sequences in three other grass species primarily by a short (49 bp) insert into the region of numerous short repeats corresponding to nucleotides 15,750/15,751, 28,728/28,729, and 27,472/27,473 in wheat, maize, and rice, respectively.

6.2 ALLOPLASMIC RYE

Lein (1949) named alloplasmic rye as those plants whose cells have wheat cytoplasm and rye nucleus, that is, chromosomes. Sasaki (1956) and Lacadena (1969) described further alloplasmic plants. The latter obtained two plants of alloplasmic rye (Cal1 and Cal2) from a *Triticum durum–S. cereale* cross (F_3 amphiploid *T. durum* × *S. cereale* × /5/*S. cereale*). Meiosis of pollen mother cells appeared to be quite normal until the formation of tetrads. However, in the young pollen cells sometimes asynchronous cycles of condensation and replication of the chromatids can be observed. The young microsporocytes show a gradual chromatid condensation cycle. Some nuclei belonging to the same tetrad showed different "mitotic" stages, that is, interphase, prophase, or prometaphase. There is a high degree of uncoiling of the centromere region due to alloplasmy. The following metaphases and anaphases of the first pollen mitosis appear quite normal. Fertility of plants is clearly influenced by the alloplasmic background (see Table 6.2).

Several other plant characteristics can also be influenced by the cytoplasm. Cold hardiness is not as strictly controlled by cytoplasm. Alloplasmic rye with *T. tauschii*

TABLE 6.2

Reciprocal Crosses of an Alloplasmic Rye Cal1 with Common Diploid Rye Var. Elbon

Cross Combination (Female × Male)	Florets Pollinated	Seeds Obtained	Fertility (%)
Cal1 × Elbon	833	395	47.4
Elbon × Cal1	40	9	22.5

Source: Lacadena, J. R., *Wheat Inf. Serv.*, 29, 21–22, 1969. With permission.

cytoplasm showed similar frost resistance as autoplasmic rye, indicating that the cold hardiness genes of rye have normal expression in the *T. tauschii* cytoplasm. Based on observations made in these two studies, it was concluded that the cytoplasm has little direct effect on cold hardiness, or on the nuclear expression of cold hardiness (Limin and Fowler 1984).

7 Breeding

In light and poor soil conditions rye is still the most economical crop of temperate or tepid latitudes. Its low specific production costs, winter hardiness (see Table 8.4), fertilizer (nitrogen) efficiency, and unpretentious growth guarantee the highest yield stability. These are good arguments for intensifying rye breeding and adequate breeding research. Independent of population or hybrid rye, the acreage is ~60% of the total rye area. Recent objectives for breeding rye are grain yield, maturity, plant height, 1000-grain weight (TGW), higher crude protein content, better dough features, frost and lodging resistance, resistance to mildew, *Rhynchosporium*, brown rust, and sprouting.

Yield has considerably varied from year to year (see Figure 7.1). On average, the yield even decreased, as a result of three extreme drought years (2003, 2007, and 2010) (see Figure 7.2). Therefore, drought resistance becomes an additional breeding objective (see Section 3.4). Extreme weather conditions have increasingly occurred over the past decade, which are reflected in more and more spring droughts. Generally, these occurrences apply to all cereal species, but the light locations, in which rye is mainly grown, have been particularly affected.

In Germany, fluctuating produce revenues have also caused the areas under rye cultivation to decline from 930,000 ha to approximately 600,000 ha. Furthermore, this decline has come at the expense of better locations for lighter locations. As with other crop species, rye reaches its limits in light soils with less nutrients and extreme weather conditions, although there is still an advantage compared with other cereal species. Rye is especially characterized by its vigorous qualities under very low temperatures, better exploitation ability of winter moisture compared with spring varieties, high water use efficiency based on a well-developed root system, and the high amount of grains per spike. Even in the future rye will continue to be the cereal crop of choice for light and medium locations.

Progress in breeding, particularly hybrid breeding with sophisticated methodology, has prevented a further negative trend. The yield level on light soils has even increased.

Breeding methods have naturally been influenced by the outbreeding nature of the crop. Rye has a gametophytic two-locus incompatibility system (Lundqvist 1956). Before 1850, landraces developed by natural selection on site (see Table 7.1). Up to around 1900 mass selection was the standard procedure of rye improvement, mainly introduced by the German breeder W. Rimpau at Langenstein, where rye is provable selected and has grown since 1732. He also described for the first time the self-sterility of rye (see Section 2.6.3) in one of the most famous commercial grain cultivars at that time, the Schlanstedt Roggen (Schlanstedt rye; Schlanstedt is a very old German village, near Magdeburg).

The next breeding approaches (1885–1921) could be best described as forms of family selection, that is, simple recurrent selections. Ferdinand von Lochow (1849–1924) at Petkus (a small old German village, near Berlin; see Figure 4.3) utilized this method

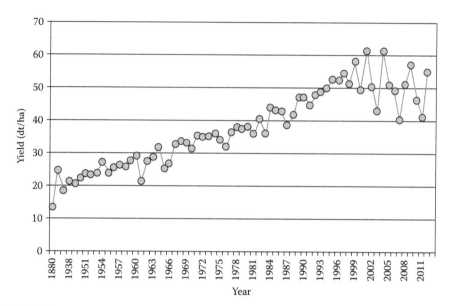

FIGURE 7.1 Average yields (dt/ha) of diploid winter rye in Germany during the past eight decades (hybrid varieties are included during the past 10 years). (After Anonymous, *DESTATIS—Statistisches Bundesamt*, Wiesbaden, Germany, 1956–2012.)

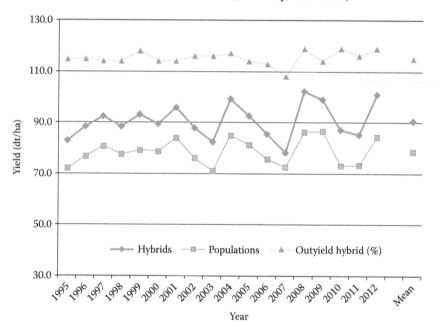

FIGURE 7.2 Grain yields (dt/ha) of best hybrid rye cultivars (Amando, Askari, Avanti, Brasetto, Fernando, Marder, Memphis, Minello, Palazzo, Treviso, Visello) in comparison with best population cultivars (Hacada, Halo, Nikita, Recrut, Conduct) in German yield testings under standard agronomical conditions (1995–2012). (Courtesy of Rentel, D., Bundessortenamt, Germany.)

TABLE 7.1
The Origin of Some European Rye (*Secale cereale*) Varieties

Landraces	Varieties Selected from Landrace	Year
Finnish landraces	Eelis-Antti, Haukipuro, Iivo, Leivonen, Mansikka-ahon	>1900
	Vihtori, Taavetti	
French shrub-rye	Himmel's Deutscher Champagnerroggen	1886
	JÄGERs Norddeutscher Champagnerroggen	1896
Saxon landraces	Saaleroggen	1911
	Pirnaer Roggen	
Probsteier landrace	Petkuser Roggen (Pirnaer × Probsteier) (at Petkus, Germany);	1887
(see Section 7.4)	Schlanstedter Roggen	
Lower Austrian landraces	Marchfelder Roggen	1925
Scandinavian landraces	Professor Heinrich Roggen	1880
	Brandt's Marien Roggen	1908
Dutch landraces	Heine's Zeeländer Roggen (at Hadmersleben, Germany)	1867
	Heine's Klosterroggen (at Hadmersleben, Germany)	ca. 1868
	Krafft's Zeeländer Roggen (at Hadmersleben, Germany)	ca. 1900
	Mette's Zeeländer Roggen (at Hadmersleben, Germany)	1852

most successfully, which is known as "Deutsches Ausleseverfahren." Subsequently, Petkuser Winterroggen (Petkus rye) became the most famous variety worldwide. F. von Lochow started his career as an agricultural civil servant (1875–1879) and he became an independent farmer during 1879–1924, which was the beginning of the rye breeding story at Petkus and worldwide. With improved genetic knowledge, more sophisticated methods appeared. Since 1920 at Petkus, Germany, W. Laube's method of overstored seeds including regulated cross-fertilization and progeny testing became the standard before hybrid breeding was introduced (see Figure 7.3). This so-called Ohio method has been used for maize since 1905. From selected single plants, only half of the seeds were sown, while the remaining half were kept for the next year's sowing. Then only seeds were sown from sister plants with positive yield characters.

A further technical development of directed pollination was achieved by the so-called separation breeding that was developed in 1910 for rye by Heribert-Nilsson (1916a) in Sweden. The remaining seeds of elite plants, whose progenies showed positive traits, were grown on isolated sites (isolation garden) allowing pollination just within single families, that is, half-sisters. Franz Frimmel at Brünn (now Brno, Czech Republic), in 1924, made another adjustment for breeding stations contaminated with big amounts of rye pollen (Frimmel and Baranek 1935). It was called "Miss Amerika Aggregat." A special pollination arrangement of parent and grandparent plants around a single family allowed selection of the "most beautiful" (plant) for the next generation. However, this method did not reach a practical acceptance.

The strongest way of directed pollination on both the female and the male parent is performed by diallelic crosses. This was proposed by T. Roemer (Fruwirth and

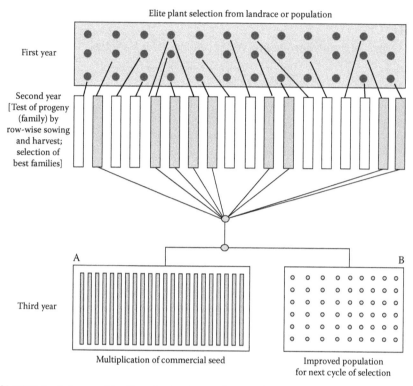

First year

Second year
[Test of progeny
(family) by
row-wise sowing
and harvest;
selection of
best families]

A   B

Third year

Multiplication of commercial seed   Improved population
for next cycle of selection

FIGURE 7.3   Scheme of family selection in rye, first applied by F. von Lochow since 1885. A, further propagation for commercial seed; B, source for new selection cycle.

Roemer 1923) in order to estimate the "Erbwert der Väter" (male heritable value) considering quantitative traits. The method of full-sib selection included pollination between one spike of the male and one spike of the female parent under an isolation bag (see Figure 7.4). Since mid-1930s, R. von Sengbusch (1940b) applied the technique at the Luckenwalde breeding station near Berlin, Germany, and called it "Pärchenkreuzung" (pairwise crosses), that is, single plants cannot cross-pollinate freely but do so pairwise in numerous crossing combinations. The following selection is then performed between full-sibling families. Parents with the best test performance are included in the improved population.

Rye shows inbreeding depression but inbred lines of acceptable vigor can be isolated and used in the construction of synthetic (and even hybrid) varieties, following suitable progeny tests for combining ability. Unlike those of wheat and barley breeding, aspects of disease resistance, at least in the beginning, have not dominated the objectives of rye breeding.

Considerable attention to improvement of grain yield, improved stability, faster growing, finer stems, resistance to diseases, and protein content together with cold tolerance and shorter straw. The main objectives of recent rye breeding are yield, early ripening, short straw, winter hardiness, lodging resistance, resistance to fungi

FIGURE 7.4 Isolation bags for controlled pollination in rye breeding and research. (Courtesy of Banaszak, Z., Choryn, Poland.)

(mainly mildew and rusts), grain size, resistance to preharvest sprouting, content of crude protein, and specific flour traits, which are described as follows:

1. *Grain yield*. It is by far the most important trait for practical rye growing. The average grain yield of rye in Germany is 50 dt/ha, ranging from 30 to 80 dt/ha. Generally, hybrids yield 10%–20% more than open-pollinated populations, sometimes up to ~40% (see Figure 7.2), together with higher stress tolerance.
2. *Tolerance*. Drought and nutrient stress resistance are an important component of yield stability because rye is widely grown on poor, sandy soils.
3. *Straw*. The breeding goals are shortness of straw and lodging resistance. Rye was and still is the tallest small grain cereal. Since the culm of rye is an important storage for carbohydrates and water, very short-strawed rye varieties may suffer from lower yield potential and lower stress tolerance. In modern agriculture, lodging of rye is often due to dense stands, insufficient potassium and copper supply, and wrong application of growth regulators. Usually, during the middle of May, the awn emergence of rye spikes starts. Up to that point, the shortening of culm must be completed (see Section 8.4).
4. *Diseases*. Resistance is needed for pink snow mold, powdery mildew, leaf rust, foot rot, head blight, and ergot. Generally, rye is less susceptible to most diseases than wheat and reacts with less yield reduction (see Chapter 8).
5. *Quality*. The milling and baking features are the main objectives of quality selection in rye. The major traits for baking rye are high kernel weight and resistance to preharvest sprouting (see Section 3.6). Caused by its low dormancy, the rye kernel may start germinating already on the head before harvest when the weather is warm and moist. This leads to deterioration of

the starch and reduces baking quality. The Hagberg falling number test is used as a measure of latent sprouting and ability to gelatinize grain starch.

The test provides an indication of α-amylase activity and depends on the action of this enzyme in reducing the viscosity of heated flour-and-water slurry. α-Amylase is an enzyme involved in the degradation of starch to sugars and is usually associated with germination. Sprouted grains give low Hagberg falling numbers. A high number is required for better bread-making quality (in wheat: <200 low; 230 medium; >250 high; ~290 very high; >310 extremely high).

Quality rye for bread making shows falling numbers higher than 140 (s). In contrast to wheat, the protein content is less important. It should not be below a certain level. Another important component of baking quality is a high pentosane content (see Section 2.5). The soluble pentosans, however, which are in much greater concentration in rye than in other cereals, are considered to be inhibitory components for feeding quality responsible for lowering feed intake, feed conversion efficacy, and growth rate in animals fed rye diets. Rye grain can be fed to swine and cattle to ~50%–60% in the diet (see Section 9.2). Besides this, rye grain has a feeding value of ~85%–90% of that of maize and contains more digestible protein and total digestible nutrients than oats or barley.

6. *Industrial utilization.* Since the 1990s, rye has increasingly provided for the production of alcohol and plastics, as well as for generation of energy. Even for such purposes specific genotypes are needed. It is currently being worked intensively, especially as the use of grains for human consumption is declining (see Chapter 9).

7.1 DIPLOID RYE

Inspired by the English cereal breeder Patrick Shireff (1791–1876), in 1868 Wilhelm Rimpau started spike selection in Probsteier (see Section 7.4) rye at Schlanstedt, Germany. However, this positive mass selection did not lead to significant rye improvement even after ~9 years of breeding work. He learned that the selection method in rye has to be modified and did the first steps toward recurrent selection associated with field isolation because of the outcrossing as well as incompatibility, that is, self-sterility character. He stated, "Roggen ist Fremdbefruchter und Windblütler" ("Rye is a cross-fertilizer and wind-pollinator"; Laube 1959).

Conventional rye variety development still takes between 8 and 12 years. Nowadays, three types of varieties are being distinguished:

1. *Population varieties.* They consist of many single genotypes selected from isolated plant sibs. They are genetically balanced and constantly yielding.
2. *Synthetic varieties.* They consist of several, genetically reproducible subpopulations, maintained by either controlled intercrossing or free open-pollination. The reproduction happens without further control of pollination. Usually after the second year of cross-pollination (Syn2), the varieties are genetically balanced and constantly yielding.

3. *Hybrid varieties.* They consist of a limited number of inbred lines that are—under controlled pollination—crossed with each other. The performance and genetic combining ability of inbreds are checked before. The hybrid seed has to be established year by year.

7.1.1 Population Breeding

7.1.1.1 Mass Selection

Directed rye breeding began with mass selection of grain types, spike shapes, and plant habits, characterized by sowing mixed samples of selected plants instead of individual offspring (Rimpau 1877, Lochow 1894, 1900). When the selected material was sown separately, for example, yellow seeds, green seeds, or brown seeds, it was called "group selection" and included cross-pollination within the group.

7.1.1.2 Family Selection

A significant improvement of rye breeding was family selection, that is, the selection of progeny families on their mean performance. In addition, the best individuals (elite plants) are usually selected from the best families. Selecting the best mother plants (P1) was already common, but the separated row-wise sowing of the progeny (P2, P3, P4, etc.) and the individual judgment was new. Heribert-Nilsson (1937) called this approach "Nivellierungszucht" (leveling breeding). Moreover, Lochow (1894) independently applied Vilmorin's principle of isolation in allogamous crop breeding onto rye. Therefore, Fruwirth (1907) described the procedure as original Deutsches Ausleseverfahren (German selection method). The method was improved stepwise and resulted in the so-called *mother pedigree selection* that F. von Lochow successfully applied for more than 30 years at the Petkus breeding station (see Figure 7.3). It is a postflowering selection where the breeding value of the paternal parent is not directly estimated. Thus, Lochow's famous rye breeding started in 1881. After 50 years of intensive work, breeding progress could be documented as shown in Table 7.2.

His initial positive mass selection finally led to a new procedure—the residue seed method or method of overstored seeds syn. half-sib progeny selection syn. remnant seed procedure syn. Ohio method (see Figure 7.5). In pedigree breeding of allogamous rye, it is the most effective method of regulating cross-fertilization. Usually, during the first year a greater number of individual plants are harvested

TABLE 7.2
Breeding Progress during Four Decades for TGW and Yield, Starting from Landraces to Elite Strains of Diploid Rye (*Secale cereale*)

Breeding Material	Thousand-Grain Weight (g)	Yield (dt/ha)
Landrace	22.8	13.2 (~1880)
Petkus elite selections	30.9	24.4 (~1920)

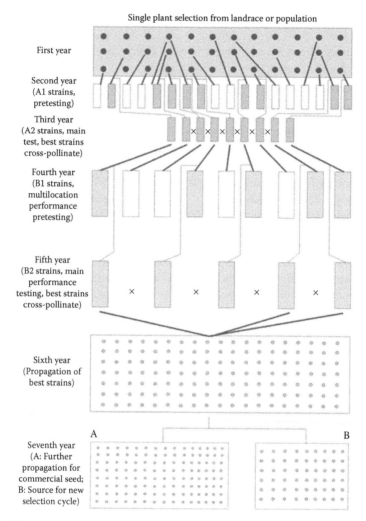

Single plant selection from landrace or population

First year

Second year
(A1 strains,
pretesting)

Third year
(A2 strains, main
test, best strains
cross-pollinate)

Fourth year
(B1 strains,
multilocation
performance
pretesting)

Fifth year
(B2 strains, main
performance
testing, best strains
cross-pollinate)

Sixth year
(Propagation of
best strains)

Seventh year
(A: Further
propagation for
commercial seed;
B: Source for new
selection cycle)

A   B

FIGURE 7.5 Method of overstored seed. (After Laube, W., and Quadt, F., *Züchtung der Getreidearten. Handbuch der Pflanzenzüchtung*, Kappert, H. et al. (eds.), Verl. P. Parey, Berlin-Hamburg, 1959.)

from a genotypic mixture of a certain population. Agronomic characters, such as spike shape, high tillering, plant length, resistance to diseases, and general vitality, are considered. Lodging resistance, tolerance to sprouting, TGW, and grain quality are equally important. During the second year, their progenies from free cross-pollination are sown as "A" families in smaller plots (mostly drill plots without replication), while half of the seeds of all elite plants are retained in reserve. These A families, which meet all the requirements, are not directly multiplied during the third year, but the remaining seeds of the corresponding elite plants are sown in the following year. Those selected so-called half-sib families are again sown in micro-plots under isolated conditions. They enter the fourth year in the so-called A family trial, that is, a multilocal yield testing with two replications. During the fifth year, a performance testing was carried out that includes propagation of best families again with residue seeds. An indirect pollen (paternal) selection occurs by this procedure because only preselected plants contribute to the pool of pollen.

In this way economically valuable traits of the A family can be definitely evaluated after maturity. The best A families determined for further breeding were already pollinated by a pollen mixture that also contained pollen of less valuable plants. Since 1919, it was applied by T. Roemer at Schlanstedt, Germany, and, since 1921, at Petkus, Germany, by W. Laube, respectively. Roemer called it "Verfahren der Hälften" (procedure of halfs).

During the following years, the family breeding and the residue seed method were supplemented by the so-called Separierungszüchtung (isolation breeding; Heribert-Nilsson 1921). It took into account the increased knowledge of pollen distribution, environmental variability, and efficient spatial isolation of plots. Cross-pollination among the selected families should be avoided by distant cropping of family plots ($\geq$50 m) and/or by a cordon of other crops around the families. The disadvantage of this approach is the need for large field space and additional labor (see Figure 7.6).

7.1.1.3 Inbreeding

Although inbreeding was never a serious breeding option in rye, it has been used for identifying homozygotes within a population and characterizing genetic variability. Homozygosity is always associated with a drop in vitality and yield performance. Nevertheless, during the 1930s, E. Baur at Müncheberg (a small village near Berlin, Germany) had the idea of combination breeding in rye by utilizing homozygous pure lines (Aust and Ossent 1941). There are genotypes in rye that show more tolerance to inbreeding than others do. Ossent (1934) had found self-fertile genotypes in several rye populations already in the 1930s. They were routinely used for developing inbred lines until recent breeding programs. For years ~50,000 spikes per year were kept under isolation bags in order to select self-fertile plants. Approximately 2000 genotypes could be collected. The selection of inbreds was focused on yield characters. Since 1935 qualitative traits were also considered, such as protein content and baking quality (see Section 9.3).

In Sweden, Heribert-Nilsson (1937) compared several inbreeding methods, such as (1) group crossings (free cross-pollination of elite strains and subsequent separation of lines), (2) elite crosses (free cross-pollination of two elite strains without subsequent separation of lines), (3) elite lines (spatial isolated cropping of single-plant progenies after free cross-pollination), (4) pairwise crosses (individual crosses of two plants with different pedigrees), and (5) inbred crosses (cross-pollination of two inbred lines). From each approach positive phenotypes were selected from both individual plants and families (see Table 7.3). However, no significant progress in selection was achieved, which W. Laube at Petkus confirmed with advanced breeding material (Roemer and Rudorf 1950). If more distinct and primitive lines are included in the adequate studies, the selection progress is better. Then heterotic effects become significant as demonstrated by Peterson (1934) and Leith and Shands (1935).

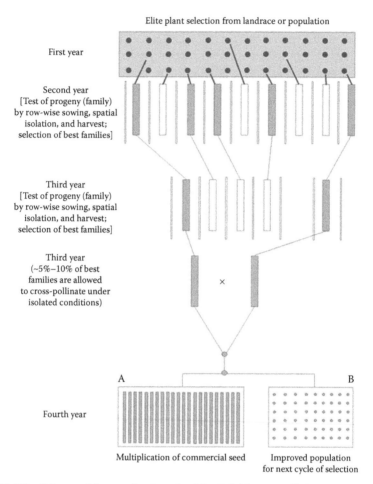

Elite plant selection from landrace or population

First year

Second year
[Test of progeny (family)
by row-wise sowing, spatial
isolation, and harvest;
selection of best families]

Third year
[Test of progeny (family)
by row-wise sowing, spatial
isolation, and harvest;
selection of best families]

Third year
(~5%–10% of best
families are allowed
to cross-pollinate under
isolated conditions)

×

A   B

Fourth year

Multiplication of commercial seed   Improved population
                                           for next cycle of selection

FIGURE 7.6 Scheme of "separation breeding" in diploid rye, applied by Heribert-Nilson since 1921. A, Further propagation for commercial seed; B, source for new selection cycle. (After Heribert-Nilson, N., *Hereditas*, 2, 364–368, 1921, applied since 1921.)

TABLE 7.3
Yield Performance of Breeding Strains in Rye after Different Inbreeding Methods

	Inbreeding Approach				
Method	1	2	3	4	5
Number of strains selected	29	31	31	34	33
Mean performance (%) compared to conventional selection of Petkus rye (100%)	97.7	97.2	89.7	84.5	76.2
Variation	79–107	85–111	79–96	61–97	53–103

Source: Heribert-Nilsson, N., *Hereditas*, 23, 236–256, 1937. With permission.

7.1.1.4 Crossbreeding

Crossbreeding in rye is not feasible. It needs homozygosity and constant inheritance of the parental varieties. This does not happen in rye. Nevertheless, crossbreeding can be applied when landraces should be improved by directed intercrossing of advanced breeding strains in order to improve the selection efficiency. Tschermak (1915) did it when he crossed Petkus rye with several Austrian (e.g., Marchfelder rye) and French landraces.

7.1.1.5 Population Breeding

Population breeding comprises the development of open-pollinated and synthetic varieties. In both cases the variety constitutes a panmictic population. The population is produced by random fertilization, at least in the final generation of seed production. Gametophytic self-incompatibility (SI) prevents self-fertilization under open pollination. It helps to avoid inbreeding depression.

In population breeding, only self-incompatible varieties have usually been produced. An open-pollinated variety is the direct outcome of population improvement. When a breeding population has reached a per se performance level comparable to that of the existing cultivars, it may be considered as a new variety. Depending on the local statutory regulations it can be released and registered as a variety.

Various selection procedures are being described for population improvement in rye (von Sengbusch 1940b, Vettel and Plarre 1955, Ferwerda 1956, Laube and Quadt 1959 Wolski 1975, Geiger 1982) (see Figure 7.5). Self-incompatible and self-fertile rye populations can be considered. Improvement of self-compatible populations aims at improving either the per se performance, the potential of the population for synthetic variety production, or both. In self-fertile materials, selection exclusively aims at improving the potential of the population for hybrid variety production (see Table 7.4).

For all approaches the intra- and/or interpopulation general combining ability of the parental units, that is, plants, clones, pairs of plants, or pairs of clones, has to be improved. In addition, for hybrid breeding the mutational load of the populations has to be diminished in order to reduce inbreeding depression in line establishment. The different selection procedures can be operated in various ways according to the experimental facilities. Moreover, the procedures can be used simultaneously or successively in a given selection scheme. It follows the generalized scheme of population improvement of Hallauer and Miranda (1981). The breeding procedure is divided into different selection cycles. Each cycle includes a parental unit (plants, clones, pairs of plants, or pairs of clones to be evaluated), a selection unit (plants, clones, or progeny that provide the data used as the basis for selection), and a recombination unit (plants, clones, or progeny that finally recombine to form the improved population).

Cloning of parental material is necessary to achieve sufficient seeds for progeny tests in multilocation yield trials, or even for microplots. The number of clones per genotype can be limited or even few when cytoplasmic male sterility (CMS) testers are involved. Selfing of heat-treated self-incompatible S_0 plants generally yields few seeds only. In order to increase the number of seeds for yield trials, the S_1 lines have to be multiplied by free pollination under isolation cages and/or in isolation (see Figure 7.7).

TABLE 7.4
Different Selection Procedures for Interpopulation Improvement in Self-Fertile Rye

Method	Parental Unit	Selection Unit	Recombination Unit	Evaluation
Testcrosses (CMS tester)	Individual plants	Testcrossing	S_1 line from self-fertile S_0 plant	Visual screening in one- or two-rowed observation plots
	Individual clones	Testcrossing	S_1 line from self-fertile S_0 plant	Regular yield testing
Inbred family selection	Individual plants	S_1 line from self-fertile S_0 plant	S_1 line from self-fertile S_0 plant	Visual screening in one- or two-rowed observation plots
	Individual clones	S_1 line from self-fertile S_0 plant	S_1 line from self-fertile S_0 plant	Microplot yield testing (1–2 m^2)
	Individual S_1 plants	S_2 line from self-fertile S_0 plant	S_2 line from self-fertile S_0 plant	Visual screening in one- or two-rowed observation plots
Combined testcrosses of S1 families	Individual clones	Testcross of S_1 line from self-fertile S_0 plant	S_1 line from self-fertile S_0 plant	Regular yield testing, and/or microplot yield testing (1–2 m^2)

Source: Geiger, H. H., *Tag. Ber. AdL*, 198, 305–332, 1982. With permission.

FIGURE 7.7 Isolation cages for controlled pollination in rye breeding. (Courtesy of Banaszak, Z., Choryn, Poland.)

Considering a wide range of experiments, it is difficult to predict an optimum number of parental, selection, and recombination units, and an optimum allocation of testing facilities. Even if just one selection procedure is taken into consideration, the optima may vary considerably depending on the underlying genetic and environmental factors.

With some modifications, over decades this standard method was successfully applied in rye breeding. (1) It starts with the selection of mother plants, equally spaced (usually 25 × 25 cm) under comparable agronomic conditions. (2) The second step includes the testing of progeny of mother plants as "half-sib families." The evaluation is made on unreplicated drill plots on two or more sites. (3) During the third cycle the remnant seed of best half-sib families is multiplied on either spatially isolated field plots and/or artificial isolation means, such as separation walls and foil cabins. (4) Finally, the advanced half-sib families are grown on ~10 m² field plots and drill rows for evaluation. Several environments should be considered.

Vegetative cloning of mother plants before the first cycle and planting clones together with drilled breeding populations may reduce reproduction time from 4 to 2 years; however, the procedure is laborious and cost-intensive.

The so-called full-sib selection was another improvement of the residue seed method (see Figure 7.8). It was even more efficient because it controlled both parents, that is, the paternal and maternal components. (1) Pairwise crosses under isolation bags are made for two-local field testing during the following year. Hand emasculation is not necessary because of SI. (2) There are sufficient seeds for the observation plots during the year. (3) From them, the best families are multiplied under pollen isolation, mostly between isolation tissues (walls) on the field, under plastic tunnels, or in greenhouses (see Figure 7.9). (4) Finally, the full-sib progeny is evaluated. Genotypes with highly heritable characters as mentioned above are included in the crosses. From residue seed advanced full-sib families can enter a new selection cycle.

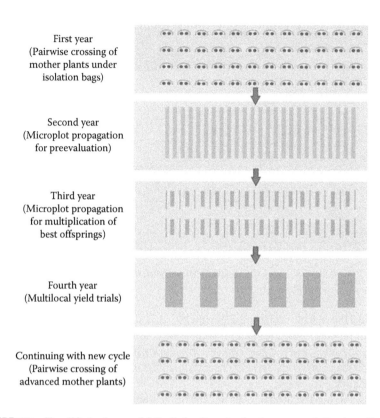

First year
(Pairwise crossing of
mother plants under
isolation bags)

Second year
(Microplot propagation
for preevaluation)

Third year
(Microplot propagation
for multiplication of
best offsprings)

Fourth year
(Multilocal yield trials)

Continuing with new cycle
(Pairwise crossing of
advanced mother plants)

FIGURE 7.8 Simplified scheme of full-sib family selection in rye population breeding.

FIGURE 7.9 Isolation walls for control of cross-pollination in rye breeding. (Courtesy of Korzun, V., KWS Lochow GmbH, Bergen, Germany.)

Recent population breeding makes additional use of subpopulation mixtures, that is, merging of genetically distant progenies. It may lead to heterotic expression of several characters in the final panmictic variety. Up to 20% yield increase is possible, just by including two components (Hepting 1978). However, during maintenance selection and common multiplication the heterosis effect will drop because of decreased heterozygosity within the population.

7.1.2 Synthetics

In rye, the term "synthetic" is used to designate varieties that are produced by crossing *inter se* a number of selected parents with subsequent multiplication by open pollination under isolation. The parents can be clones, inbred families, or other genotypes. Long-term maintenance of clones is not feasible under practical conditions. However, their gametic arrays can be reproduced from S_1 families, which can be partially established by selfing under heat treatment (see Section 2.6.3).

Its general combining ability indicates the potential of a genotype as a constituent of a synthetic variety. Population improvement and the formation of synthetics ought to be organized as an integrated program. When a high testing accuracy is practiced in population improvement, a selected fraction of the recombined material may serve as varietal parent without additional testing. The optimal number of parents is determined by almost the same parameters as the optimum population size in recurrent selection. Since synthetic breeding is not directed on selection with varieties but on creating better parents, intravarietal genetic variance is of minor concern, and the number of varietal parents can be chosen to be smaller than that in long-term population improvement. Studies on rye revealed that the genetic variance among synthetics sharply decreases as the number of parents increases (Geiger et al. 1981). If several unrelated, well-combining, and high-performing plant populations are at the breeder's disposal, the question arises whether a synthetic variety should be composed of parents from one single or a certain number of such populations. Only in case a population is outstanding in both per se performance and variance of general combining ability, it is likely to be better suited as source for synthetic than any population set. In all other situations, the optimum has to be determined by predicting the expected performance of the best synthetic of each set of populations. Frequently, two or more populations turn out to be more promising than just one.

In this complex theoretical and practical situation, synthetic rye varieties have not gained much acceptance in the seed market. If improvement of both hybrid and population varieties is conducted at the same breeding company, an integrated approach would be desirable to make better use of genetic resources, labor supply, and technical equipment, a general scheme for simultaneous development of hybrid and synthetic varieties in rye was proposed by Geiger (1982) (see Figure 7.10).

Currently, Diekmann Seed Ltd., Germany, produces synthetic varieties for poorer cropping areas, such as Herakles or Kapitän, with good yield performance and resistance.

The idea of synthetic varieties is rather old. Hayes and Garber (1919) first mentioned it. The release of so-called Maultier-Roggen (mule rye) in Germany probably was the first attempt to produce a synthetic variety. It was also called "Bastard-Roggen"

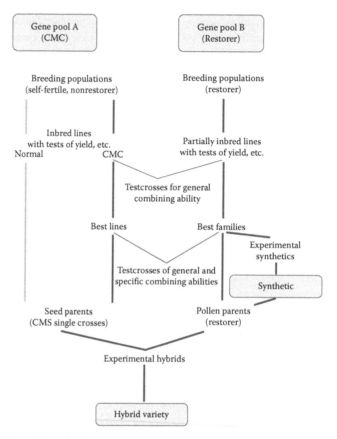

FIGURE 7.10 Schematic drawing of an integrated breeding program for simultaneous development of hybrid and synthetic varieties in rye. (After Geiger, H. H., *Tag. Ber. AdL*, 198, 305–332, 1982; Geiger, H. H., and Miedaner, T., *Handbook of Plant Breeding*, Carena, M. J. (ed.), Springer Publication, New York, 2009.)

(hybrid rye) and was the result of cross-pollination of two accepted varieties for production of heterosis seed. Actually, it was not a breeding method rather than a certain method of seed multiplication. During 10 years of testing several variety combinations for synthetics, Bredemann and Heuser (1931) classified the so-called years of heterosis and years of nonheterosis, that is, heterosis effects showed strong fluctuation. Nevertheless, the positive heterosis effects in some years compensated the low effects in other years, and even exceeded them in average of all F_1 populations. The F_2 populations behaved as the F_1, respectively, however, with somewhat lower yields. Even F_3 and F_4 populations were still more vital than the parental varieties.

Later, Sprague and Jenkins (1943) described synthetic varieties as populations that are derived from more than four lines after open pollination. They can not only be grown as F_1 population but also the next populations (F2–F4). S_1 lines of high combining ability may be combined into synthetic varieties for fringe-area production and as a stopgap improvement measure as proposed by Lonnquist and McGill (1956).

Based on testcross progeny performance, a number of lines can be selected to be included in the synthetic variety. The synthetic variety is generally formed by compositing equal amounts of seed of all possible single crosses between the selected S_1 lines. In subsequent generations, mass selection may be practiced to maintain the synthetic variety.

Although heterotic effects are used, the method is related to population breeding (Schnell 1982). Synthetic varieties are maintained as populations. A differentiation within the population breeding is given by the maintenance breeding of synthetics. While multiplication of open-pollinating varieties is made directly with the population, in synthetics the single components (lines, plants, or families) are maintained.

7.1.3 HYBRID BREEDING

A substantial reason for the difference in selection progress between conventional and hybrid rye is the impact of heterosis, from which hybrid breeding is strongly profiting. Accordingly, the divergent parental gene pools and traits are optimally used and the traits desired can be selectively combined, thus enabling an increase in yield. Due to their broad genetic spectrum, hybrid varieties adapt very well. This allows for utilizing favorable environmental conditions and also buffering unfavorable ones.

Over the years heavy investments have been made in hybrid rye breeding, which additionally accelerate the great progress made in breeding with the system described here. In addition, royalty income from conventional rye varieties is clearly lower that is forcing breeders to limit their investments in this type of variety.

In 1877, C. Kleinert, a German gardener, at Gruschen near Schlichtingshausen (near Poznan now Poland) reported about rye crosses between Swedish snow rye and Correns rye. He found better tillering and grain formation in F_1 as compared to Probsteier population rye (see Section 7.4). Obviously, he demonstrated heterosis effects in rye for the first time, which is long before the phenomenon was discovered in maize and other crop plants (Rimpau 1883). Another early report about hybrid seed production was published by Frimmel and Baranek (1935) when they discussed the utilization of xenia in rye breeding (see Section 2.5.1).

There was also a cytogenetic approach for hybrid seed production. As in barley, de Vries and Sybenga (1989) proposed the utilization of the so-called telotertiary compensating trisomics. However, as in barley, the experimental stage was not exceeded.

As in maize, in allogamous rye selfing results in strong inbreeding depression, that is, hybrid genotypes display heterosis. Major heterotic effects are found for grain yield, sometimes up to 210% when the parental mean is considered. As hybrid breeding in rye progressed, line performance increased and relative midparent heterosis for grain yield decreased to ~140%. This is comparable to maize and much higher than in self-fertilizing small grains, such as barley, wheat, and rice, in which usually 5%–10% of midparent heterosis is found (Geiger and Miedaner 1999). In addition, hybrid breeding takes the most advantage of general and specific combination ability in the final variety.

According to maize nomenclature, all rye hybrid varieties so far released are top-cross hybrids (see Schlegel 2009) following the formula: $(A_{CMS} \times B) \times Syn_{Rf}$ (A_{CMS}, population A (cytoplasmically male-sterile); B, population B; Syn_{Rf}, synthetic (fertility restorer)), that is, a CMS single-cross hybrid as seed parent and restorer synthetic as second pollinator (Geiger and Miedaner 2009). The single-cross parents derive from either one or two different heterotic gene pools; however, both are unrelated to that of pollinator. Because of the high degree of inbred depression in rye, single hybrids as in maize or other crops are not yet feasible.

The main goal of hybrid breeding is a stable grain yield. This includes tolerance to drought (see Section 3.4) and nutrient stress because rye is widely grown on poor, sandy soils where it has a higher relative performance than wheat and triticale. Caused by hybrid breeding, rye can now compete with these cereals even on more fertile, productive soils. Since hybrids are more genetically uniform than population varieties, breeding for disease resistance needs more attention, especially for those diseases that cannot be prevented by chemical means (Miedaner et al. 1995, Miedaner and Geiger 1999).

Systematic hybrid studies in rye started in 1970 at the University of Hohenheim (near Stuttgart, Germany) with the discovery of the so-called P-cytoplasm (Geiger and Schnell 1970; see Table 6.1) in a *Secale cereale* ssp. *tsitsinii* var. *multicaule* genotype (see Table 2.3). As for many crops, CMS became the cornerstone of hybrid production. Stepwise, CMS could be transferred to inbred lines by subsequent backcrosses, resulting in male-fertile lines with normal cytoplasm and male-sterile lines with mutated cytoplasm while the nuclear complements are largely identical.

From the beginning, efficient conversion of elite lines to CMS was a challenge. Therefore, the latest marker-assisted backcrossing is one of the main breeding and research tasks. In contrast to gene introgression, for which donor segments around target genes have to be considered, background selection for CMS conversion focuses solely on recovery of the recurrent parent genome. The optimal selection strategies for CMS conversion differ from those for gene introgression and need to be improved. CMS conversion requires fewer resources than gene introgression with respect to population size, marker data points, and number of backcross generations. Combining high-throughput assays in early backcross generations with single-marker assays in advanced backcross generations may increase the cost efficiency of CMS conversion for a broad range of cost ratios. The maintenance of CMS lines is organized by controlled crossing with the fertile analogous line.

In order to improve the seed parent by new traits (see Figure 7.10) continuously, new crosses and subsequent backcrosses are required, again resulting in male-sterile lines. They carry the "maintainer" genes located within the genome. The improvement of the pollen parent happens in a similar way. However, permanent control of restoration ability has to be checked.

The production of those inbreds needs many years. It is exclusively achieved by selfing. There is no alternative yet. As in other crops, a doubled-haploid method (see Section 3.5.4) is not routinely available. Because of inbred depression, the selfed progeny on average shows 30%–50% lower performance than the heterozygous donors. Therefore, the performance and combining ability of inbreds are always proofed in two steps. The male-sterile seed parents are tested by a top-cross design. They are grown in an isolated area and/or place and surrounded by the pollinator, while the inbreds of the pollen parent are placed between the isolation walls, each together

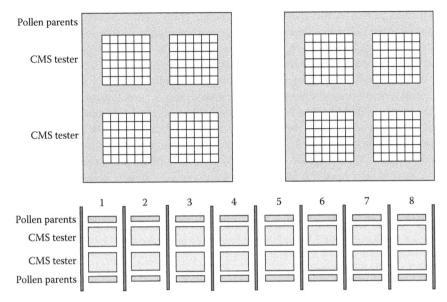

FIGURE 7.11 Testing designs of performance and combining ability of seed (top) and pollen (bottom) parents during the process of hybrid development.

with the pollinator. Thus cross-pollination is strongly reduced. Usually, the female parents are simple-cross progenies of the gene pool A (see Figures 7.10 and 7.11). During the following year, both sorts of cross progenies are grown for yield analysis. This biannual procedure is then repeated with advanced inbred candidates under stronger selection parameters.

Before submission of the hybrid candidate to the official state trials, it is finally tested again as the so-called experimental hybrid variety. A 3-year official testing of the variety candidate is done before it is released as new hybrid cultivar.

7.1.3.1 Production of Seed Parents

One of the essential prerequisites of hybrid breeding is the full male sterility of the seed parent over a wide range of environments and years. It has to be genetically uniform in order to control its breeding contribution during all the selection processes. It has to be appropriately vigorous for normal propagation. The cross of unrelated and more or less homozygous lines best accomplishes the requirements mentioned earlier.

Systematic search for gene pools with maximal heterosis revealed that two German populations Petkus and Carsten's were particularly well matched. Inbred lines from the two pools were used successfully for the development of hybrids. A simplified scheme of hybrid rye breeding can be taken from Figure 7.11. Seed–parent lines are developed from Petkus and pollinator lines from Carsten's gene pool. Intensive selection for line performance is practiced in selfing generations S_1 and S_2. Selfing is done by hand under isolation bags. After one or two stages of selection, the seed–parent lines are transferred into the CMS-inducing P-cytoplasm or G-cytoplasm (see Table 6.1) by repeated backcrossing. Resulting BC_1 and BC_2 are subsequently crossed with parents from the opposite pool to select for testcross performance, that is, the ability to combine for grain yield. The reverse procedure is used for the pollinator lines. They are grown between isolation walls in adequate plots and crossed to CMS single crosses as testers. The testcrosses are evaluated in multienvironmental trials with two replications.

In detail, from pairwise intercrosses of preselected inbred lines ($I^1 \times I^2$), the progeny is selfed twice, that is, during 2 years. Their seeds (S_0, S_1) are already tested for self-performance in multilocal trials and the best lines are selected. It follows the cross with CMS lines and selfing under bags. In this way a backcross progeny (BC_0) and a selfing progeny (S_3) can be harvested. During the following year, the BC_0 is again backcrossed when it was fully male sterile, while the S_3 progeny is multilocally checked for yield performance. The resulting BC_1 is tested for general combining ability in top-cross trials (see Figure 7.12). Two other backcrosses with parallel performance trials are the following: BC_3 and S_7. This procedure takes ~10 years. Marker-assisted selection and the gametocides for production of male-sterile lines could and should improve the complex production of CMS and maintainer pools in the future.

7.1.3.2   Pollen Parents

Only selfed progenies with complete seed setting enter the next selection cycle. This procedure can be repeated twice or more, each time accompanied by performance testing. It ends up with pollen parents showing the best restoration ability in crosses with the CMS partner. Compared to the development of seed parents, the production of pollen parents is somewhat shorter. Approximately seven years are needed. When some plant generations are realized under greenhouse conditions, the production of pollen parents can be even shorter. So, during the seventh year, inbreds can already be used for the composition of synthetic populations.

Recent progress in restorer line development allows constricting the genetic variability of the pollinator. It can be achieved by the combination of sister lines or the utilization of partial inbreds. Therefore, specific combining ability might be exploited better and the period of pollen shed be prolonged.

7.1.3.3   Restorer

The restoration of fertility in the CMS lines is achieved by the so-called restorer genes that are found in the nuclear genome. First restorer genes were described by Geiger (1972) in the line, for example, L18, from a European gene pool. However, their restoration was incomplete. Since 2005, a new restorer gene is used which induces almost complete pollen shedding. For example, the hybrid cultivar Pollino carried this gene for the first time.

Pollen fertility in the hybrids is restored by using dominant nuclear-coded restorer genes (*Rfc*) from either European or exotic populations. Generally, the breeding of rye hybrids is based on the P- (CMS-P) or G-sterility-inducing cytoplasm (CMS-G). The use of an alternative type of sterility-inducing cytoplasm in practice is strongly recommended to reduce risks arising from plasmotype uniformity (Morgenstern and Geiger 1982, Adolf and Winkel 1985, Geiger et al. 1995). Another potentially applicable sources of CMS in rye is the C-cytoplasm (CMS-C) discovered by Lapinski (1972) in the old Polish cultivar Smolickie. Considering the low frequency of sterility alleles in populations of cultivated rye, development of

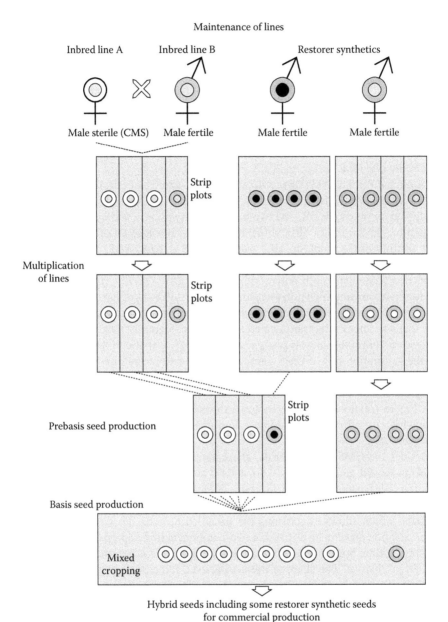

FIGURE 7.12   General schematic drawing of hybrid seed production in rye.

new nonrestorer lines in the CMS-C system can be significantly facilitated by the application of advanced molecular markers (see Table 6.1). The *Rfc1* gene controls restoration of male fertility with C-cytoplasm. Diversity Arrays Technology (DArT) markers were mapped on chromosome arm 4RL in the close vicinity of the *Rfc1* gene. Numerous other markers are tightly linked with the restorer gene, that is,

91 DArT markers and 3 sequence-characterized regions (SCARs). All these markers were mapped with the distance not exceeding 6 cM. A second mapping population resulted in nine DArT markers located also close to the *Rfc1* gene (Hackauf et al. 2009a, Stojalowski et al. 2011).

Another effective restorer gene (*Rfp1*) for the P-cytoplasm was isolated from the Iranian primitive rye population Iran IX (also on chromosome arm 4RL) and from an Argentinean landrace Pico Gentario (*Rfp2*) (Stracke et al. 2003). The closest markers were found to be 2.9 cM for *Rfp1* and 5.2 cM for *Rfp2*, respectively. Recently, a conserved ortholog set (COS) marker was localized even closer to *Rfp1*, that is, within a 0.7 cM interval (Hackauf et al. 2012).

Adolf and Winkel (1985) detected cytoplasmic–genic male sterility caused by the G-type cytoplasm (Gülzow). Restorers were found by the recessive gene, *ms1*, again located on chromosome 4R, and two modifying genes, *ms2* and *ms3*. The latter are located on chromosomes 3R and 6R, respectively. Because all of them restore the male fertility of G-cytoplasm, they were renamed as *Rfg1*, *Rfg2*, and *Rfg3* (Börner et al. 1998). The target gene, *Rfg1*, was mapped on chromosome 4RL distal to three restriction fragment length polymorphisms (RFLPs) (*Xpsr119*, *Xpsr167*, *Xpsr899*) and four random amplified polymorphic DNAs (RAPDs) (*Xp01*, *Xap05*, *Xr11*, *Xs10*) loci. *Xpsr167* and *Xpsr899* are known to be located on the segment of chromosome 4RL that was ancestrally translocated and is homeologous to the distal end of other Triticeae 6S chromosomes. It is suggested that *Rfg1* is allelic to the gene determining the restoration of rye CMS caused by the P-cytoplasm and to the gene *Rfc4* that lies on rye addition lines of chromosome 4RL and restores the male fertility of hexaploid wheat with *Triticum timopheevii* cytoplasm. Homeoallelism to two loci for CMS restoration on chromosomes 6AS and 6BS in hexaploid wheat is likely. Meanwhile, on all rye chromosomes, dominant pollen restorer genes were found (see Table 7.5).

Besides the CMS-based hybrids, alternative mechanisms were tested. England (1974) proposed gametophytic SI for hybrid seed production. In rye, Wricke and Wehling (1985) studied at the University of Hannover, Germany, the manipulation of gametophytic SI by heat stress during the flowering period. SI can be temporarily inhibited which allows establishing the so-called S_1 lines. They can be multiplied under isolation and can be utilized in two-component mixed propagation for hybrid seed production. Since partial inbreeding occurs in this system, the hybrid varieties are less productive. The share of inbreeding even increases by spontaneous introgression of SI alleles during line propagation. Therefore, the proposed SI method for hybrid seed production in rye did not succeed in practical breeding.

7.1.3.4 Production of Hybrid Seed

Commercial hybrids are produced between CMS single crosses as seed parents and restorer synthetics as pollinator parents. The latter are mostly composed of two inbred lines crossed by hand and further multiplied by random open pollination. This complex type of hybrid (top-cross hybrid) needs several stages or years for production (see Figure 7.11). The main advantages are a cheaper and more stable seed production with less risk for the breeder, a higher vigor of the hybrid seed, and an extended period of pollen shedding by the pollinator synthetic. Substantial

TABLE 7.5

Chromosomal Localization of Dominant Pollen Restorer Genes in Rye with Major (+++,++) and Moderate Effects (+) according to the Pampa Male Sterility-Inducing Cytoplasm

Chromosome/Chromosome Arm	Effect
1R, 1RS	++
2R	
3R, 3RL	+
4R, 4RL	+++
5R	+
6R	++
7R	

Source: Compiled from Genotypes, such as Inbred 18 (L18), Inbred 161, Iran IX, Massaux, and Pico Gentario, and from different sources.

progress in grain yield has also been achieved. In 1999, the best hybrid surpassed the best open-pollinated check by ~20% in the official German trials. Progress in hybrid breeding steadily increased this superiority since the 1990s (Geiger 1990). This clearly reflects that hybrid breeding is genetically the most efficient breeding method. Hybrids became more attractive to farmers than the traditional populations.

After establishment of methodological basis at the University of Hohenheim, Germany, hybrid breeding was taken over by Lochow-Petkus Ltd., Germany, and its daughter company Hybro (1978). The first three hybrid varieties Forte, Aktion, and Akkord were released in Germany in 1984, followed by Cero (1985), Amando (1987), Rapid (1989), Marlo (1991), Gambit (1992), Dino and Uso (1995), Gamet (2000), Askari (2003), Rasant, Fugato, Festus (2004), Agronom, Amato (2005), Minello (2008), and so on. A second company, Diekmann Seed Ltd., Nienstädt, Germany, produces successful hybrid seeds by utilization of the G-cytoplasm (Melz et al. 2001). Its top sellers are Helltop and Hellvus.

Following the first breakthrough, more than 40 hybrid varieties were bred during the period between 1985 and 2012 occupying more than 60% of the total rye acreage. Some of these hybrids also are registered and distributed in Austria, Denmark, France, the United Kingdom, Scandinavian countries, and the Netherlands. In Sweden and Poland, one and two independent hybrid rye breeding programs, respectively, are running at present with the first released Polish hybrids in 1999. In addition, in Russia, several programs are conducted in different areas of the country. Outside Europe, the only hybrid rye breeding program is situated at the University of Sydney, Australia.

Commercial hybrid seed production is more complex than standard elite seed production. It requires additional experience, extra precise seed processing, and

FIGURE 7.13 Stripwise cropping of cytoplasmic male-sterile mother plants with restorer lines for hybrid seed production in rye under field conditions. (Courtesy of Melz, G., Dieckmann Seeds, Germany.)

multifaceted logistic over years. Spatial cropping and space planting of inbreds decrease normal seed reproduction and have thus to be taken into account. The multiplication of CMS lines and their maintainer requires at least 2 years. CMS lines and their fertile analogs are stripwise grown in isolated fields, and are nationally or internationally distributed (see Figure 7.13).

Nonrye cropping areas are favored. Optimal climatic conditions improve seed setting of characteristically sensitive inbreds. Off-types have to be rouged. For sufficient cross-pollination, the ratio of crossing partners is 2:1, and sometimes 3:1 (see Figures 7.2 and 7.14). Nonrestorer B lines and their male-fertile maintainer are treated similarly. The female:male ratio can be 4:1 depending on the field size. In order to minimize seed admixing of the maintainer and CMS material, the maintainer are harvested shortly after flowering. What was described for A, B, and CMS lines applies to the restorer lines and their synthetics as well.

Certified seed for commercial distribution is produced in fields by mixing CMS seed parents and restorer (and/or synthetic) parents. The ratio is ~95:1, respectively. In this way female and male plants cannot be separated as is done in stripwise cultivation. Therefore, the ~5% of seeds formed on the pollinator cannot be mechanically eliminated from the true hybrid seeds of the CMS parent. However, the cost savings from the mixed cultivation offset the small loss of hybrid yield performance.

Presently, more than 40 hybrid varieties are on the commercial list in Germany (see Table 7.6), occupying >60% of the total rye acreage, and sometimes 90%, depending on the region. Some of these hybrids are registered and distributed in Austria, Denmark, France, the United Kingdom, Scandinavia, the Netherlands, and Australia. An additional yield of 14%–20% can be expected by the hybrids, although population breeding progressed as well (see Figure 7.14).

At the beginning of hybrid breeding an increased ergot contamination was observed. The hybrid variety Rasant can be taken as an example. Later, when the so-called PollenPlus® technology (see Section 8.3.2) was applied and the introduction of the CMS–genic male sterility of the G-type hybrids was successful, ergot infection could be significantly reduced. It is achieved by increased pollen shed and higher stress tolerance. Moreover, hybrid seed is not mixed with common rye anymore as it was done at the beginning of hybrid breeding in rye. The varieties Askari (P-type) and Helltop (G-type) are examples of recent progress in breeding ergot-tolerant hybrids.

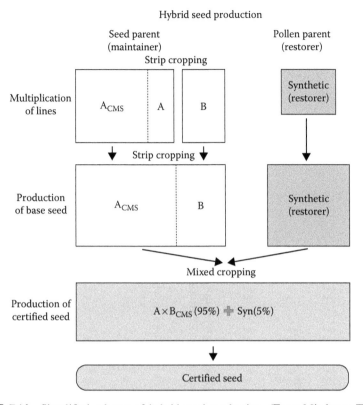

FIGURE 7.14 Simplified scheme of hybrid seed production. (From Miedaner, T. et al., *Zwischen Tradition und Globalisierung 1*. Zikeli, S. et al. (eds.), Köster Verlag, Berlin, Germany, 2007.)

Looking at the general yield development of rye during the past eight decades in practice, a story of success can be noticed (see Figure 7.2). It started with level of ~12 dt/ha in 1880, realized by landraces. The first jump in yield occurred from the years 1880 to 1920 when targeted mass selection methods were used. Since 1935, there was further continuous increase in yield reinforced by improved selection technology. What is true for Germany is also true for Poland as a neighboring country, which confirms a general trend. The past two decades brought an increase in the use of the hybrid method. A substantial contribution of research and targeted selection can always be assumed. This can make a breeder happy! The strong fluctuation in yield during the twenty-first century is obvious too. Is this a consequence of the worldwide climatic change that is displayed in rye?

7.1.4 MOLECULAR BREEDING

Currently there is great interest in developing fast and secure methodologies for DNA analysis. A DNA marker can precisely reveal genetic differences between individual genotypes of rye. Different genetic marker systems such as RFLPs,

TABLE 7.6
A Compilation of International Breeding Activities and Varieties of Rye

Country	Location	Breeder	Aims	Varieties
Afghanistan				
Joudar (L), Kalhek (L) (spring type)				
Argentina				
Choique (P) <1989, Don Enrique (donor Pampa CMS), Forrajero Klein (P), Forrajero Massaux (P), Insave (P), Forragero Insave (P), Greenfold (P), Harriet (L), Harriet's Centeno, Klein, Manfredi Suguia (P) (spring type), Pastoreo (P), Pico (P), Pico Genterio (P) 1957, Remeco (P), Safico (P), Rio Cuarto (P), San Jose (P), Tropero (P)				
Australia				
Black winter (P) <1946; Kenija 12-31-2 (P), Weethalle (P)	South West Australia	University of Adelaide	Selected from semidwarf spring types	Bevy, a composite of mostly (80%) semidwarf plants with 15% of plants as tall as SA Commercial and 5% being very short. When mature, heads also range in length.
Austria				
Brandroggen (L), Brucker Harrach, Chrysanth Hanseroggen, Edelhofer, Edelhofer neu (P) <1971, Eho Kurz (P) (Edelhof, Eiskorn, Elego, Hanna, Harrach Universal, Heertvelder, Heinefelder (L) <1938, Hohenauer (L), Jogo (P), Kaltenberger (L), Kärtner, Kärtner 1 (P), Kefermarkter (L) <1938, Lassaer (L) <1938, Loosdorfer Reform (P) <1938, Loosdorfer Tetra (P) tetraploid, Lungauer Tauern, Marienhofer (L) <1955, Melker (L) <1955, Mühlenviertler (L), Oberkämtner, Otterbacher, Panmers Hohenauer (P) <1938, Petroneller Tymauer (L) >1953, Radstadter Bergland (L), Ritselhofer (L) <1938, Stalker (P) <1938, Tiroler Tschermak's Marchfelder Roggen (1926), Schlägler (L) <1938 (very winter hardy), Prof. Tschermak Roggen (1927), Tymauer Roggen (L) <1938, Universal, Universal Neu, Wienerwald (L) <1938, Wieselburger (L) <1955				
Azerbaijan				
Somopolevaya Chavdar (P)				

Belarus

Benyakonskaya (L) <1937, Bruktinskaya (L) <1940, Jaselda (P), Kalinka (P), Lippovskaya (L) <1940, Plavunetc (P) <1940, Radzima (P), Shatckaya (P), Sosnovskaya (L) <1940, Volostyanskaya (L) <1940, Vysokolitovskaya (L) <1940, Zubrowka (P)

Minsk, Shodino

Belorusskaya (L), Belorusskaya 23, Belta (P) <1972 tetraploid, Charkovskaya 55, Charkovskaya 60, Charkovskaya 194, Drushba (P) <1972, Partizanskaya (P) <1953, Radzima (P), Shvedskaya (P) <1940, Zazerskaya (P) <1953, Zhabchitckaya (L)

Belgium

Celestyner, Court de Flandre (P) <1946, Feniks (P), Fliämischen Landroggen (L) (Conckelaer, H. Proot, 1903), Galma (P, Geant de Flandre (P) <1946, Jobaro, Waregem (L) <1946

Brazil

Bagé (P), Bar mario (P), Blanco (P) 1987, source for Al tolerance, BR1 (P), Centeno 52 (P), Centeno 53 (P), Centeno 54 (P), Centeno de La Estanzuela (P), Common (P), Garcia (P), Gefulio Vorgos (P), Guarapnovo (P), Passo Fundo (P), JN Vernandor (P), Sao Paulo (P), Serrano (P) 2005

Bulgaria

Bul Milenium, Joegewa, Kolarovka (P) 1971, Kolarovka 1 (P), Losien 14 (P)

Sadovo
Plovdiv
Dobrich

Sadovo (P) <1947
Nizkostebelnaya (P)
Tosevschi (L)

Canada

Antelope (P) 1963, Cad II (P), Cavdar, Centrene 2831, Chaupon, Explorer (P), Gator (P), Horton (P) <1949, Prima (P), Saskatoon (P), Star (P) <1930, Swift Current, Standard (P) <1949, Standard 1 (P) <1950, Standard 2 (P) <1950, Standard 3 (P) <1950, Storm (P), Swift current (P), Turkey

Manitoba

University of Manitoba

From an open-pollinated selection in a composite cross of European and Canadian varieties

Cougar (P) 1967 (medium yield, winter-hardy, late, medium height, fair lodging resistance; small green and tan seed, medium test weight); Puma (P) 1972 sel. from Dominant (medium yield, good winter hardiness, medium late, medium height, poor lodging resistance, small green seed, medium test weight); winter-hardy selected from Dominant

(Continued)

TABLE 7.6

(Continued) A Compilation of International Breeding Activities and Varieties of Rye

Country	Location	Breeder	Aims	Varieties
	Saskatchewan	University of Saskatchewan		Caribou (P) <1969, Frontier (P) 1983, Gazelle (P) 1974 (spring type), Prima (P), Prolific (P) <1940
		Semiarid Prairie Agricultural Research Centre, Saskatchewan		Hazlet (P), Remington (P) 2000, Rifle (P) 1996
		Agricore United		Dakota (P)
				Gauthier (P), Musketeer (P) 1983, Prima (P) 1986
		SeCan Association		Kodiak (P) 1973
	Alberta	Viterra		Kodiak (P) 1973

China

Dongmu 70 (P), Gansu (P) (spring type), Ging Zhouzao (L) (spring type), Gong (P) (spring type), Jilin (spring type), Jinzhou (P), Kye-shan-hye-mai (L) <1953 (spring type), Lanzhou (P), Nixiajy (P), Nato (P), Qingling (P), Weining, White rye, Wugong (spring type), Xinjiang (spring type)

Croatia

Karlovacka zupaniya (L)

Czech Republic

Albedo (P) 1991, Aventine, Beskyd (P) 1991 tetraploid, Breno (P), Bruno, Ceske (P) 1949, Chlumecke (L), Conduct (H), Debrett (P), Detenicke, Detenicke Kratkostebelne (P), Dobřenické krmné (P), Glapskie, Karkovske, Kerkovske (L), Kribice Krajove Nalzovske (P) 1971, Krmne Zito (L), Lochovo Blatenske (P), Predborske, Pudmericke (L), Puselske (P), Ratborskie (P) 1925, Recrut (P) 2002, Rodosinsky Rekord (P), Selgo, Slapske, Tesovske (L) (spring type), Těšovské Jarní (P), Valticke, Viglasske (P), Vschi, Zenit (P) 1948, Zidlochovicke Panis	Stupice Kwassitz Prague, Ruzyne	Dr. Ritter von Proskovetz-Kwassitz	Stupicke (P) <1951, Hanna (P) 1897 sel. from Bohemian landrace Hanna (P) 1897 sel. from Bohemian landrace Detenicke (L) <1947, Detenicke kratkostebelne (P), Dobrovicke (P) <1971, Doupovske (P) <1947, Dregerovo (L) <1947, Dvorskeho (L) <1947, Laznicke (L) <1947, Nalzorske (L) <1947, Napajedelcke (L) <1947, Ratborske (P) <1947, Slancke (L) <1947, Terrasol (P) <1947, Valticke (L) <1947, Zdanicke rane (P) <1947, Zidlochhvicke rane (P) <1947

Denmark

Borris Perlerug (P) (spring type), Brattingsborg (L), Rogo (P) 1969 (spring type), Rotari, Sejet Kaernerug (L)

Ecuador

Trigo nacional (L)

Estonia

Elvi (P), Estonskaya (L) <1940, Getera (P), Jogewa 1 <1946, (P), Jogewa 2 (P), Jogewa 1 2 (P) <1972, Sangaste[a] (P) <1946, Tulvi (P), Vambo (P) <1975, Viku (P)

Finland

Aitta (P), Akusti (P) 1992, Anna (P), Auvinen (P) (spring type), Belhjoa (P), Ensi (P) <1937, Fin, Härmä, Harmanruis, Hankkijas Jussi (M) 1975, Hännilä (L) (spring type), Harmanruis (P) <1937, Harmanzen's, Jussi (P), Juuso, Kartano (P), Kausala (P), Kelpo (P), Koylio (P), Korhosen (P), Monarüvesi (P) <1946, Myttäälä (L) <1937, Niemisjärvi (L) <1937, Oiva (L) <1937, Onni (L), Onni II (P) <1937, Pekka (P) <1946, Ponsi, Puhoksen (L) <1937, Puhoksen wis (P) <1937, Pvole oline wert (L) <1937, Reetta, Riihi 1999, Sampo (P), Slarna (L) <1937, Sulikava (P), Toivo (P) <1937, Uusimaa (L), Vaasa (L), Vatia (P), Voima (P) 1966, Visa (P) <1960

(Continued)

TABLE 7.6
(Continued) A Compilation of International Breeding Activities and Varieties of Rye

Country	Location	Breeder	Aims	Varieties
France				
	Crouelle, Marne Institut national de la recherche agronomique (INRA) Corsica	A. Berbigier		Beaulieu (P) (der. from Petkus Kurz), Grand Crouelle Quenza (P) <1955
Cilion (S) 2000, Champagner (L) <1890, Clou, Elect (P), Everest (P) <1966				
Germany				
	Nienstädt	Dieckmann GmbH and Co. KG, took over the breeding program from Carsten and PZG		Cantor (P) 2007, Herakles (S) 2008, Helltop 2009 (H, G-cms), Hellvus 2007 (H, G-cms), Kapitän (S) 2008, Marcello (P), 2007, Novus (H)
	Einbeck, Bergen	KWS Lochow GmbH		Arantes (P) (spring type), Avanti (H) 1997, Balistic (H) 2006, Bellami (H) 2008, Bonapart (H), Brasetto (H, PollenPlus®) 2009, Conduct (P) 2006, Esprit (H) 1995, Evolo (H) 2006, Fernando (H) 1998, Goliath (H), Gonello (H) 2009, Guttino (H, PollenPlus®) 2009, Locarno
Amalanclor, Baro (P), Bendelebener, Bergroggen, Born (P) 1998, Calypso, Canovus, Borfuro, Dakko (P), Eche Kurz, Esto (P) pale-grained, Erzgebirgsroggen (L), Deutscher Ringroggen, Dohlauer (L) <1945, Dompfaffer Roggen (L), Düppeler Roggen (L), Epos, Gaiton (P), Goliat, Halo (P) 1977, Hess-dorfer Johannis, Hessischer Staudenroggen, Janos (P), Jubiläumsroggen, Kampiner Roggen (L), Kapitän, Karlshulder Winterroggen (P), Karlshulder Sommer-roggen (P) 1968 (spring type), Lekkeroggen (P) 1971, Lischower (L), Lukas (P), Marienroggen, Merkator (P) 1980, Oberambacher, Ovid, Padana (P) 1971,				

Station	Breeder	Cultivars
Petkus, Belzig	F. von Lochow and successor VVB Saat-und Pflanzgut (1945–1990)	(H), Marcello 2007, Marder (H), Minello (H) 2008, Nikita (P) 1998, Palazzo (H, PollenPlus®) 2009, Petkus Ameliorant (P) 1971, Picasso (H) 1999, Placido (H) 2007, Pollino (H) 2005, Progas (H) for biogas, Recrut (H) 2002, Resonanz (H) 2004, Visello (H, PollenPlus®) 2006, Sorom (P) 1980 (spring type), Treviso (H) 2001, Ursus (H) 1997. Paleschker Biesenroggen (L), Pettenser Roggen (L), Pluto (P) 1979, Protector, Purmitoi 1971, Rümkers (P), Sellino, Schmidt Roggen, Sturmroggen, Sächsischer Schwäbischer (L), Ostpreussischer Johannisroggen, Göttinger, Heidenreich's Riesen, Kuckucks Champagner, Maultier-Roggen, Nomaro, Norddeutscher Champagner (P) ca. 1900, Perolo (P), Quadriga, Tero (P) tetraploid, Waldecker Stauden (L), Uso, Vitallo (F)
Gülzow	Inst. Pflanzenzüchtung (<1990)	Danae (P) 1968, Donar (M), Lübnitzer (P) 1998 sel. from Lübnitzer landrace, Muro (M), Pekuro (P), Petka (P), Petkuser Moorroggen, Petkuser Winterroggen, Petkuser Sommerroggen, Petkuser Moorroggen, Pollux (M)
Hadmersleben		Kustro (P) 1973, Nomaro (P)
Schenkenberg	Hybro Saatzucht	Heines Hellkorn (P) Acronym 2005, Agronom (H) 2005, Allawi (H) 2011, Amando (H) 1987, Amato (H) 2005, Askari (H, good ergot res.) 2003, Dino (H) 1995, Drive (H) 2012, Dukat (P) 2008, Festus (H) 2004, Fugato (H) 2004, Gambit (H), Garnet (H) 2000, Marlo (H) 1991, Mephisto (H) 2011, Plato (P) 2001, Santini 2012, Satellit 2012, Stakkato 2012, Rasant (H, low ergut resistance) 2004, Rapid (H) 1989, Timo (H), Uso (H)

(Continued)

TABLE 7.6
(Continued) A Compilation of International Breeding Activities and Varieties of Rye

Country	Location	Breeder	Aims	Varieties
	Grundhof	Petersen Saatzucht		Apart (H) 1997, Dominator (P), Matador (P) 2001, Ovid (P) 1995 (spring type), Weidmannsdank (P)
	Steinach	Saatzucht Steinach		Boresto (P) 2000, Borellus (P)
	Schwartau b. Kiel	Saatzucht Carsten		Caroass (S) 2002, Caro Kurz (P) 1926, Carotop (S) 2002, Carotrumpf (S) 2003, Carsten's Kurzstroh (P)
	Aschersleben	Terra-AG Samenzüchterei		Askanischer Riesenroggen (P) 1900 sel. from Zeeländer Roggen
	Aderstedt	Rüben-/Getreidesa-menzüchterei Ltd.		Aderstedter Roggen (P) 1894 sel. from Probsteier
	Hadmersleben	Ferdinand Heine		Heine's Hadmerslebener Klosterroggen (P) 1895 sel. from Zeeländer, Heine's Zeeländerroggen (P) 1869 sel. from Zeeländer Roggen (Holland)
	Schlanstedt	Wilhelm Rimpau		Rimpau's Schlanstedter (P) 1892 sel. from Probsteier
	Vienau b. Braunau (Altmark)	Rudolf von Kalben		Kalben's Vienauer Roggen (P) 1903 sel. from Pommer landrace
	Pfiffelbach and Sundhausen n. Gotha	Kirschen		Kirschen's Roggen (P) 1890 sel. from Zeeländer × Schlanstedter, Kirsches Stahler (P)

Halle/S.	Landw. Institut	Roemer's Emsroggen (P), Roemer's Erzgebirgsroggen (P), Saaleroggen (P) 1911 sel. from Saxonian landrace
Friedrichswerth	A. Eduard Meyer–Friedrichswerth	Friedrichswerther (P) 1902 sel. from Petkuser
Pirna	Zucht-/Verkaufsgenossenschaft	Pirnaer (P) 1896 sel. from Saxoian landrace
Könkendorf n. Sadenbeck	A. V. Jäger	Jäger's Norddeutscher Champagner (P) 1895
Schladen	Otto Breustedt	Deutscher Ringroggen (P), Harzer Viktoria (P) 1896 sel. from Probsteier
Quedlinburg	Himmel	Himmel's Champagner (P) 1901 sel. from Champagner
		Probsteier (P) 1863
Schönberg	Zucht- und Verkaufs-Genossenschaft	
Buhlendorf-Lindau	J. Sperling	Buhlendorfer (P)1895 sel. from Petkuser
Rostock	Professor Heinrich	Heinrich Roggen (M) 1908 sel. from Swedish mutant source
Erfurt	N. L. Chrestensen Samen-/Pflanzenzucht	Chrestensen's Riesenroggen (P) sel. from Propsteier; Kolbenroggen (P) 1907
Toitenwinkel b. Rostock	H. Brandt	Brandt's Marien syn. Mecklenburger Marienroggen (P) 1920–1961

(Continued)

TABLE 7.6
(Continued) A Compilation of International Breeding Activities and Varieties of Rye

Country	Location	Breeder	Aims	Varieties
	Getreidezüchtungsforschung Darzau	Dr. Karl-Josef Müller		LiKoRo (Lichtkornroggen) (P) 2011 (sel. from Schmidt Roggen, strain 1988/406)
	Salzmünde	Groetzner Saatzucht		Rye Food (P), Varda (P), Waldi (P) sel. from *S. montanum*

Greece

Hania (P), Kastoria (P)

Hungary

Country	Location	Breeder	Aims	Varieties
Beka (P), Duna-Tiszaközi feher (L) <1955, Frosz (P), Huszaj (P), Kisvárdai (L) <1954, Kisvárdai 1 (P), Kisvárdai Alacsony (P), Kisvárdai Legelő, Kriszta (P) 1998 perennial, Lovászpatonai (P) <1952 *Lr8* gene, Lovaszpatonai 2 (P), Nagycenki (L) <1954, Perenne (P) perennial	Heviz			Fleischmann (P) <1937
	Kecskemet			Kecskemeti, Kecskemeti Törpe (P) 1961
	Gyor-Sopron			Ovari, Soprohonchonpasci (P)

Italy

Abruzzi (P) >1900, Cerasi (P), Cinquecento (P), Fasto, Forestal, Gambarie, Germano, Maly, Matese, Musco-Jermano, Primizia, San Giovanni (P), Secina, Sito 70, Tetti Bartola (L), Tetti Bariau (L)

Japan

Aokei 3 (P), Autuma, Dongmu (spring type), JNK (L), Kouhai Shuudan, Mantenboshi (P), Petokudza (P), Senbon-Hadaka (P), Shikoku-Hadaka (P), Shikoku-Hadaka 83 (P)

Korea

Paek am Dzong, Paldanghomil (P)

Kazakhstan

Belagagskaya (L) <1948, Dlinnokolosaya (P), Dolinskaya (L) >1947, Gibrid (P), Gibridnaya 173, Mnogokoloska (P) <1946, Pamirskaya (L), Sornopolevaya (P), Vetvistaya (P), Vetvistostebelnaya (P), Vysokogornaya (P) <1952, Zima (P) <1948

Kyrgyzstan

Kirgiszkaya 1

Latvia

Arupe (P). Joniai, Priekuli, Priekulskaya II (P), Priekulskaya (L) <1946, Priekulskaya 1069 (P), Stende, Stedskaya II (P) <1952, Vaive Tetra (P) tetraploid, Virgiai

Lithuania

Baltiya (P), Dotnuvskaya Auksteyn (P) <1946, Duoniai (P), Joniai (P) resistant to powdery mildew, Kombaininyai (P), Lietuvos 3 (P), Lietuvos 3 (P), Litovskaya 3 (P), Rukai (P)

Macedonia

Berovo (L) <1947, Kogani (L) <1947, Shtip (L) <1947, Tebreliya (L) <1947, Valandovo (L) <1947

Mexico

Mexico (P), Snoopy "S" (P) (spring type)

The Netherlands

Admiraal (P), Akkerroggen, Animo (P) 1976, Bandalier (P), Bonfire (P), Celestijner (L), Durmitor, Eldera (P), Heertvelder (P) <1968, Humbolt, Kruiproggen, Leuvenumseroggen (L), Norderas (P), Ottersumseroggen, Parana (P), Pearl syn. Perl (P), Rekka, Royal (P), Speedogreen, Staal, Wageninger, Zeelander (L), Zelder (P) <1966

Cebeco Seed Ltd.　　　Dominant (P) <1968

New Zealand

Rahu (P) (spring type)

(Continued)

TABLE 7.6
(Continued) A Compilation of International Breeding Activities and Varieties of Rye

Country	Location	Breeder	Aims	Varieties
Norway				
Norsk Graerug, Refsum (L) <1930, Vågones vårrug (P) (spring type)				
Poland				
	Choryń, Dankow	Danko Hodowla Roślin		Altpaleschkener Riesenstau-denroggen (P)1882, Danko (P) 1980, Dańkowskie Amber (P) 2010, Dańkowskie Diament (P) 2005, Dańkowskie Nowe (P) 1978, Dańkowskie Selekcyjne (P) ca. 1810, Dańkowskie Złote (P) 1968, Warko (P) 1991, Zielonkowe Dańkowskie
	Modrow-Gwisdzyn			
	Poznan	Poznańska Hodowla Roślin		Motto (P) 1991, Motto Nawid (P), Nawid (H) 1998, Puławskie (P), Pulawskie Wczeswe (P)
	Puławy			Krzyca Pulawska (P), Pulawskie (L) <1940, Smolickie, Smolickie Krotkie (P) 1967
	Smolice	Hodowla Roślin Smolice		Abago (P) 1999 (spring type), Bosmo (P) 2001, Hegro (P) 1999, Klawo (H) 2000, Luco (H) 1999, Ludowe Smolickie (P), Wibro (P) 1994, Zduno (P) 1996
	Lublin	Stanisław Ramenda		Noteckie Tetra (P) tetraploid
Agricolo (P) 2003, Amilo (P) 1994, Antoninskie, Arant (P) 1993, Bojko (P), Borkowskie Tetra (P) tetraploid, Chodan (P), Czeslawicki Tetra (P) tetraploid, Chrobre (P), Daran (P), Domir (P) 2008, Foivd, Gradan (P), Garczyna, Garczynskie, Garczynskie ludowe, Golskie (L) <1937, Granum, Herakles (P), Instytuckie Wczesne (P), Jelenieckie (L), Kazimierskie (L), Kfiszulskie zelazna (L) <1937, Kier (P) 1999, Konto (P), Kortowo (L), Ludowe (P), Madar (P), Miku-kickie (L) <1940, Moytto (P), Noszanovskie, Nowe, Oltarzewskie (L) <1937, Pancerne (L), Pastar (P), Pastewne pozne (P), Putza (L) <1937, Rogalinskie (L) <1936, Rostockie (P) Saskie, Slowiańskie (P) 2004, Skaltio, Stach, Stanko, Strze-kecinskie (spring type), Szklane (L) 1835, Tempo (P), Tetra Grozow (P) tetraploid, Tetra Lubelskie (P) tetraploid, Turbo (P), `Universalne (P), Walet (P) 1999, Warko (P) 1991, Wielkopolskie, Wierzbienskie (L) <1937, Włodko (P) 2002, Włoszanowskie (P) <1940, Wloszanowskie Nowe (P) <1968, Wojcieszyckie				

Portugal

Alvão (L), Atilho (L), Barroso 5040 (P), Brigada de Mirandela (P), Castello Branco (P) <1971, Centenico (L), Centenico do Alto (L), Cepeda (L), Edral (L), Estacao Agrar (P), Forragero Klein (P), Gimonde (L), Gremio da Lavoura de Felqu (L), Guarda (P), Lamego (L), Lamas de Ôlo (L), Macos, Malhadas (L), Miranda do Corvo (L), Montalegre (L), Morgade (L), Padrela (L), Porto (P), Vilar Douro (L), Vila Pouca (L)

Romania

Brasov 200-N (P), Bukovina 1 (P), Bukovina 2 (P), Bukovina 3 (P), Dragomirna (L), Dorna (P), Fundulea, Glapski (L), Gloria (P), Radminski (L), Tosevschi (L), Secara de Primavera (P) <1953 (spring type), Secara de Toamna (P) <1954 (spring type), Suceveana, Suceava 25

Russia (42 breeding stations and 63 varieties in 2012)

Arkhangelsk	Gnezdovka (L) <1930
Ingushetia	Burunnaya (L)
Michurinsk	Michurinka (P)
Kaluga	Kaluzhskaya 45 (P)
Promorye, Amur	Spasskaya (L)
Kaliningrad	Slavyanskaya (P) <1954
Dagestan	Akulskaya (L) <1950, Akushinskaya (L) <1950, Dikaya (L) (spring type)
Sakhalin region	Dzyanantina (L) <1949 (spring type), Hokaidodzairai (L) <1949 (spring type), Karafuto (L) <1949 (spring type)
Stavropol, Bakasan	Buyanova (L), Gibridnaya mnogoletnya derzhavina (P) <1948, Kolmykovskaya (L) <1938
Kazan	Kazanskaya (L) <1948, Kazanskaya 5 (P) <1948, Kazanskaya 6 (P) <1948

Alpha (P) 1999, Antares (P) 2002, Baltay (P) 1999, Benjakonskaya, Burunaya, Bylina (P) 2000, Cherkovskaya (L), Chitkinskaya (L) 1939, Chniven (P) 1993, Davydovskaya (L), Era (P), Gran (P) 2011, Irina (P) 2004, Iset (P) 2005, Iskra (P), Jaroslawna (P) 1976

LriO gene, Kalvskaya 45 (P), Kitmodovskaya, Kompus (P), short-straw type, tetraploid, Lygena 112 (P), Narkowskaya 194, Shitomirskaya, Snezhana (P), Spaskaya (L), Stendskaya 2, derv. from Tschulpan >1989 tetraploid, Verzbinskaya, Vlada (P) 2007, Vólchova (P) 1989, Voschod 2 (P), Vympel (P), Weiwe, Zaria

(Continued)

TABLE 7.6

(Continued) A Compilation of International Breeding Activities and Varieties of Rye

Country	Location	Breeder	Aims	Varieties
	Yekaterinburg			Gri(P) <1948, Samoseika (P) <1950, Rzanka (P), Tavrinskaya (L) <1948
	Orel			Lisitcyna (L) >1947, Orlovski
	Voronesh			Alpiiskaya (P) <1950, Kuznetckaya (P) <1951, Mup (P) <1948, Savala (P), Talovskaya 12 (P), Talovskaya 33 (P), Voronezkaya (L) >1947, Voronezhskaya shi (P) 1947
	Samara, Penza	N. M. Tulaykov Samara Research Institute of Agriculture		Bezencukskaya (L) <1948, Bezencukskaya 3 (P), Bezencukskaya 23 (P), Bezencukskaya 87 (P) 1993, Petrovskaya 7 (P) <1950
	Jaroslavl; Kirov	Stavropol Research Institute of Agriculture		Dymka (P) 1993, Falenskaya (L) <1946, Golubka (P), Kirovskaya 89 (P), Teleginskaya (I) <1941, 1993, Vyatka (L) <1946, Vyatka 2 (P) 1950, Vyatka-kustovka (P), Zvezdochka (P)
	Novgorod			Shpanskaya (L) <1938
	Rostov			Manycskaya (L) <1946, Tatsinskaya golubaya (P)

Ufa	Urals Research Institute of Agriculture, Bashkortostan	Bashkierski karlik (P), Birskaya 2 (P), Cisminskaya 2 (P), Cisminskaya 3 (P). Parom (P) 2008. Saratovka chishmiskaya (P) <1953, Tschulpan syn. Chulpan syn. Zulpan (P) 1975 from short-strawed crosses, Tschulpan 3 (P) *Lr6* gene
Orenburg	Tatarstan Research Institute of Agriculture	Avangard (P), Avangard 2 (P) <1948, Estaveta tatarstana (P), Ioldyz (P), Sharlykskaya (L) <1940, Tatarstana (P) 1999
Irkutsk, Novosibirsk	Siberian Research Institute of Agriculture	Buriatskaya (L) <1946, Chernyshevskaya (P) <1952, Erzhavinskaya (L) <1947, Estafeta erzhavinskaya 2 (P), Korotkostebelnaya 69 (P) 1985, Novosibirka (P) <1946, Onohoyskaya (P) <1946 (spring type), Severnaya (P) (spring type) (Yakutia), Sibirskaya (P), Tulunskaya selenozernaya (P) <1948, Udinskaya (L) <1947, Zarya (P), Zitkinskaya (L) <1946
Barnaul, Altai		Elowskaya (P) <1948, Kamenskaya (L) <1940, Kytmanovskaya (L) <1940

(Continued)

TABLE 7.6

(Continued) A Compilation of International Breeding Activities and Varieties of Rye

Country	Location	Breeder	Aims	Varieties
	St. Petersburg	VIR (42 varieties)		Banlig (P), Bant (P), Banvos (P), Chernigovskaya 3 (P) *Lr4* gene, Dyuimovochka (L), Fioletovozernaya (P) <1952, Getera 2 (P), Gibridnaya (P) <1952 (spring type), Ilmen (P) 1993, Imerig 2 (P), Immunaya 1 (P) *Lr5* gene, Immunaya 4 (P) *Lr6* gene, Izumrudnaya (P) <1948, Karbezanli (P), Karbezanvos (P), Korichnevokolosaya (P) <1952, Leningradskaya tetraploidnaya (P), Luga (P), Malysz 72 (P), Monstrozum 2 (P), Neva (P), Onega (P), Orgib (P), Pigmei (P) (spring type), Pionerskaya gibridnaya (P) <1952 (spring type), Rossiyanka (P), Rossul (P) *Sr1* gene, Russkaya (P) <1950, Sanim (P) *Lr4* gene, Selenga (P) (spring type), Suida (P), Trostnikovaya (P), Volhova (P), Zelenostebelnaya 2 (P), Zheltozernaya (L) <1945 (Karelia), Zheltozele-nolistnaya (P)

Krasnoyarsk	Institute of Agriculture	Aprelka kubanskaya (P), Jeniseyka (P) 1993, Kamalinskaya (L), Kamalinskaya 4 (P), Kamalinskaya 13 (P), Kamalinskaya 327 (P) <1945, Krasnoyarskaya (L) <1934, Krivonozhka (L9 <1940, Mininskaya (P), Skorospelka (P) <1940, Sitnikovskaya (L) <1946
Bryansk		Feya (P), Krupnozernaya (P) 1975, Novozybkovskaya 4 (P) 1962 *Lr7* gene, Zhatva (P)
Kaluga		Kaluzskaya 45 (P)
Nemcinovka		Nemcinovskaya, Nemcinovskaya 50 (P) <1975
Omsk, Tomsk		Atrachinka (P) <1939, Garantiya (P) <1946, Omka (P) <1946, Kustarka (P) <1939, Novoderevenskaya (L) >1941, Taezhnaya (P), Tevrizskaya (L) <1941, Ust-ishimka (P) <1941
Orel		Lisicyna (P), Orlovskii gibrid (P)
Saratov		Pervenec (P), Rzhaksinskaya (L), Soratovskaya (L) (spring type), Soratovskaya 1 (P) <1947, Soratovskaya 4 (P), Saratovskaya krupnozernaya, Volzhanka (P) <1946, Wolchova syn. Volkhova (P) >1989

(Continued)

TABLE 7.6
(Continued) A Compilation of International Breeding Activities and Varieties of Rye

Country	Location	Breeder	Aims	Varieties
	Moscow, Nizhniy Novgorod			Dikaya tetraploidnaya (L) tetraploid, Gibridnaya 2 (P) <1947, Kormovaya 61 (P), Krona (P) 1989, Moskovskaya kartikovaya (P) (recessive short straw by ct allele), Muraveika (L) <1930, Purga (P), Tulskaya 1 (P), Vetvistaya (P) <1952, Vetvistaya Tetra (P) tetraploid, Vyatka, Voshod 1 (P)
	Volgda region			Kustovka (P)
	Smolensk			Smolenskaya (L) <1940
	Tjumen			Chibrik (P) <1941, Ishimka (P) <1941, Chelyuskinka (P) <1941, Krivorozhka (P) <1941, Vagaika (P) <1941
Serbia				
Maksimirska (P)				
Slovakia				
Beskyd				
South Africa				
Polko (P)				
Spain				
Ailes (P), Altobar (P), Centeno de La Estanzuela (L), Cereco, Del Pais (L), Granja, Huesca (L), Llevante (L) syn. *S. villosum* <1750, Mansilla, Piniella, Riodeva (I), Villarrobledo (L)				

Sweden

Aslan, Björn (P) 1940, Björn 101 (P) 1940, Caspian, Gimli, Fljam (P) <1937, Fordland Wasa II, Floured (P) tetraploid, Gotlands (L), Ivan, Johannes (L), Kaskelott, Kungs, Kungs II syn. King II (P) <1937, Kungsrug, Maelm, Malm (P) <1937, Malt (P), Midsummer rye (L), Norderas, Orebro (P), Olle, Ötsgöta (L), Ostgota grey (L) syn. Ostgota grarag, Ottarp, Otello (P) 1970 sel. from Kungs II, Potter, Rorik, Sjalvbindar (L) <1937, Slarma, Slash-and-burn rye (L), Stal syn. Stalrag (P) 1911 syn. Steel-rye sel. from Stärnrag, Stjärnrag (L) <1937, Swedish green rye; Svalof Fourex (P) 1960 (spring type), Svalöf's star, Tello, Värnerag (L), Värne 64 (P), Wasa (L) <1930, Wasa II <1937

Switzerland

Adliker (L) <1928, Avantgard, Cadi (P), Gadi (P), Rothenbrunner (L) <1928, Schwand (L) <1928

Turkey

Ankora (P), Aslim (P), Turkey 77 (L) *Dn7* gene

Turkmenistan

Karagal (L) <1949

UK

Ashill Pearl (P), Rhyader (P), Rheidol (L)

Ukraine

Beregovskaya (L) <1952, Bylina (P), Charkovskaya (L) <1935, Charkovskaya 55 (P) *Lr4* and *Sr1* genes, Charkovskaya 60 (P) <1972, Charkovskaya 194 (P) <1948, Desnyanka, Desnyanka 2 (P) 1972, Donetckaya (L) <1951, Glybokskaya 2 (P), Gnom (P), Inka (P), Kievskaya 2 (P), Kievskaya 80 (P), Kormovaya 1 (P), Kormovaya 51 (P) 1992 tetraploid, Kustovka- kievskaya (P), Lvovskaya (L), Melodiya (P), Mikulinskaya (L) <1955, Mikulitckaya radchanskaya (P) <1955, Nedoboevskaya (L), Nemyshlyanskaya 953 (P) <1948, Nika (P), Niva (P), Odesskaya 1 (P), Olimpiada 80 (P), Pallada (P), Polesskaya (L) <1947, Sasskaya (P), Tarascanskaya 2 (P) <1947, Tarascanskaya 4 (P) <1947, Turkovskaya (L), Uluchshennaya (L) <1940, Vega (P), Veselopolyanska (L) <1947, Vetvistaya (L) <1935, Zhitomirskaya (L)

Uruguay

Centeno de La Estanzuela, Chica, Pico Mag

(Continued)

TABLE 7.6

(Continued) A Compilation of International Breeding Activities and Varieties of Rye

Country	Location	Breeder	Aims	Varieties
USA				
Acco WR-811 (P) 1975, Akusti (P), Elovs (P), Emory (P), Gurley Graser, Weser (P)	Dade City, Florida	Hancock Seed	Selected from Balbo	Aroostook 1981 (low yield, good winter hardiness, very early, tall, poor lodging resistance; small seed of brown and tan color, low test weight), Florida (P) (spring type), Florida 401 (P) 1991, Florida Black (P) 1969, Florida dwarf (P) (spring type), Gator (P) <1967, Hancock (P)
	New York	USDA, Cornell University and Maine Department of Agriculture		Dakold (P) 1902, Pierre (P)
		North Dakota Agricultural Experiment Station	Selected from von Lochow Petkus	Elk (P), Frederick 1984 (medium yield, good winter hardiness, medium late, medium height, poor lodging resistance, medium-size seed, predominantly tan color, high test weight)
		South Dakota Agricultural Experiment Station		Metzi (medium yield, medium late, tall with average lodging resistance, seed predominantly green, high test weight)
		Nutriseed Co.	Selected from Lochow-Petkus Ltd. (1958)	Von Lochow 1964 (medium yield, fair to poor winter hardiness, medium late, medium height, good lodging resistance,

	Institution	Cultivars/Description
	Minnesota Agricultural Experiment Station	large seed, predominantly green color, high test weight), Dean (L) ca. 1900, Emerald (P) <1969, Minnesota, Minnesota 2, Minnesota 90 (yellow grained), Minnesota 91 (pale-green grained), Minnesota 92 (dark-green grained), Rymin (P)
	University of Georgia	Abruzzi (P) ca. 1900 (origin Italy), Athens Abruzzi (P), Athens Bonel (P) 1968, Emory (P), Hazel (P) 1991, Weser (P) 1970, Wintergrazer 70 (P) 1970, Wrens Abruzzi (P) 1970, Wrens 96 (P) (spring type) 1996
Michigan	University of Michigan	Rosenb (P), Wheeler (P) 1971
	University of Mississippi	Explorer (P) 1966, Mississippi Abruzzi (P) 1969
	University of Tennessee	Balbo (P) 1967, Hiwassee (P) 1972
	Wisconsin Agric. Exp. Stat.	Adams (P) 1953, Coloma (P) 1991, Imperial (P) <1967 sel. from Schlanstedter, *SrR* gene, Wisconsin
	California Agric. Exp. Stat.	Merced (P) 1969
	Idaho Agric. Exp. Stat.	Vern (P) 1904
College Station	Texas A&M AgriLife Research	Wintermore (P) 1975

(Continued)

TABLE 7.6
(Continued) A Compilation of International Breeding Activities and Varieties of Rye

Country	Location	Breeder	Aims	Varieties
	Ardmore	Samuel Roberts Noble Foundation, Oklahoma State University	Forage, grain rye	Bates (P) 1995, Bonel (P) 1966, Elbon (P) 1956, Maton (P) 1975, Maton II (P) (spring type) 2007, Okema (P) 1970, Oklon (P) 1993

Origin unknown

F, fodder rye; I, inbred; Adar, Alfa, Bandelier, Deprimavare (spring type), Ergo, Fontane, Hacada (P) 1993, Harlan, Japayedelske, Kirszkayh, Massaux, Oktavian (P), Orizont, Pabrenicke Kyune, Perkow, Peros, Priaborschi, Protektor, Sofia 59, Sydar, Talowskaya H, hybrid varieties; L, landrace; M, mutation variety; P, potulation variety; S, synthetic variety.

Note: Most varieties are winter types, and spring-type growth habit and time of release are given in the table.

a Friedrich Georg Magnus von Berg (1845–1938) was the breeder at Sangaste (Sagnitz) mansion in Estonia. In 1875, a local population "Sangaste" was developed, which was characterized by winter hardiness and large kernels. He started formation of an initial material for plant breeding in 1868. Over 40 crossing combinations between several varieties of rye were made. Mainly Probstei rye contributed to the first release. Currently, the variety Sangaste is considered as the oldest cultivated rye variety in the world. At the World's Fair held in Paris in 1889, Sangaste rye was awarded a gold medal and at the world exhibition in Chicago in 1893, first prize was given to that variety. Silver medals were awarded at the All-Russian Exhibitions held in Kharkiv in 1888 and at Tsarksoselskaya Anniversary Exhibition in 1911. Count Berg immediately started varietal improvement and conservationary breeding of Sangaste. He selected heavier rye spikes, which also assessed larger number of grains and grain weight per spike. Spike length, color, and density were determined in subsequent selection as important traits as well. By the instrumentality of special sowing desk, seeds from selected spikes were manually planted in remote fields. Seeds were mixed on selected plots and planted in wide-row plots for multiplication purposes. The yield of multiplication plots was harvested and cleaned with spin dryer and sorting machine personally designed by Count Berg (Anonymous 2012).

b Joseph A. Rosen (1877–1949), born in Moscow and studied agronomic sciences in Russia and Germany, gained international renown in the field of agriculture after he developed the new variety of winter rye named Rosen's rye after him, which was widely used by the US farmers. When he was a Russian student at Michigan Agricultural College in 1908, he was sent to bring the original rye seed from his father back in his homeland. Successive selections of desirable plants were made from this population by F. A. Spragg, and, in 1912, the first bushel of Rosen's rye went from Michigan station to Carleton Herren of Albion. At that time, probably the most famous of Rosen's rye growers were George and Louis Hutzler of South Manitou Island in northern Lake Michigan. Here, in perfectly isolated island fields, these growers were working with Rosen's rye of the finest, purest stock. Other growers could renew their supplies when cross-pollination contaminated their seed crop. During the beginning of the twentieth century, Rosen's rye was grown worldwide. First in 1934, rye from Poland was imported again to the United States in order to widen the spectrum of rye varieties and the gene pool of cropped rye (Schlegel 2007).

RAPDs, sequence-tagged sites (STSs), amplified fragment length polymorphisms (AFLPs), single-sequence repeats (SSRs), microsatellites, and single-nucleotide polymorphisms (SNPs) were developed and applied. RFLPs, SSRs, and AFLPs were most effective in rye. In 2012, Tomita and Seno (2012) offered new markers for rye mapping, based on chromosome-specific polymerase chain reaction (PCR) from restriction enzyme "Eco0109I" recognition site. These markers were sequenced and converted to STS primers (see Table 5.5).

When developing markers for precision breeding, marker-assisted selection, or molecular breeding—the terms are different but the content is similar—one should identify the gene that affects the traits having breeder's interest and design markers within the gene or flanking regions. Until now more than 10,000 markers have been identified and mapped in certain chromosomes of rye, using special cytogenetic approaches and linkage maps (Schlegel and Korzun 2013). The selective marker must be diagnostic in the target population to be effective in selection. In addition, linkage equilibrium should be maintained between the target gene and the marker for utilization in breeding. Moreover, the effect of quantitative trait loci (QTL) and the candidate marker for the character require validation in other genetic backgrounds. To increase adoption of the marker-assisted selection by breeders, more reliable markers are needed that can be done by utilization of greater number of mapping populations. Phenotypic characterization of plant individuals has to be precise, even when different genetic backgrounds are considered.

The breeding of ergot-resistant hybrids was the first practical example of successful application of molecular genetics in actual rye breeding. Among others, it makes use of tagging the restorer gene *Rfp1*. The restorer gene shows close genetic linkage to the RFLP marker *Xiac76* (<0.8 cM). This molecular marker offered the chance of "smart screening" of the restorer gene during the selection process (see Section 7.1.3). Further approaches on this issue can be expected soon.

7.1.5 Mutation Breeding

The wide genetic variability in rye is already described by Rimpau (1899), characterizing extreme morphological deviations in growth habit. The search for spontaneous mutations that are not recessive segregants is extremely difficult in rye because they are often masked by heterozygosity.

Spikes of square-head type were described by Amend (1919) and awnless spikes by Duckart (1927). Antropov and Antropova (1929) and Nowikow (1953) in Russia, cited after Fedorov et al. (1970a), described the first short-strawed mutants. Before them Benno Martiny (1870) mentioned about a mutant with forked spikes. It was found in 1757 (see Figures 7.15 and 7.16). In 1935, T. Roemer (1883–1951; Martin Luther University of Halle-Wittenberg, Germany) founded a Department of Mutation Research at his Institute, which, in 1937, was extended to a Department of Cytogenetics with R. Freisleben as the first head. Spontaneous and induced mutations should be evaluated for cereal breeding. In rye, targeted experiments began during the 1960s (Schmidt 1980).

Eine dreigablige Roggenähre i. J. 1779 aufgefunden, jah ich bei Professor Koch in Berlin. Dieselbe war etwa 6¹/₂ Cm. lang und hatte auf der einen Seite zwei Aehrenauswüchse zu etwa 2 Cm. lang, von denen der eine etwa 1¹/₂, der andere etwa 3 Cm. über dem Grunde der Hauptähre von dieser ausgingen. Außerdem hatte Herr Prof. Koch die Güte, mir die Abbildung einer Roggenähre mit 17 derartigen Aehrenenden zu zeigen, die im Jahre 1757 bei Struppen im Königreich Sachsen gefunden worden sein soll.

FIGURE 7.15 Facsimile of "Der mehrblütige Roggen—Eine Pflanzenkulturstudie." (From Martiny, B., *Schriften Naturforsch. Gesell.*, Verlag A. W. Kasemann, Danzig, Poland, 1870. With permission.)

FIGURE 7.16 Rye with forked spike as evidence for morphological mutations, detected in former times. (Modified from Weinmann, J. W., *Phytanthoza—Iconographia Oder Eigentliche Vorstellung Etlicher Tausend, So wohl Einheimisch- als Ausländischer, aus allen vier Welt-Theilen, in Verlauf vieler Jahre gesammelter Pflantzen, Bäume, Stauden, Kräuter, Blumen, Früchten und Schwämme*, Regensburg, Germany, 1742.)

In 1964, Müntzing and Bose (1969) treated dry seeds of three different inbred lines and population plants of the rye variety Stalrag with ethyl methanesulfonate (EMS) solution in concentrations that ranged from 0.5% to 2.5%. After treatment for 3 h at 18°C and ~20 h of rest, the seeds were sown in the field. Only seeds treated with 0.5% and 1.0% of EMS germinated. Seedlings from stronger treatment did not survive the winter. The treatments, including two more inbred lines, were repeated in the following year. This time the seeds were sown immediately after treatment, and in this way germination, seedling vigor, and survival were sufficiently good to result in a larger number of M1 plants and M2 progenies; that is, strong concentrations of

EMS reduced germination significantly. In M1 seed set was reduced, the degree of sterility again varying with the strength of the EMS treatment. The stronger concentrations of EMS resulted in obvious decreases of plant weight in M1. In contrast, there was an indication that plant weight was stimulated by weak concentrations. M2 seeds derived from the strongest EMS treatments germinated less well than those from weaker concentrations. This was found to be true of the germination of the treated seeds as well as seedling height and survival. The inbred lines, which could be considered perfectly homozygous after a long period of inbreeding, were found to differ in their frequencies of spontaneous chlorophyll mutations. They also reacted differently to the EMS treatments with regard to primary effects as well as to the frequencies of induced mutations. Tiwari and Sisodia (1978) were able to produce an amber-white seed mutant through chemical mutagenesis.

Several physical and chemical mutagens were used for inducing genetic changes in rye (see Table 7.7). For better monitoring of mutational effects, Grzesik and Nalepa (1974), Sturm (1974), and Kuckuck and Peters (1970) included inbred lines in their experiments. The latter authors found new genetic variation for self-fertility, straw length, and resorcinol content. They also regarded the utilization of self-fertile and inbreeding-tolerant rye strains as mandatory, knowing that inbreds are less vital. Gacek (1972) was able to select several short-strawed mutants (50–60 cm) useful for breeding. Aastveit (1967) and Pillipcuk (1970) demonstrated an increased genetic variability in tetraploid rye too. They described in M3 changes of plant height and

TABLE 7.7
A Literature Compilation of Some Mutation Experiments in Rye

Mutagen	Rye	Notes	Reference
Fast neutrons	Autotetraploid strains	Modification of quantitative traits and length culm	Aastveit (1967)
X-rays	Diploid strains	Short-strawed mutants	Batygin (1965)
Nitrosomethyl urea (NMH) and γ-rays	Diploid strains	Short-strawed mutants	Bogomolov and Cyganok (1971)
^{60}Co γ-rays	Tetraploid strains	Chromosome aberrations, short-strawed mutants	Gielo and Starzycki (1979)
Ethyl methanesulfonate (EMS), NMH, and γ-rays	Diploid strains and self-fertile	Chlorophyll mutants	Joshi et al. (1962)
Isopropyl methanesulfonate (isoPMS)	Diploid and tetraploid strains	Increased morphological variability	Müller (1975)
2,4-Butane sultone	Diploid strains	Cellular aberrations	Singh et al. (1972)
Diethylamine (DEA) and EMS	Diploid strains	Amber-white and chlorophyll mutations	Tiwari and Sisodia (1978)

seed setting after γ-ray treatment in Norderas and Wasa II varieties. The mutations were heritable.

The comprehensive study by Schmidt (1980) of M2 generations revealed a significant increase in variability for most of the agronomic traits compared to the control populations. Dominant and recessive mutations were described. Generally, the efficient mutagens induce the same effects in rye as found before in autogamous crops. Population rye was clearly more resistant to EMS treatments than inbred lines, as were more tetraploids than diploids, respectively (Schmidt 1980, Wölbing 1980).

Several varieties of rye derived from mutation breeding: Heinrich Roggen, Donar, Muro (*Mu*tations*roggen*) or Pollux (Germany), and Hankkijas Jussi (Finland) (see Table 7.6). The successful breeding scheme is given in Table 7.8.

Akgün and Tosun (2004) showed that perennial rye seeds were exposed to different doses of γ-rays (^{60}Co) (0, 2, 4, 6, 8, 10, 12, 14, 16, 20, 25, and 30 krad). The lethal dose (LD50) was 18–20 krad and the growth reduction dose (GR_{50}) was 12 krad. In M1 generation, as doses of γ-rays increased, the frequency of metaphase I with univalent and the frequency of irregular anaphase I cells increased. Anaphase I irregularities were especially seen as chromatin bridge. However, the frequency of anaphase I cells with laggards was higher than that of anaphase I cells with bridge at 20 and 25 krad doses of γ-rays.

Molecular approaches allowed production of specific mutations (Ohnuma et al. 2004). For example, a complementary DNA (cDNA) encoding rye seed chitinase (RSC) was cloned by rapid amplification of cDNA ends and PCR procedures. It consisted of 1191 nucleotides, and encodes an open reading frame of 321 amino acid residues. Recombinant RSC (rRSC) was produced in the oxidative cytoplasm of *Escherichia coli* in a soluble form by inducing bacteria at a low temperature (20°C). Purified rRSC showed properties similar to the original enzyme from rye seeds in terms of chitinase activity toward a soluble substrate, glycolchitin, and an insoluble substrate, chitin beads; in terms of chitin-binding ability to chitin; and in terms of antifungal activity against *Trichoderma* sp. *in vitro*. rRSC mutants were subsequently produced and purified by the same procedures as those for rRSC. Mutation of Trp23 to Ala decreased chitinase activity toward both substrates and impaired chitin-binding ability. Furthermore, the antifungal activity of this mutant was weakened with increase in the NaCl concentration in the culture medium. Complete abolishment of both activities was observed upon the mutation of Glu126 to Gln.

Frequencies of mutations were studied by Figueiras et al. (1991). Cytogenetic analysis of the rye cultivar Ailés and its derivatives resulted in high chromosome (2.05×10^{-2}) and gene (9.3×10^{-3}) mutation frequencies. Highly repetitive DNA extremely varies (Cuadrado et al. 1995). It was demonstrated by the highly repetitive rye DNA sequences pScl19.2, pSc74, and pSc34, and the probes pTa71 and pSc794 containing the 25S–5.8S–18S ribosomal DNA (rDNA) nucleolus organizer region (NOR) and the 5S rDNA multigene families, respectively. The authors suggested the existence of a transposon system responsible for this instability. Recently, Liu et al. (2008) published a new sequence, $pSaD15_{940}$, derived from *S. africanum* and showing no homology to any of the known clones. It is highly diagnostic for detection of *S. africanum* DNA in certain wheat background.

TABLE 7.8
Simplified Scheme of Mutation Breeding in Diploid Rye

Year	Generation	Activities			
1	M1	Treatment of advanced breeding strains or population varieties with mutagen, e.g., isoPMS			
2	M2	Selection for target character, e.g., plant height			
3	M3	First cross-pollination among mutant lines	Line selection		
4	M4	Second cross-pollination among mutant lines	Line maintenance		
5	M5	Growing mutant lines on isolated plots for single plant selection			
6		Strain formation A1		Strain formation 2	Strain formation
		Approach 1		Approach 2	
7		A strains		A strains	A strains
8		A' strains		multiple topcross	
9		B strains		Offspring	
10		Elites		Elites	

Source: Wölbing, L., Erzeugung und Verwendung künstlich induzierter Mutanten in der Winterroggenzüchtung. PhD thesis, Akad. Landw. DDR, Berlin, 1980. With permission.

7.1.6 INTERSPECIFIC AND INTERGENERIC CROSSES

7.1.6.1 Interspecific Crosses

7.1.6.1.1 Secale cereale × primitive rye

Genetic diversity of elite breeding material can be increased by introgression of exotic germplasm to ensure long-term selection response. Starting from a cross between inbred line L2053-N (recurrent parent) and a heterozygous Iranian primitive population Altevogt 14160 (donor), two backcross and three selfing generations were performed to establish molecular introgression libraries A and B. AFLPs and SSRs were employed to select and characterize candidate introgression lines [preisogenic lines (ILs)] from BC_1 to BC_2S_3. Among others, several pre-ILs were selected showing significantly higher pentosane content. Two pre-ILs showed significantly higher starch content than the elite recurrent parent. The results indicated that exotic genetic resources in rye carry favorable alleles for quality traits of baking, which can be exploited for improving elite breeding material, predominantly by marker-assisted selection. Introgression libraries can substantially foster breeding programs and provide a promising opportunity to proceed toward functional genomics (Falke et al. 2008).

7.1.6.1.2 Secale silvestre

This species and *S. vavilovii* were used as the source of short-strawed genes. Since the 1970s, it became the main breeding goal to reduce plant height in rye. Besides natural and induced mutations, interspecific crosses seemed to be an alternative (Nürnberg-Krüger 1974). Several derivatives with short stature were isolated from these crosses. However, their general yield performance was weak. They could not be included in advanced breeding programs.

Secale silvestre is also a species that shows grain characteristics that are of interest for breeders (Shang et al. 2005). Three low-molecular-weight (LMW) glutenin-like genes (designated as *Ssy1*, *Ssy2*, and *Ssy3*) could be isolated and characterized. The three genes consist of a predicted highly conservative signal peptide with 20 amino acids, a short N-terminal region with 13 amino acids, a highly variable repetitive domain, and a less variable C-terminal domain. The deduced amino acid sequences of the three genes were the LMW-m type due to a methionine residue at the N-terminus. Phylogenetic analysis indicated that the prolamin genes could be perfectly clustered into five groups, including high-molecular-weight (HMW)-glutenin subunit (GS), LMW-GS, and α/β-, γ-, and κ-prolamin. The LMW glutenin-like genes of *S. silvestre* are more orthologous to the LMW-GS genes of wheat and B-hordein genes of barley, which was confirmed by homology analysis with the LMW-GS of wheat at *Glu-A3*, *Glu-B3*, and *Glu-D3* loci. The results indicated that a *Glu-R3*-related locus might be located on the R^{sil} genome with functions similar to the *Glu-3* locus in wheat and its related species.

7.1.6.1.3 Secale africanum

This species also serves as a valuable source of increasing the diversity of cultivated rye (*S. cereale* L.) and provides novel genes for wheat improvement (Lei et al. 2012). Schiemann and Nürnberg-Krüger (1952) carried out intensive crosses with *S. africanum*. Viable seeds are formed when *S. cereale* is the seed parent. No viable seeds develop from the reciprocal cross. Histological comparisons of developing *S. africanum* × *S. cereale* seeds with those of *S. africanum* indicated that the ultimate cause of seed failure was abnormal endosperm development. The endosperm became cellular and contained small starch grains in a few cells. The endosperm cells progressively collapsed and deteriorated. No abnormalities in maternal tissues were observed. When excised embryos were cultured in nutrient-supplemented agar medium, some grew into vigorous hybrid plants when transferred to soil, which thus indicated that the embryos from this cross are potentially viable (Price 1955, 1975).

7.1.6.1.4 Secale montanum *(in German: Waldroggen,*
Staudenroggen, Johannisroggen)

Rye appears to be at least as promising a candidate for perennialization. Its chromosomes are homologous with those of its direct perennial ancestor, *S. montanum* (see Section 2.2). Both species are diploid and cross-pollinated. They are winter-hardy, well adapted for grazing, and useful in weed control because of their allelopathic properties. Russian breeders started programs for perennial varieties during, 1940s. In the North Caucasus region, perennial dual-purpose fodder/grain rye was common for decades. It was based on introgression of *S. kuprijanovii*. There are reports that Circassians took their perennial rye seeds to Turkey when they emigrated in 1905.

Cytologically, a chain of translocations involving three chromosomes separates the two species. Gene(s) of *S. montanum* governing perenniality are located on one or more of the translocated chromosomes. Because meiosis in plants heterozygous for one or more translocations produces many inviable gametes with duplications or deficiencies of chromosomal segments, plants in interspecific rye populations fall into one of three categories:

- Homozygous for the *S. cereale* chromosomal arrangement (fertile, annual)
- Heterozygous for one or more of the translocations (highly sterile)
- Homozygous for the *S. montanum* arrangement (perennial, fertile)

Plants of the last category would seem to answer the breeder's need. However, they are rare and a large portion of their genomic content is derived from *S. montanum*. This reduces their agronomic desirability and spike fertility. According to Reimann-Philipp (1995), seed set in *S. montanum* itself is low (~80%). He selected intensively for the *S. cereale* phenotype within an interspecific, perennial population. This population was presumed to be homozygous for the three *S. montanum*-derived chromosomes involved in the translocations, identified as 4R^mon, 6R^mon, and 7R^mon by Koller and Zeller (1976) but referred to as 2R^mon, 6R^mon, and 7R^mon by Reimann-Philipp (1995). He was attempting to keep these chromosomes fixed while completely restoring the other four chromosome pairs from *S. cereale* through recombination and selection. However, he could not achieve a TGW greater than 15 g (compared with typical values of 40 g for annual rye under those conditions).

Dierks and Reimann-Philipp (1966) had postulated that a single gene that lay approximately 10 crossover units from one of the break points governed perenniality. Selection for perenniality would be routine, but selection for the *S. cereale* chromosomal constitution would require either a laborious testcross procedure or a cytological test. A morphological difference between chromosomes 6R and 6R^{mon} (Reimann-Philipp and Rhode 1968) did not prove satisfactory for this purpose. Today, the extensive genetic maps of rye would allow marker-assisted selection for the *S. cereale* arrangement.

Systematic breeding for perennial varieties from *S. cereale* × *S. montanum* crosses was initiated as early as 1906 (Tschermak 1906, Derzhavin 1938). The Russian perennial grains program included a large effort in rye (Derzhavin 1960), and they produced some weakly perennial genotypes that were used in limited production (Wagoner 1990). In Hungary, G. Kotvics (1970) began hybridization experiments during the 1950s. Later, a decade-long effort to breed perennial rye in Germany met with only partial success (Reimann-Philipp 1995, see Section 7.1.6.1).

L. F. Myers and R. J. Kirchner used another breeding strategy in the breeding of Black Mountain perennial rye in Australia (Freer et al. 1997). They focused their activities on backcrossing the interspecific hybrid twice with the *S. montanum* parent. Perennialism was quickly restored by backcrossing. However, their cultivar was intended primarily as a forage grass, with *S. cereale* donating genes for non-shattering rachis and improved seed production. Oram (1996) practiced six cycles of half-sib family selection for grain and forage yield in Black Mountain, achieving gains in both traits while maintaining a low level of shattering.

Recently a perennial rye cultivar, Perenne, was released in Hungary for grain and forage production (Hodosne-Kotvics et al. 1999). The crude protein content of 18% in Perenne grains was higher than that in annual varieties (~12%), and contained more crude fiber, crude fat, and ash than in the annuals. Regarding the quantity and composition of amino acids, Perenne showed values between *S. montanum* (high amino acid content) and *S. cereale* (lower amino acid content).

While its farinograph softening value was inferior to those of the annual varieties, its flour, mixed with wheat flour, outperformed them. In terms of other properties (falling number, farinograph water absorption capacity, baking test), the Perenne flour, either in mixture or in pure form, was not left behind in the annual varieties.

Pure perennials can also be used for bread making since they have great grain composition (Füle et al. 2005). However, as a new crop perennials have some problems with agronomical technology. Compared to common rye, germination of seeds is very less (50%–70%).

The development of a new forage mixture was initiated with perennial rye (designated as *S. cereanum*) and perennial alfalfa in the Research Center of Nyiregyhaza, Hungary. The two plants with different fodder values in the mixture are complementary to each other from the point of view of dietetics; the mixture produces great mass fodder with digestible fiber and high quantity of crude protein. In the second year, perennial rye produced 30–35 t/ha green mass, but the mixed stand yielded 78–85 t/ha green mass (Halasz and Sipos 2007). The variety Kriszta, that was developed in 1998 by J. Kruppa (Debrecen) and used in the experiment, shows good general performance.

Owing to its high green productivity, low seed productivity, and perennial type, it is utilized as a green fodder, grazing land, or grassland. This variety can be in culture 3–5 years; after this time the stand grows thinner and thinner, and gets weedy. Its seed production is ~500–1000 kg per year. The plant is 150–180 cm high and the 1000-kernel weight is 12–15 g. It can be sown with 3–3.5 million seeds in September and in springtime in March too. Its drought resistance is very good and this perennial crop adapts well to the disadvantageous soil and weather conditions, so it can be sown also in weak habitats and loose sandy soils. It covers the fields the whole year. Attributable to the fast shooting and excellent stooling (the number of its shoots can reach 100), this rye is instrumental in protecting loose sandy soils from erosion and deflation. It fixes the surface of the land. Expectedly, its role will extend to game farming too, because it is exceedingly suitable as winter pasture. It can be grazed also in winter and it tillers fast early in spring.

All breeding activities with perennial rye are faced with the problem of inter-crossing with normal rye. Reimann-Philipp (1995) warned of a hazard when grow-ing diploid perennial rye with the *S. montanum* chromosomal arrangement on a field scale. If pollen from the perennial drifts into seed production fields or breeding nurseries of annual rye, the resulting translocation heterozygosity would seriously and irreversibly degrade fertility in subsequent generations. The utilization of peren-nial rye remains restricted to niches of agriculture.

7.1.6.1.5   Secale vavilovii

Fragile rachis, tiny seeds, very weak straw, and susceptibility to all *Puccinia* pathotypes as well as powdery mildew are the detrimental characteristics of this annual wild species. However, its self-fertility was of interest for breeders. Kuckuck (1973) collected several ecotypes in Iran. Some of them were included as males in crosses with *S. cereale* var. Heines Hellkorn. After about seven inbreeding generations, he was able to select self-fertile stocks showing the "cultivated" habit. Later, the author proposed *S. vavilovii* for broadening the genetic variability of *S. cereale* and fixing genetic stocks. Even for the production of self-fertile tetraploid rye, *S. vavilovii* was considered (Kuckuck and Peters 1977). Additional breeding work of Friedt (1979) led to promising tetraploid strains. Nevertheless, their straw length and fertility remained insufficient (see Section 7.2).

7.1.6.2   Intergeneric Crosses

Compared to polyploids, diploid rye is less suitable for alien gene introgression by crossing means. Nevertheless, there are reports of intergeneric hybridization with rye. By studying embryo culture methods, Brink et al. (1944) crossed *Hordeum jubatum* and *S. cereale*. They macerated yeast in water, which was then boiled for 30 min, and after cooling, the starchy material was removed by centrifugation. It served as a culture medium.

In recent times, Apolinarska et al. (2010) reported *Aegilops* and *Secale* crosses. The valuable genes of *Aegilops biuncialis* (UUMMR^{cer}R^{cer}), *A. ovata* (UUMMR^{cer}R^{cer}), *A. kotschyi* (UUSSR^{cer}R^{cer}), *A. variabilis* (UUSSR^{cer}R^{cer}), and *A. tauschii* (DDR^{cer}R^{cer}) were transferred to rye, by crossing *Aegilops*–rye amphi-ploids with tetraploid and diploid substitution rye. The C-banded karyotypes of the BC$_1$ and BC$_2$ generations of amphiploids with tetraploid substitution rye and BC$_1$ with diploid substitution rye revealed great variation in chromosome number and composition of hybrids. Partial plant sterility in all crossing combinations of amphiploids with tetraploid and diploid substitution rye resulted in elimination of donor genomes. Nevertheless, a few introgression lines could be recovered, demonstrating both *A. variabilis*–rye substitutions and translocations. Tetraploid ploidy level compensates introgression better than the diploid ploidy level (Kwiatek et al. 2012).

Nevertheless, Schlegel and Kynast (1988) were able to produce stable introgressions, for example, a translocation with chromosome 6B of hexaploid wheat. Later, Apolinarska (2008) was able to confirm the result by using hexaploid triticale as a donor, and Lukaszewski and Kopecky (2010) by introducing chromosome 5B of wheat into the diploid rye recipient.

7.1.6.2.1   Aegilops tauschii × Secale silvestre

Cytology and gene expression of an amphiploid between *A. tauschii*, native to China, and *S. silvestre* revealed high frequencies of aneuploids in the progenies of the amphiploid hybrid ($2n = 4x = 28$), indicating genomic incompatibility. The nucleolus organizer region of chromosome 5D of *A. tauschii* is visible in the amphiploid, but is not recognizable as that of chromosome $1R^{sil}$ of *S. silvestre*. The resistance genes of *S. silvestre* against stripe rust and powdery mildew are also suppressed in the amphidiploid and *A. tauschii* background. However, the analysis of endosperm storage proteins indicated gliadin and glutenin genes from both parents, thus clearly expressed in the rye endosperm (Yang et al. 2001, 2008b).

7.1.6.2.2   Triticum durum × Secale africanum

Another amphiploid between *Tritium durum*, native to Sichuan, China, and *S. africanum* was produced and evaluated by cytological observation, electrophoretic analysis of seed storage protein, and surveys of disease resistance. Feulgen staining and Giemsa C-banding of somatic metaphases indicated that the nucleoli from *S. africanum* were frequently suppressed in the amphiploid. Electrophoretic studies showed that most gliadin and glutenin of both parents were observed in the endosperm of the amphiploid with codominant expression. Inoculated by the stripe rust and powdery mildew isolates, the amphiploid totally expressed the resistance from *S. africanum* (Yang et al. 2000).

7.1.6.2.3   Hordeum pubiflorum × Secale africanum

A low frequency (0.86) of intergeneric rod bivalents was observed at metaphase I of the meiosis in the F_1 hybrid between diploid *H. pubiflorum* and *S. africanum* (Fedak 1985). It confirms true chromosome pairing, chiasma formation, and recombination between the species. Seeds were obtained from colchicine-doubled sectors on the hybrid. An average of 2.49 univalents plus 12.77 bivalents per cell characterized meiosis in the amphiploid. The univalent and bivalent frequencies among *Hordeum* and *Secale* chromosomes were approximately equal, although a high proportion of ring bivalents were formed by the *Hordeum* chromosomes. Over 90% of the pollen was stainable, but only 40% seed set was obtained on the amphiploid. Although the seeds were badly shriveled, some were viable.

7.1.6.2.4   Thinopyrum distichum × Secale cereale

Marais et al. (1998a) made attempts to determine whether the salt-tolerant genes of *Thinopyrum distichum* were expressed in the presence of the rye genomes in order to breed a salt-tolerant fodder crop. *Thinopyrum distichum* florets were pollinated with diploid and tetraploid rye, but hybrids were obtained only with tetraploid rye. The F_1 plants ($2n = 4x = 28$) expressed high levels of salt tolerance. Colchicine treatment failed but some F_1 plants produced apomictic seeds (see Section 2.6.4). The F_2 plants showed 20–22 chromosomes and retained two complete or near-complete rye genomes. In F_2, plants responded better to colchicine treatment; however, the doubled hybrids remained male-sterile. Again, a single F_2 plant produced many apomictic seeds with chromosome numbers ranging from 14 to 16 (primarily rye chromosomes). The mechanism that caused the preferential elimination of *Thinopyrum* chromosomes remains unknown. The F_3 phenotypically resembled rye, had primarily reductional male meiosis, and produced virtually functional pollen, but were self-sterile. Fedak and Armstrong (1986) found a total of 16 seeds from 250 pollinating florets of *S. cereale* × *T. intermedium* and only 5 of them contained embryo whereas strikingly only two germinated to produce seedlings.

7.1.6.2.5   Secale cereale × Triticum ssp.

In many taxa, introgression is a more or less frequent phenomenon, depending on the spontaneous tendency to interspecific hybridization. As hybridization played an important role in the evolution of plant speciation, it has been suggested that interspecific crosses will offer better results in taxa in which such evolutionary mechanisms have been common. Only few experiments have been successful with diploid crop plants, although the improvement of their performance is as important as in polyploids.

By crossing hexaploid wheat and diploid rye and subsequent backcrossing, several monosomic and monotelosomic rye–wheat chromosome additions were produced (Schlegel 1982b, Schlegel et al. 1986a, Apolinarska 2008) using rye cytoplasm as background. However, during the first experiments also alloplasmic rye–wheat additions were available, when, in crosses with rye, the hexaploid spring wheat Chinese Spring was used as female.

Monosomic additions were maintained by backcrossing to the euploid rye parent and subsequent microscopic screening. Because of missing male transmission of the extra chromosome, self-pollination did not result in disomic addition lines. Characteristic morphological and physiological distinctness could be detected depending on the alien chromosome added to rye. Since the population size was too small, no agronomic performance testing was possible.

In order to transfer wheat chromosomal segments and/or single genes into the rye complement, experiments were designed to take advantage of gametic selection after spike irradiation, which would facilitate the recovery of desirable translocations. Thus, premeiotic spikes of several addition lines were exposed to X-rays of 1000 rad dosage.

The resulting pollen grains were used to fertilize emasculated euplasmic spring rye plants. The resulting F_1 seedlings were screened by chromosomal banding techniques for the presence of alien chromatin and further characterized by meiotic analysis. A successful example of gene transfer could be demonstrated involving chromosome 6B of hexaploid wheat, var. Chinese Spring. Thirty-six treated spikes used for pollination gave 364 seeds for cytological screening. Since male transmission of the intact extra chromosome could not be fully excluded, all of the progenies were checked for somatic chromosome number. Among 188 plants analyzed, 169 (90%) individuals had 14 chromosomes, 1 (0.5%) 14 + t^{wheat} chromosomes, 1 (0.5%) 13 chromosomes, 1 (0.5%) 28 chromosomes, and 16 (8.5%) 14 + t^{wheat} chromosomes. Thus, the added wheat chromosome can be transmitted, at least, to some extent through the pollen cells, although previous studies of selfed alloplasmic 6B additions never yielded disomic progeny. By application of differential chromosome staining, all 15-chromosome plants were classified as rye plus one wheat chromosome 6B. However, among the 14-chromosome progeny, two types were found to have novel sites of N bands compared to rye. One of the two plantlets showed in all somatic cells, in addition to the 13 normal rye chromosomes, one prominent chromosome with a heavy N band situated near the centromere. This band did not match any of the rye-specific N bands. Its location and size resemble the heterochromatic block of chromosome 6B. More studies that are detailed revealed a dicentric structure of the rye–wheat translocation. The results provide clear evidence of transfer of alien genetic material to diploid rye. It is tolerated not only in the embryo but also in adult plants. It also demonstrates that wheat genes can compensate for the loss of rye genetic information (Schlegel and Kynast 1988).

Meanwhile, other introgression lines were produced, such as 1RL.1DL (the introgressed wheat segment contains locus *Glu-D1* with allele *d*, encoding HMW-GS 5+10). The introgression substitutes the rye locus *Sec-3*, encoding secalin. The introgression does not affect plant morphology or fertility but may lower test weight. The introgression increases the sodium dodecyl sulfate (SDS)-sedimentation volume by ~75% and improves bread loaf volume in tests performed using protocols for wheat–rye blends. Parameters of bread-making quality tested using procedures for rye flour (see Section 9.3) are not affected by the introgression (Lukaszewski et al. 2000).

7.1.6.2.6   Secale cereale × Haynaldia villosa

Gruszecka and Piettrusiak (2001) reported about rye–*Haynaldia villosa* (syn. *Dasypyrum villosum*) chromosomal additions and translocation lines. Modified morphological characters of backcross progenies demonstrated the introgression, but no cytogenetic evidence was provided.

7.1.6.2.7   Secale cereale × Hordeum ssp.

Several workers have obtained *Secale* × *Hordeum* hybrids. Brink et al. (1944) reported a first cross between *S. cereale*, var. Imperial, and *H. jubatum*. Later Wagenaar (1959) described another one. In hybrids between *H. depressum* and species of *Secale*, which have chromosomes considerably larger than the *Hordeum* chromosomes, almost complete pairing of the two sets of chromosomes of *H. depressum* was observed (Schooler 1967, Petersen 1991). In the hybrids between diploid species of *Secale* and tetraploid species of *Hordeum*, believed to be segmental allotetraploids (i.e., modified H genomes), it can be directly observed that only the chromosomes from the *Hordeum* genomes are involved in chiasma formation.

Fedak (1978) reported a first viable hybrid of *H. vulgare* × *S. cereale* when he injected GA$_3$ to the pollinated florets. Later he obtained a hybrid between *H. vulgare*, var. Betzes ($2n = 2x = 14$), and *S. vavilovii* ($2n = 2x = 14$) in which chromosome instability was observed in somatic and meiotic tissues. In somatic tissue, the chromosomes per cell varied from 7 to 24 with a mean of 19.7 (Fedak and Nakamura 1982). Similarly, in meiotic tissue, the chromosome number varied from 14 to 26 with a mean of 18.3. The mean chiasma frequency was 12.9 consisting of an average M1 configuration of $0.02^{IV} + 0.3^{III} + 6.68^{II} + 3.92^{I}$. It was concluded that the hybrid was derived from the union of an unreduced gamete from Betzes barley with a normal gamete from *S. vavilovii*.

7.2   TETRAPLOID RYE

With the 1937 discovery of the colchicine method for inducing polyploidy, development of polyploid crops was regarded as an unconventional technique to penetrate yield and other barriers in plant breeding. The two universal effects of chromosomal doubling are increased cell size and decreased fertility. Consequently, crops that benefit most from increased cell size and suffer least from reduced fertility are inherently predisposed to benefit from polyploid breeding. Crops most amenable to improvement through chromosome doubling should (1) have a low chromosome number, (2) be harvested primarily for their vegetative parts, and (3) be cross-pollinating. Two other conditions, the perennial habit and vegetative reproduction, have a bearing on the success of polyploid breeding by reducing a crop's dependence on seed production. In the 1950s–1970s, induced autopolyploidy in rye was considered to be an important breeding method. Rye showed good prerequisites for an autopolyploid crop. Russia, Poland, Germany, and Sweden spent considerable attention to the production of tetraploid rye. It resulted in varieties such as Belta (Belarus) or Tero (Germany). However, the common aneuploidy of ~10%–30% of the tetraploids could not be substantially reduced. Interchromosomal substitutions as a source of irregular chromosome pairing and aneuploidy, that is, aneusomy, could be excluded as well (Schlegel et al. 1985).

In Germany (Institute of Plant Breeding, Hohenthurm, near Halle (Saale)), a breeding program of pale-grained tetraploid rye led to suitable strains but the milling industry and parallel cropping of diploid and tetraploid varieties prevented the official release of those tetraploids. Already in 1949, Plarre (1954) started an approach with tetraploid rye, which had been obtained from the Plant Breeding Station at Petkus where R. von Sengbusch had produced the tetraploids by colchicine treatment in 1938. The seed set of tetraploid rye averaged 71%, while diploid controls showed ~83%. Nearly 90% of the tetraploids were found to have meiotic irregularities of different kind. A weak positive correlation between the increased seed set and the decreased irregularity in metaphase I was observed. W. Plarre used these data for cytological preselection in an experiment with cloned individuals as well as controlled pollination and fertilization. The effect of intercrossing of cytologically preselected plants amounted to 5%–8% higher seed set. Walther (1959) used the same tetraploid material at Hohenthurm. However, he did not confirm the correlation between cytology and fertility. He considered physiological reasons for partial sterility. Müller (1969) initiated another approach targeted toward tetraploids with reduced plant height. The diploid rye Danae was doubled by means of colchicine in 1961. In 1966, the plant height was compared between the tetraploid rye Hohenthurm St. A (a strain from Plarre and Walther, Hohenthurm) and the tetraploid Danae. The plant height was 135 and 103 cm, respectively, while the records for seed set were 65% and 60%, respectively.

The idea of breeding pale-grained rye met the increasingly changing consumer taste for white bread already during the 1950s (Roemer and Rudorf 1950). Actually, it was a smart idea, that is, to combine the higher nutritional value of rye flour with the brighter color; instead to use pure wheat bread, at least in countries preferring rye bread.

In order to overcome low fertility, seed shriveling, and aneuploid offspring—all three features were believed to be influenced by irregular meiotic chromosome pairing—several attempts were made. Four major approaches of chromosomal pairing regulation were investigated:

1. The research group belonging to H. Rees at Aberystwyth (UK) (Hazarika and Rees 1967) favored an increased quadrivalent formation with convergent or parallel centromere coorientation as a mean of reduced aneuploidy and consequently improved fertility. A "disjunction index" [the number of pollen mother cells (PMCs) without univalents and trivalents divided by the total number of PMCs] was used as a measure for meiotic pairing regularity. They even demonstrated a positive correlation between the disjunction index and the fertility.

2. Alternatively, Sybenga (1964) proposed a preferential bivalent pairing in order to reduce meiotic irregularities by induced allopolyploidization of an autotetraploid. Besides a genetic control of diploid-like chromosome pairing, a complex system of experimentally induced reciprocal translocations was intensively discussed. As Reimann-Philipp and Eichhorn-Rhode (1970) and Chapman (1984) examined F_1 hybrids between *S. cereale* and *S. montanum* where tetraploidy had been induced by the use of colchicine, the frequency of ring bivalents was higher, and the average size of multivalents was lower than a model assuming random pairing of chromosome arms predicts. The author interpreted it as evidence for preferential chromosome pairing. Recent observation shows that the *Ph1* locus on wheat chromosome 5B, which enforces strictly bivalent pairing in polyploid wheat, may also control chromosome pairing in autotetraploid rye by apparently restricting chiasma formation between dissimilar homologs. Unlike in wheat, the effect appears to be dosage dependent, which may be a reflection of an interaction between *Ph1* and the rye chromosome pairing control system. With two doses of *Ph1* present, chiasmatic pairing was severely restricted resulting in a significantly higher number of univalents and bivalents per cell than in the controls (Lukaszewski and Kopecky 2010).

3. Schlegel (1976) introduced a method for mass production of autotetraploid rye by the so-called valence crosses. Under microplot isolators, nonemasculated tetraploid genotypes (clonal plants) were crossed by spontaneous

pollination with diploid genotypes (clonal plants). The tetraploids used as mother plants showed a recessive controlled pale grain character, while the diploids used as male plants showed dominantly green seeds. In this way, green xenia could be selected among the pale grains after harvest of the mother plants. All the green xenia must be hybrids, either triploid or tetraploid. The triploids result from fusion of a diploid female gamete and a haploid male gamete, while the tetraploids are derived from a diploid female and an unreduced male gamete (Schlegel and Mettin 1975c). Microscopic chromosome counting can separate triploids from tetraploids.

In this way, diploid genotypes showing high chiasma frequencies per PMC (>17 Xta per PMC) and other useful agronomic characters were transferred into the so-called half-meiotic tetraploids. The method proved to be efficient for broadening genetic variability of tetraploid breeding material without deleterious effects of colchicine treatment. Until this time, all tetraploids available worldwide were based on colchicine-induced genotypes. It was a very limited number and genetically confined. Moreover, by utilizing diploid genotypes with useful agronomic and cytogenetic characters the resulting tetraploids could be modified according to the breeder's need. The application of diploid males with high chiasma frequency for tetraploidization was done with the intention to stabilize chromosome pairing of the resulting tetraploids.

After numerous experiments, the higher chiasma frequency of the diploid parent did not significantly influence the number of multivalents of the tetraploids, despite the missing correlation between quadrivalent frequency, aneuploid frequency, and fertility, at least in advanced breeding strains. However, depending on the karyological structure of the diploid parent, a more or less strong modification of the bivalent frequencies could be observed, that is, the bigger the differentiation of the chromosome structure between the parental genotypes the higher was the number of bivalents per PMC. Although just minor changes of the karyotype have been considered, they contributed to a preferential bivalent pairing, obviously within the parental diploid genomes of both female and male donors (see Figure 7.17). In some way, it confirmed the theory of directed manipulation of tetraploids' chromosome pairing as proposed by Sybenga (1964).

4. Another approach was proposed by Apolinarska (2003). She produced tetraploid rye with single-substitution wheat chromosomes—1A, 2A, 5A, 6A, 7A, 3B, 5B, and 7B. The study showed that the rye genome is closer to the wheat genome A in relation than to genome B, respectively. She also found that the wheat chromosomes influenced the pairing of rye genome chromosomes, as well as the frequency of ring and rod bivalents and tri- and quadrivalents.

Summarizing the data of over 60 years of research, neither an increased chiasma frequency and/or higher frequency of quadrivalents per PMC nor diploidization

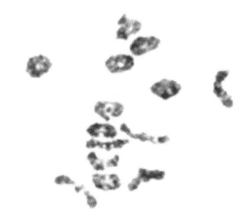

FIGURE 7.17 Metaphase I chromosome pairing of autotetraploid rye, $2n = 4x = 28$ (*Secale cereale* L.) after Feulgen staining, showing preferential bivalent associations (14^{II}).

mechanisms substantially contributed to the breeding progress of tetraploids. Gradual selection for yield, fertility, and low degree of shriveling are more successful than experimentally induced genetic or cytological changes.

7.2.1 Perennial Tetraploids

In an effort to improve the kernel weight of tetraploid rye, Reimann-Philipp (1995) used colchicine to double the chromosome number of a perennial diploid *S. cereale* × *S. montanum* population homozygous for the $4R^{mon}$, $6R^{mon}$, and $7R^{mon}$ translocated chromosomes. The resulting tetraploid, named Permontra, had a 1000-kernel weight of approximately 30 g (double that of the diploid), and first-year grain yields over 20 dt/ha when grown in Germany. Yields declined in subsequent years (Reimann-Philipp, 1986). Permontra became attractive to the steppe regions of southwest Asia rather than to Europe. Actually, this hybrid variety was developed in Germany for use as a cereal crop for human consumption or as a perennial forage or hay crop. It was first planted in the United States in 1990, with the first harvest in 1991. Permontra achieved a similar grain yield at the Land Institute's plots in Kansas, but only 15%–20% of the plants regrew in the next season (Piper 1993). After a first-year harvest in the hot, dry summer of 2001, a stand of Permontra died out completely by September. Another problem with Permontra was its poor seed set—as common in annual tetraploid rye. The formation of numerous multivalent associations led to production of gametes with extra or missing chromosomes.

Further selection was thought to improve the seed set of the material. Bremer and Bremer-Reinders (1954) and Reimann-Philipp (1995) pointed out that selection for seed set and meiotic stability could be practiced much more effectively in a perennial, by screening phenotypically or cytologically in one flowering cycle and intercrossing selected plants in the next. This author demonstrated that phenotypic selection greatly improved the seed set in the perennial spring rye

Soperta, which was derived from seven Permontra plants. Soperta did not require vernalization in order to flower (see Section 3.7), that is, it showed spring-type growth habit.

Tetraploid perennials may provide additional advantages beyond increased kernel weight. Reimann-Philipp (1995) found in one study that Permontra had greater heat and drought tolerance (see Section 3.4) and a much more extensive root system than diploid or tetraploid annuals (Gordon-Werner and Dorffling 1988). The latest approach of introducing perennial tetraploids was reported by Akgün et al. (2011). By applying 0.2% colchicine solution at 24°C, they produced sufficient tetraploids for selection. The low seed set rate was their main breeding problem. However, by subsequent assortment, fertility could be increased from 30% at C1 to 61% at C4.

As a consequence, both diploid (see Section 7.1.6.1) and perennial tetraploids of rye remain restricted to niches in agriculture.

7.3 DUAL-PURPOSE RYE

Under extensive growth conditions in the United States and other countries, rye was tested as a dual-purpose crop. In the southern country of Brazil, the cropping area has decreased substantially in the last five decades, but rye still presents an important potential in cereal production, mainly as pasture or soil cover and for human food. Breeding efforts were aimed at yield improvement (grain and green forage), resistance, and wide adaptation. In 2008, a new variety BRS Serrano was derived from a cross made at Embrapa Trigo during the winter of 1998 between Garcia and Bagé rye populations. These populations were selected in a field trial of colonial rye genotypes for agronomic and forage evaluation. After three cycles of open pollination, the population of BRS Serrano was established. The genetic seed multiplication process began in 2000 and continued until 2001. Between 2002 and 2004, the population was described for "distinctness, uniformity and stability" according to Union internationale pour la protection des obtentions végétales (UPOV) and evaluated in field trials for dual purpose (forage and grain production) in seven distinct environments. During this period, BRS Serrano produced 10.7 t/ha dry matter, 30% higher than the yield of rye BR1 (check cultivar), and the potential was higher than 12.0 t/ha. BRS Serrano is diploid and a spring type, has a high stature, and is medium–late ear emergence and maturity. In addition, it is highly tolerant to soil aluminum toxicity (see Section 3.4.2), resistant to leaf rust, powdery mildew, spot blotch, *Septoria* leaf blotch, and barley yellow dwarf virus (BYDV) as well as moderately resistant to scab and grain shedding but susceptible to stem rust and lodging.

7.4 BREEDING ACTIVITIES, VARIETIES, AND INSTITUTIONS WORLDWIDE

European and Russian breeding activities are already outlined in Sections 7.1–7.3 (see also Table 7.6). Poland as a main grower is paid individual attention. In the fifth century BC, rye was brought to Poland. Therefore, rye improvement by breeding has a long tradition. The first known Polish cultivar Szklane was grown in 1835 on

Suchorzewski farm near Wrzesnia in western Poland (Arseniuk and Oleksiak 2003). So far, over 80 cultivars were under production. The first breeding companies that also were involved in rye breeding were established in 1880 by Wladyslaw Zelenski in Grodkowice nearby Cracow and by Aleksander Janasz in Dankow. Cultivars with an epithet Dańkowskie are known for about a century. Dańkowskie Selekcyjne was the first cultivar released after a purposeful crossing of parental forms. This cultivar was grown commercially over 90 years, and despite of its withdrawal from the Polish register in 1988, it is still being met on farmers' fields in Poland. The other cultivar Dańkowskie Złote was released in 1968 and is being cultivated. Its shares in commercial and certified seed production in 1999 and 2000 were 49.4% and 71%, respectively. Up to 1995 solely population cultivars were released. Poznan Plant Breeding Co. Ltd., Danko Plant Breeding Co. Ltd., and IHAR/Smolice Plant Breeding Co. Ltd. are the main breeders and competitors over the past half a century. In 2000, 21 rye cultivars were officially registered in Poland (see Table 7.6). The first hybrid cultivar was registered by German breeders in 1995 under the name Marder. Afterward, other hybrid cultivars have been appearing in the Polish market—Esprit (1996), Nawid (1998), Luco (1999), and Klawo and Ursus (2000).

Although rye is grown worldwide, its genetic diversity is rather limited. Europe was the main region where rye was subsequently developed to advanced varieties. When the genetic variation and diversity were studied for three improved varieties and 18 landraces of *S. cereale*, originating from northern Europe, by starch gel electrophoresis, most of the genetic diversity was found within the accessions (Persson and Bothmer 2002). Landraces from Germany and Norway show low genetic variation compared to landraces from Sweden and Finland. In dendrograms, Finnish landraces and 11 of the Swedish are grouped together with a very small diversity index. Thus they can almost be considered as part of the same accession. This is probably because a high number of Finish immigrants arrived in Sweden during the seventeenth century bringing their own rye material—obviously the same as the early Finnish did.

Extended genetic studies among 20 spring and 22 winter accessions of agronomically different ryes from 14 countries by employing RAPD markers surprisingly revealed just two main groups of rye: spring and winter populations. Within the spring group, cultivars fell into a North European and an American–Chinese group (Ma et al. 2001, 2004b). Cultivars of winter rye fell into four groups: Northern European, Russian, American, and Chinese populations. An Unweighted Pair Group Method with Arithmetic mean (UPGMA) dendrogram based on genetic distances of cultivars of rye within the winter and spring groups showed that the clusters corresponded well to their geographical locations. It indicates that isolation has played an important role in the evolution of cultivated rye, and that temporal isolation has influenced the genetic diversity of rye more than geographical isolation (Rui et al. 2004).

Research on rye at the United States probably started in 1863 when the US Department of Agriculture (USDA) assigned the area in Washington, DC, between 12th and 14th streets as an experimental tract by the Commissioner of Public Buildings (Moseman et al. 1993). Among the five crop investigation leaders, G. F. Sprague (maize and sorghum), G. A. Wiebe (barley), H. C. Murphy (oats), and C. R. Adair (rice), L. P. Reitz was responsible for wheat and rye. In the fall of 1865, part of the land was plowed, fertilized, and planted with 346 cultivars including 62 cultivars of winter wheat, mostly from France, Russia, Prussia, the United Kingdom, Chile, and China. In the spring of 1866, 66 cultivars of spring wheat (including *Arnautka durum*), 17 of oats, 13 of barley (including Oderbrucher), 17 of rye, 19 of maize, and 4 of sorghum were planted. In 1867, the cereal cultivars in the plots included 43 winter wheats, 66 spring wheats, 5 winter ryes, 16 spring ryes, 21 barleys, 20 oats, 10 maizes, and 3 sorghums. In the autumn of 1889, red kafir corn, several other cultivars of sorghum, and 40 acres of Arctic rye were sown.

The coordination of the research in the USDA on wheat and rye was initiated when the Bureau of Plant Industry (BPI) was organized in 1901. M. A. Carleton was designated cerealist in charge of the Cereal Laboratory in the Division of Vegetation, Physiology and Pathology. He was directly in charge of all wheat, rye, spelt, and emmer experiments from 1901 until 1913. D. D. Morey at the Coastal Plains Experiment Station at Tifton collaborated with L. P. Reitz in the growing, testing, and breeding of rye from ca. 1955 until the 1972 reorganization. He developed and released some diploid and tetraploid rye cultivars that were grown in the southeast of the United States.

From 1924 to 1928, G. F. Sprague, and from 1929 to 1932, N. E. Jodon were Cereal Investigation employees at North Platte. They were responsible for research on the production and testing of cereal crops including wheat, rye, spelt, and emmer. From 1995 to 1972, D. D. Morey conducted agronomic, production, and breeding research on rye at Tifton, GA; H. H. McKinney (1919–1925) was working on the physiology and resistance to ergot. The University of Missouri, CO, was an important center for cytogenetic and interspecific hybridization research on wheat and rye between 1936 and 1972. In the late 1920s, L. J. Stadler used both diploid and polyploid species in his pioneering X-ray mutation experiments. Cytogenetics research was started ca. 1932 by Luther Smith when he was a PhD student under Stadler. In 1936, J. G. O'Mara and E. R. Sears joined the projects. They studied procedures for chromosome doubling and behavior in diploid and amphidiploid hybrids. O'Mara pioneered the production of wheat–rye addition lines and developed the first systematic method for producing such testers. He also produced the first hexaploid triticale before leaving the project for war-related service in 1942. In 1937, Sears began producing aneuploids and exploiting them in genetic analysis. W. J. Sando (1921–1955) at the Agricultural Research Center, Beltsville, Washington, DC, successfully crossed wheat with *Agropyron*, rye, and other related genera and species. He also developed a tetraploid rye that contained a high percentage of rutin. Rutin is used to treat capillary fragility, a condition that may result in stroke.

8 Rye Cropping

Cultivated rye is derived from the wild species *Secale montanum* and *S. vavilovii* that are endemic to forest clearings and field margins of southwest Asia, up to 2500 m above sea level. Its growth at this height is the cause of the winter hardiness of rye. This contributed to its wide distribution in Central Europe, particularly during the Middle Ages. Its agronomic features such as tough rachis, three florets, lower husk wrapping, and larger seeds were indirectly selected according to the selection among the cultivars of wheat and barley. The first pure cropping of rye in Anatolia, Turkey, can be dated back to 4000 BC. Further selection in Europe led to landraces; after 1850, conscious selection began.

By learning from several inaccessibilities of early agricultural experiments, new types of precise field testing were proposed. Among them, long-term experiments of different fertilization effects on yield, processing performance, or continuous crop rotation were introduced. Julius Kühn (1825–1910) began such an experiment with rye in the nineteenth century at the University of Halle/S., Germany, which lasted until recent days (Eternal Rye Cropping). It was established in 1878 together with the first Agricultural Experiment Station at the University of Halle/S., after an earlier approach by J. B. Lawes (1814–1900) who established the Rothamsted Experimental Station on his farm in England in 1841.

Systematic field experiments began in 1845 and continued to the present day (Schlegel 2007). J. Kühn (see Figure 8.1) designed the long-term experiment as a rye monoculture in order to detect the long-term effects of mineral and manure fertilization (Böttcher et al. 2000). Since the past 135 years this rye field, of ~6000 m², continuously produces ~2.5 t/ha yield of grain, that is, about half of the fertilized plots, although year by year potassium, phosphorus, and nitrogen are taken away from the soil!

In this long-term experiment, winter rye (*S. cereale*) is continuously cropped on a parabraunerde-tschernosem derived from sandy loess. The variants of treatment are (1) manure 1, (2) -PK, (3) NPK, (4) NPK + manure since 1991, previously "N-," (5) unfertilized, and (6) manure 2 till 1952, then unfertilized. After the subdivision of the initial plots in 1962, rye was also cropped in rotation with potatoes. This was initiated by Karl Schmalfuss (1904–1976).

Recent studies showed that on the plots without any fertilization (5) where the N-input was confined to the atmospheric deposition (40 kg N/ha/year), the yield in these years amounted (in accordance with the long-term trends) to only 40% and 48%, respectively, of that on NPK plots. A similar response was observed with treatments 2 and 6. Crop rotation increased the grain yield on an average of all treatments by 9.4 dt/ha compared to continuous cropping (with 42.2 dt/ha). There were no marked differences between the treatments regarding quality, relating neither to 1000-kernel weight nor to the tested milling properties. The same was true with the processing criteria of the flour (falling number, pasting properties of starch) and with the quality

FIGURE 8.1 Julius Kühn (1825–1910), one of the founders of modern agriculture, designed the long-term experiment "Eternal Rye Cropping"; picture as young scientist in 1871. (From http://de.wikipedia.org/wiki/Julius_K%C3%BChn.)

parameters of the baked bread (bread volume, pore space continuity, crumb elasticity). Differences between the 2 years regarding pentosane content, falling number, and pasting properties of starch have likely been caused by climatic conditions. The observation that the remarkable yield differences resulting from differentiated fertilization are accompanied by only small differences in quality might be because these differences in yield are due mainly to differences in the number of kernels per square meter without any significant change in 1000-kernel weight (Böttcher et al. 2000).

Physiologically rye shows a C3 photosynthetic pathway and may serve as grain, hay, pasture, cover crop, green fodder, and green manure. Several traits have to be considered not only in breeding but also for technology, depending on utilization (see Table 8.1).

Rye is a good pioneer crop for sterile soils as well. When used as a cover crop, it is grown for erosion control, to add organic matter, to enhance soil life, and for weed suppression. It may also stabilize and prevent leaching of excess soil or manure nitrogen. It has been used to protect soil from wind erosion in exposed areas and, with its tall stature, may be of some value in providing windbreaks. It is a good green manure because it produces large quantities of organic matter. It should be used only in rotation with row crops because other grain crops are graded down in the market if they contain rye seed. Depending on the end-use, several traits have to be considered in cropping (see Table 8.2).

Rye may also contribute to nitrate (NO_3) reduction in surface water (Kaspar et al. 2007). A significant portion of the NO_3 from agricultural fields which contaminates surface waters in the Midwest of the United States is transported to streams or rivers by subsurface drainage systems (tiles). Previous research has shown that nitrogen fertilizer management alone is not sufficient for reducing NO_3 concentrations in subsurface drainage to acceptable levels. Therefore, additional approaches were tested.

TABLE 8.1

Utilization of Breeding Methods Considering the Reproduction System and Origin of Population

Reproduction		Origin of Population		Breeding Method	
Autogamous	—	Crossing	**	Mass selection	—
Allogamous	**	Wild origin	—	Pedigree method	—
Anemophilous	**	Polyploidization	**	Combination breeding	**
Self-fertile	—	Induced mutation	**	Residue method	**
Self-sterile	**			Hybrid breeding	**

Note: **, true; —, not applicable.

TABLE 8.2

Quality Characteristics of Rye in Dependence of the Type of Utilization

	Type of Utilization			
Traits	Baking	Animal Feeding	Alcohol Production	Industrial Usage
Starch content	Neutral	Neutral	Important	Important
Protein content	Neutral	Important	Neutral	Neutral
α-Amylase	Negative	Neutral	Important	Important
Pentosan content	Important	Negative	Negative	Important

In one treatment, eastern gamagrass (*Tripsacum dactyloides*) was grown in permanent 3.05-m-wide strips above the tiles. For the second treatment, a rye winter cover crop was seeded over the entire plot area each year near harvest and chemically killed before planting the following spring. Twelve subsurface-drained field plots were established in 1999 with an automated system for measuring tile flow and collecting flow-weighted samples. Both treatments and a control were initiated in 2000 and conducted from 2002 through 2005. The rye cover crop treatment significantly reduced subsurface drainage water flow-weighted NO_3 concentrations and NO_3 loads in all 4 years. The rye cover crop treatment did not significantly reduce cumulative annual drainage. Averaged over 4 years, the rye cover crop reduced flow-weighted NO_3 concentrations by 59% and loads by 61%. The gamagrass strips did not significantly reduce cumulative drainage. Thus, rye winter cover crops grown after maize and soybean have the potential to reduce NO_3 concentrations and loads delivered to surface waters by subsurface drainage systems.

8.1 NO-TILL RYE

When rye is cropped a lot of leaf residues remain on the soil surface. Therefore, no-till agriculture can be an effective way to avoid erosion and help control weeds. Mowing and using a burn-down herbicide are the two common methods of killing a rye cover crop for no-till plantings. Killing rye by mowing should be done at the flowering stage when the anthers are extended and the pollen fall from the seed heads when shaken. If mowing is done earlier, the rye simply grows back. In order to make no-till planting effective, a good stand of crop is a prerequisite.

8.2 SEEDING

Rye crop can be established when seeded as late as October 1. It grows better in cooler weather. Minimal temperatures for germinating rye seed have variously been given as 1°C to 5°C. Soil fertility influences germination. Soil fertility is the ability of a soil to supply plant nutrients. When there is sufficient nutrient supply for the crop tillers well (see Section 2.3). Rye is sown in closed drills or ordinary drills. Per hectare, 4.5 to 6 million of germinable seeds are sown. A general formula for seed density that can be applied is given as follows:

$$\text{Seed amount per hectare} = \text{1000-grain weight (g)} \times \text{germability of seeds (\%)} \times \text{seeds density (germable seeds/m}^2)$$

Rye seeds are planted at a depth of 4–6 cm. The recommended seeding depths include 2.5–5.0 cm. The suggested seeding rates are 70–110 kg/ha. The optimal seed rate for hybrid material is ~150 seeds per m^2. When rye is seeded late, the rate should be increased up to 330 kg/ha to achieve rapid and complete vegetation cover and reduce erosion. To minimize erosion, a leaf area index of 1.0, that is, complete cover, may be necessary. In general, a spike density of ~500 spikes per m^2 is sought.

As a general recommendation, the following parameters should be considered for optimal yields in rye (see Table 8.3).

TABLE 8.3
Recommendation for an Optimal Yield Structure of Rye Depending on Target Yields

Target Yield (dt/ha)	50–60	70–90	90–110
Soil fertility (C_t content, %)	25–35	30–40	40–55
Seedlings per m^2	120–140	150	200
Culms per plant	3–4	4–5	5–6
Tillering culms per m^2	500–650	600–900	1000–1100
Spikes per m^2	300–450	400–500	500–600
Seeds per m^2	12,000–13,000	15,000–17,000	20,000–24,000
TGW (g)	35–42	38–46	42–48
Spike yield (g)	1.3–1.5	1.4–1.8	1.6–2.0

Source: Dennert, J., and Fischbeck, G., *Pflanzenbauwiss*, 2, 2–8, 1999. With permission.

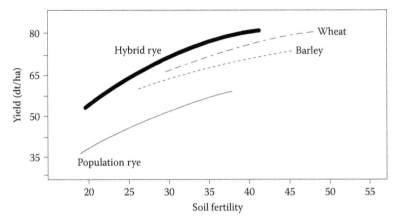

FIGURE 8.2 Yield expectation of population rye, hybrid rye, winter wheat, and winter barley under different soil conditions in Central Europe. (Courtesy of Anonymous, Landesforschungs-sanstalt, Mecklenburg-Vorpommern, Germany, http://www.landwirtschaft-mv.de/.)

8.2.1 SOIL TYPE

Rye grows on different kinds of soil except waterlogged soils. It thrives in black earth and grows best in well-drained loam or clay loam soils, but even heavy clays, light sands, and infertile or poorly drained soils are feasible. It will grow in soils too poor to produce other grains or clover. In light sandy soils rye can produce more than 8 t/ha in several years when hybrid varieties are grown (see Figure 8.2). The yield potential is even higher compared to winter wheat and winter barley when soil fertility is lower. Wheat is a serious competitor in medium and higher soil fertility. In general, it is tolerant of different soil types, and thus shows highest yield stability under varying environmental conditions.

Rye is well known to tolerate acidic soil. The range of best suitability is pH 5.0–7.0, but tolerance is between 4.5 and 8.0. It does best at a pH of 6.0–6.5. Therefore, rye also shows high aluminum tolerance (Gallego et al. 1998; see Section 3.4.2). It even gives good yields in poor, sandy soils, and does better than oat in sandy soil.

8.2.2 WATER

Rye grows best with ample moisture; but in general, it does better in low rainfall regions than do legumes and it can outyield other cereals on droughty, sandy, infertile soils (see Section 3.4). Its extensive root system enables it to be the most drought-tolerant cereal crop, and its maturation date can alter based on moisture availability. The structure of the rye plant enables it to capture and hold protective snow cover, which enhances winter hardiness. This snow retention might also be expected to enhance water availability. However, under intensive condition of rye production, a dense stand of plants in autumn can promote the growth of snow mold. Diploid varieties are more drought tolerant than those that are tetraploid. Rye requires ~20%–30% less water than wheat per unit of dry matter formation (see Section 3.4; Table 8.4).

TABLE 8.4
Water Requirement and Minimum Germination Temperature in Some Cereal Crops

Crop	Rye	Wheat	Triticale	Barley
Temperature of germination (°C)	1–2	2–4	1–3	2–4
Minimum				
Optimal	20–25	20–25	20–25	20–25
Growth of seedling (°C)	2–3	3–5	3–4	3–5
Vernalization	0 to +3	0 to +3	0 to +3	0 to +3
Temperature (°C)				
Duration (days)	30–50	40–70	35–60	20–40
Frost resistance[a] (°C)	−25	−20	−18	−15
Water requirement (l/kg DM)	250–300	300–400	280–380	250–350

Source: Geisler, G., *Pflanzenbau: Biologische Grundlagen und Technik der Pflanzenproduktion.* P. Parey Verlag, Berlin, Germany, 1980. With permission.

Note: DM, dry matter.

[a] Without snow cover.

TABLE 8.5
Average of a 4-Year Trial (1995–1998) on Yield and Nitrogen Efficiency (kg Yield per 1 kg N) of Winter Rye, Winter Wheat, and Winter Barley under Optimal and Suboptimal (Dressing and Plant Protection according to the Location) Agronomic Conditions in Central Europe

	Yield (dt/ha)		Yield (kg) per Nitrogen (1 kg)	
Crop	Optimal	Suboptimal	Optimal	Suboptimal
Wheat	56.2	50.2	26.9	33.1
Barley	51.3	43.3	30.1	43.6
Rye	75.5	67.4	43.1	56.8

Source: Adam, L., *Roggen, Anbau und Vermarktung*, Jahresber. Landesamt fur Verbraucherschutz, Landwirtschaft, Flurneuordnung, Brandenburg, Germany, 2004. With permission.

8.2.3 NUTRITION AND DRESSING

Rye even produces high yield under poor soil conditions and without or limited extra dressing. Nevertheless, it responds well to fertilizer. When it is grown after other cereals and maize, the residual soil nitrogen left from the precrop is low. When wheat, barley, and rye are compared, the highest nitrogen efficiency is shown in rye (see Table 8.5).

Intensive systems of rye production require additional nitrogen during the starting phase of vegetation. It promotes tillering and spikelet fertility. Depending on the soil and the variety, up to 30–50 kg/ha of nitrogen can be recommended. Depending on the sowing date and the sowing density, 20–30 kg/ha of nitrogen is recommended which is still applied in autumn. The application of nitrogen during the period of shoot emergence and/or two-node stage influences stand density and spike fertility (number of seeds per spike). However, dense stands are susceptible to lodging. In order to utilize the yield potential, a third portion of nitrogen can be applied during the stage of spike emergence, however, when sufficient water resources are available.

In fertile soils with a yield forecast of 45–55 dt/ha, the utilization of stabilized nitrogen fertilizers at the beginning of the vegetation period is the most efficient way of nitrogen dressing.

The application of sludge is possible, however, in limited amounts. Precise calculations of mineral fertilizers have to be considered.

8.2.3.1 Micronutrients

Zinc and copper are two of the essential nutrients needed for the normal growth and development of cereal crops. For example, copper is removed in the grain of cereal crops at the rate of ~22 g/0.4 ha/yr compared to 4.5–45 kg/0.4 ha/yr for major nutrients such as nitrogen, phosphate, potash, and sulfur. If straw is taken from a field, an additional 10–20 g/0.4 ha of copper may be removed. Organic peat soils are very prone to copper deficiency. For instance, in Western Canada, research has shown that organic soils deeper than 45 cm often respond dramatically to copper fertilization. Copper deficiency should be correctly diagnosed before copper fertilizer is applied. Response of rye to copper fertilizer in copper-deficient soils is low. In deficient soils, symptoms generally occur in irregular patches. Deficiency is often first noticed on rye as "leaf melanosis." This is a browning discoloration of plants associated with reduced yields and ergot infestations.

The mineral nutrition of cereals has been studied over a period of several decades. However, directed selection for improvement of micronutritional efficiency was only initiated by few breeding programs. Worldwide, a high percentage of arable land suffers from severe microelement deficiencies, such as Zn, Fe, Cu, and Mn. A targeted breeding for more efficient varieties is required not only from the viewpoint of cost efficiency but also from the viewpoint of sustainability. This includes the specific characterization of parental lines for new crossing combinations, a search for highly efficient genotypes among the related species of cereal crops, and detailed studies of the inheritance of micronutritional traits (see Section 3.4.2).

8.2.4 Temperature

Rye is one of the best crops where soil fertility is low and winter temperatures are extreme. It is grown in the cool temperate zones or at high altitudes. Vegetative growth for rye requires a temperature of at least 4°C. It is the most winter hardy of all small cereal grains. Its cold tolerance exceeds that of wheat, including the most winter hardy of wheat varieties, and it is seldom injured by cold weather (see Section 3.2). Winter hardiness is a complex feature, which includes cold resistance and resistance to damping, which is often related to resistance to snow mold and resistance to ice crust and lifting as well. Therefore, frost tolerance may be increased by land treatment (melioration, high quality of tillage, and timely sowing). Compared to other cereals, rye can also be spring sown, when agronomic conditions are difficult.

8.2.4.1   Harvesting

The crop can be harvested and threshed in one operation with a combine, or swathed and later threshed. To reduce shatter loss when direct combining, harvest is begun at ~22% moisture followed by drying. Moisture content needs to be below 15% to avoid discounts at elevators. Rye threshes very easily. Under dry threshing conditions, care must be taken to adjust the concave setting and/or cylinder speed to minimize cracking. Care in picking up the swathe is also necessary to avoid undue shattering. Sprout damage can occur during harvest or storage in some years. Some buyers have used the "falling number" test to check for sprouted grain and have discounted grain with a low value.

Rye at 14% or lower kernel moisture is considered dry and safe for storage. At 13% moisture, loss of condition due to molds or mites is unlikely.

8.3   DISEASES, SUSCEPTIBILITY, AND RESISTANCE

Rye is considered a healthy crop. In comparison with wheat and barley, rye is among the less susceptible cereals, but some fungi and viruses can cause heavy yield losses. It is particularly endangered by snow mold (*Microdochium nivale*), the eyespot disease (*Pseudocercosporella herpotrichoides*), and take-all (*Gaeumannomyces graminis*).

Leaf diseases are mostly caused by the pathogens of powdery mildew (*Erysiphe graminis*), the *Rhynchosporium* leaf spots (*Rhynchosporium secalis*), the rye leaf rust (brown rust), *Puccinia recondita,* as well as the stem rust (*P. graminis*). However, ergot (*Claviceps purpurea*) remains an important disease in rye (see Section 8.3.1). With the introduction of hybrid varieties, it became even more important since hybrids were often kept for long flowering. New hybrids had better cope with ergot (see Section 7.1.3).

Rye possesses slight susceptibility to head blight (*Fusarium culmorum*). Flag smut (*Urocystis occulta*) is rare as well. Presently, soil-borne viruses are the largest threat in rye cropping, mainly caused by wheat spindle streak mosaic virus (WSSMV), soil-borne cereal mosaic virus (SBCMV) as well as soil-borne wheat mosaic virus (SBWMV). The compiled data according to resistance genes are given in Figure 8.3.

8.3.1   Bacterial Diseases

8.3.1.1   Bacterial Streak (*Xanthomonas campestris*)

In rye, the bacterial streak (black chaff), *Xanthomonas campestris* pv. *translucens* syn. pv. *secalis*, can be found. It is a seed-borne pathogen. The transmission rate is very low, but ensures serious outbreaks in the field under suitable conditions.

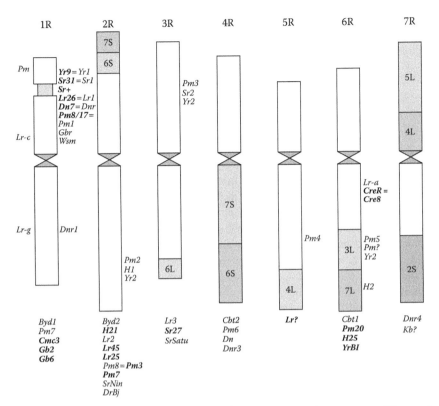

FIGURE 8.3 Schematic presentation of rye resistance genes on chromosomes of *S. cereale*, according to Schlegel and Melz (1996), Schlegel and Korzun (2013), Korzun et al. (2001), and Wehling et al. (2003); wheat gene nomenclature is given in bold letters. *Byd,* barley yellow dwarf virus; *Cbt,* karnal bunt; *Dn, Dnr, Gbr,* green bug aphid; *H,* Hessian fly; *Hm,* lethal leaf blight and ear mold disease; *Lr,* leaf rust; *Pm,* resistance gene to powdery mildew; *Sr,* stripe rust; *Wsm,* wheat streak mosaic virus; *Xct, Xanthomonas campestris* pv. *translucens*; *Yr,* yellow rust. Genes with unknown chromosomal location are *Lr4, Lr5, Lr6, Lr-b, LrSatu,* and *Xct4*. Bold letters represents wheat gene nomenclature, which differentiates the gene designation made either in wheat or rye genetics. (From Tyrka, M., and Chekowski, J., *J. Appl. Genet.*, 42, 283–295, 2004. With permission.)

The pathogen is disseminated by seed on a large scale. Infected leaves show narrow, water-soaked streaks, and yellowish to rust-colored margins. Bacterial slime exudes and dries to a thin scale-like layer, which can be flaked off. Seedlings hardly show any symptoms. Glumes and seeds show black chaff symptoms, with purple-black discoloration of the surface. The main hosts of *X. translucens* are barley, rye, wheat, and triticale. It has also been recorded on some grasses (*Bromus* sp., *Phalaris* sp., *Elymus repens*) and other Poaceae by inoculation. Its distribution is almost worldwide: Europe (Belgium, Bulgaria, France, Romania, Russia, Spain, Sweden, Ukraine); Asia (Azerbaijan, China, Georgia, India, Iran, Israel, Japan, Kazakhstan, Malaysia, Pakistan, Syria, Turkey, Yemen); South Africa; and the United States (Canada, Mexico). No known control measures exist for the disease in the field. Chemical control focuses on seed treatments. Organomercurial seed treatments were used in the past, and were mostly considered effective.

8.3.1.2 Halo Blight (*Pseudomonas coronafaciens* pv. *coronafaciens*)

Several types of symptoms were caused by *Pseudomonas coronafaciens* syn. *P. syringae* pv. *coronafaciens* in Georgia. The disease occurs quite commonly in North America, in which cases it is usually diagnosed as winterkilling: infections just before panicle emergence can reduce yield. A recent report was from Japan as well. Early season infections, however, have little impact on yield. Elliptical to linear, brown lesions, surrounded by mild chlorosis, and often in association with frost injury, were observed on leaves during the winter when temperatures rarely exceeded 15°C. Halo symptoms were prominent in early spring when conditions were mild (maximum 15–22°C). Chlorotic lesions were less common after heading. Halo lesions became buff within 1–2 days, giving the lesions a "scalded" appearance. Scald symptoms were most common from flowering until maturity, but easily recognizable halo symptoms persisted on leaf sheaths almost until maturity. Minute halos were sometimes found on the glumes, but more commonly the entire floret was blighted and dried, and became white. The toxin-induced chlorosis is influenced by irradiance. Distinct and faint halo blight symptoms appear in 3–4 days in full light and 58% shade, respectively. No symptoms or only faint symptoms appear after 7 days at 86% shade. When plants kept in 58% and 86% shade are moved to full light 5 days after inoculation, lesion size and chlorosis increase rapidly during the next 2 days. On the seventh day after inoculation, the size of the lesions from the 58% and 86% shade treatments exceeds those in full light by 2.5 and 5 times, respectively (Cunfer and Youmans 1979). Control procedures that are used where the disease is severe enough to merit them include planting-resistant varieties, crop rotation, field sanitation, and seed treatment.

8.3.2 FUNGI

8.3.2.1 Ergot (*Claviceps purpurea*)

Rye is the principal economic host of ergot. It is more sensitive to ergot than other cereals. It refers to a group of fungi of the genus *Claviceps*. The most prominent member of this group is *Clav. purpurea*. This fungus grows on rye and related plants, and produces alkaloids that can cause ergotism in humans and other mammals who consume grains contaminated with its fruiting structure. The so-called sclerotium is a giant and compact mass of hardened fungal mycelium containing food reserves. It shows black to purple color and can be easily identified on spikes. *Claviceps* includes ~50 known species, mostly in the tropical regions. Economically significant species include *Clav. purpurea* (parasitic on grasses and cereals), *Clav. fusiformis* (on pearl millet, buffalo grass), *Clav. paspali* (on dallis grass), and *Clav. africana* (on sorghum). *Claviceps purpurea* most commonly affects outcrossing species such as rye (its most common host), as well as triticale, wheat, and barley. It affects oats only rarely. *Claviceps purpurea* has at least three races or varieties, which differ in their host specificity: G1—land grasses of open meadows and fields; G2—grasses

FIGURE 8.4 Ergot kernels in a rye spike. (After Pabst, G., *Köhler's Medizinal-Pflanzen*, Verl. F. E. Köhler, Gera, Germany, 1887.)

from moist, forest, and mountain habitats; G3 (*Clav. purpurea* var. *spartinae*)—salt marsh grasses (*Spartina* sp., *Distichlis* sp.). An ergot kernel develops when a spore of fungal species of the genus *Claviceps* infects a floret of flowering plant (see Figure 8.4). The kernels are generally one to five times larger than the host seed. Thus, the largest ergots (1–5 cm) are found in large-seeded plants such as cereal rye. The infection process mimics a pollen grain growing into an ovary during fertilization. Infection requires that the fungal spore have access to the stigma, so plants are infected showing pronounced open-flowering. The proliferating fungal mycelium then destroys the plant ovary and connects with the vascular bundle originally intended for seed nutrition. The first stage of ergot infection manifests itself as whitish soft tissue (sphacelia) producing sugary honeydew, which often drops out of the infected florets. This honeydew contains millions of asexual spores (conidia), through which insects disperse to other florets. Later, the sphacelia convert into a hard dry sclerotium inside the husk of the floret. At this stage, alkaloids and lipids accumulate in the sclerotium.

When rye contains 0.5% or more of ergot, it is considered unfit for food or feed. Although the ergot of rye causes yield reductions, the significance of the disease is primarily related to the various toxic alkaloids present in the ergots (sclerotia). The ergot sclerotium contains high concentrations (<2% of dry mass) of the alkaloid ergotamine, a complex molecule consisting of a tripeptide-derived cyclol-lactam ring connected via amide linkage to a lysergic acid (ergoline) moiety and other alkaloids of the ergoline group that are biosynthesized by the fungus. The alkaloids have a wide range of biological activities including effects on circulation and neurotransmission. The alkaloids can cause severe health problems in both humans and animals. Before this disease was understood, the ergots were ground up along with rye grains and ingested when the flour was used for baking. In the Middle Ages, this led to a frightening disease of humans known as holy fire, Great Fear,

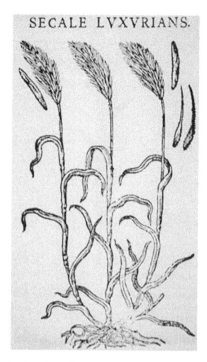

FIGURE 8.5 Botanical drawing from early description of rye showing ergots. (After Bauhin, C., *Theatri Botanici*, Basel, 1658.)

or St. Anthony's fire. Even historical events, such as French Revolution in 1789, have been linked to ergot poisoning (Matossian 1989). Ergot was mentioned very early in the literature as *S. cornutum* (=*Sclerotium clavus* = *Clav. purpurea*) (see Figure 8.5). Ergot is the French word for "spur" (Richard 1824). Long ago, people in France noted some resemblance between the sclerotia and the spurs on rooster legs. The size of the sclerotium is host plant dependent.

Today, ergot poisoning is mostly a concern in animals that may be given contaminated feed or graze where wild grasses are heavily infected with ergot. *Claviceps purpurea* infects only the ovary of cereal and grass plants; no other part of the plant is infected. In the early stages of disease development, a sticky exudate (honeydew) consisting of host sap and conidia often appears. The infected ovary (kernel in the rye spikelet) is replaced by a purplish-black sclerotium, commonly referred to as an ergot. Ergot bodies overwinter in the field, or with the seed in storage, and germinate under favorable conditions in the spring.

Because sclerotia do not usually survive for more than a year, rotation with a nonsusceptible host plant is a viable management tactic for annual crops. Deep plowing buries sclerotia, so ascospores from stromata cannot be discharged into the air. Although deep plowing buries the ergots, many cereal crops are now grown with "no-till" practices in which new crops are seeded directly into the stubble from the previous crop to reduce soil erosion. Crop rotation is even more important when deep plowing is no longer an option.

Ergot is one of the most important diseases in grass seed production. Because many of the grass species are perennial, tillage and crop rotation are not management options. Postharvest field burning has been practiced in parts of Europe and the United States to manage ergot and other diseases and pests since the 1940s, but environmental concerns have resulted in legislative restrictions to burning.

The disease can be partly controlled by sowing ergot-free seed or year-old seed on land where rye has not been grown for 1 or 2 years previously. The mowing of ergot-infested grasses adjacent to rye fields is also helpful. Ergot bodies can be removed by immersing infested rye in a 20% salt solution. The grain is stirred, and ergot bodies float to the surface where they can be skimmed off. The salt must be washed from the seed, and the seed partly dried before it is sown. Other chemicals have been applied to seed or soil to inhibit production of ascospores from sclerotia, but are not economical. Recently, sterol-inhibiting fungicides have been applied to flowers to prevent infection in grass seed production, but such treatments are not usually economical in cereal production. Resistant varieties are not yet available. Therefore, interest in genetic resistance to ergot has increased since the 1970s with the creation of male-sterile lines for hybrid seed production. Effective male sterility systems provide the opportunity for hybrid seed production; however, male-sterile plants flower longer and remain susceptible until they are fertilized. A recent study revealed rich reservoirs of genetic variation for resistance that could be used in hybrid breeding. By using ~50 full-sib families of five open-pollinated winter rye populations of highly diverse origin (Dańkowskie Selekcyjne, Poland; Charkovskaya, Ukraine; NEM4, Russia; Halo and Caro kurz, both from Germany), significant genotypic variation due to genotype and genotype – environment interaction could be determined. Genotypic variation within all populations was higher than that among the five populations. All populations showed high estimates of heritability (0.72–0.89). Recurrent selection to improve ergot resistance should be successful (Mirdita et al. 2008).

The German company, KWS Lochow Ltd., developed new approaches from 1996 to 2003 to minimize the specific risk from ergot, which caused progress in the yield of hybrid varieties. While population varieties over the years show almost the same contamination with ergot (0%–0.5%), the first hybrids (1991–1998) were heavily infected (up to 0.4%). The next generation of hybrids (1999–2006) showed moderate infection. The hybrid seed was mixed with 20% of population rye. After 2007, hybrids with so-called PollenPlus® technology were established. With the seed, a very effective protection against ergot infection is achieved. It uses an efficient restorer gene *Rfp1* on chromosome 4RL taken from an extensive Iranian rye source. The restorer gene shows a close genetic linkage (<0.8 cM) to a molecular restriction fragment length polymorphism (RFLP) marker *Xiac76*. This marker offers the chance of smart screening of the character during selection process (see Section 7.1.4).

8.3.2.2 Leaf Rust (Brown Rust)

Leaf rust or brown rust [*P. recondita* f. sp. *secalis* syn. *Aecidium clematidis* (anamorph)] is one of the most common airborne pathogens of rye and appears regularly in all rye-growing areas (Miedaner and Sperling 2004). Rust brown, oblong–oval uredo (summer) spores are visible. They become black at the end of the growing season. The lower leaf surface shows deposits of teleuto (winter) spores covered by the epidermis.

There are effective fungicides that can control the disease. Apart from the application of fungicides, breeding for resistance is the most effective protection against the disease. Resistance can be controlled quantitatively or qualitatively. There is sufficient variability between rye populations with more or less good heritability. Selection can be efficiently integrated in breeding programs.

By a recent screening of ~900 pustule isolates from Germany, Poland, and Czech Republic, just 9 of these lines exhibited virulence frequencies over 50% for the German, Polish, and Czech isolates, suggesting low or even no resistance to leaf rust. No dominant pathotypes were found. The pathotype structure proved to be relatively stable over the years (Klocke 2004). Complex pathotypes with a high number of virulence genes possessed an advantage because they are virulent to the majority of the differential lines. In contrast, pathotypes with a small number of virulence genes can attack only few differential lines.

Genetic analysis led to the identification of two dominant resistance genes, *Pr1* and *Pr2*. Both genes proved to be effective against a local leaf rust population as well as a subset of single-pustule isolates, the latter of which comprised single-pustule isolates with very high virulence complexity. Resistance conferred by *Pr1* and *Pr2* was expressed in detached-leaf tests of seedlings as well as in field tests of adult plants. Molecular marker analysis allowed us to map *Pr1* in the proximal part of rye chromosome 6RL, whereas *Pr2* was assigned to the distal part of chromosome 7RL (Wehling et al. 2003). Studies that are more recent led to the identification of four resistance genes. The genes of the lines H54/9 and S4084 are localized on chromosome 1RS and the genes of the lines H26 and 94107 on chromosome 4R. Linkage marker could be determined. A survey shows the present status of genes controlling leaf rust resistance in rye (see Table 8.6).

In the United States, severe infections of leaf rust are largely confined to the southern range of rye production and cause a reduction of tillering and grain yield. The disease overwinters in the leaves of rye as dormant mycelium. Destruction of volunteer rye in stubble fields will aid in the control of this disease. A study in Poland demonstrated that disease intensity is not significantly affected by sowing rate. Fungicidal protection of rye plants, applied at the fourth node stage (Flusilazol and Tridemorph), effectively reduced the incidence of brown rust (Szemplinski et al. 2005).

In Russia, since 1967, more than 2500 rye (*S. cereale* and *S. montanum*) populations have been studied at the N. I. Vavilov Research Institute of Plant Industry, St. Petersburg, in order to determine genetic diversity for leaf rust resistance. Plants possessing race-specific resistance to leaf rust were found in 51 accessions (cultivars, landraces, and wild species). In 2000, a study of 420 rye accessions revealed stem rustresistant genotypes (*P. graminis* Pers. f. sp. *secalis* Erikss. et Henn.) in 69 of them. Leaf rust resistance was found to be dominant monogenic in 44 accessions and digenic in cultivar Tschulpan 2. In some accessions, for example, Avangard 2, Novozybkovskaya 4-2, and Derzhavinskaya 2, leaf rust resistance of individual plants was determined by one dominant gene, whereas in other plants of the same accessions it was determined by two dominant genes. In most resistance sources (Sanim, Chernigovskaya 3, Kharkovskaya 55), genetic control of the character is determined

TABLE 8.6
**A Compiled List of Resistance Genes against Leaf Rust (*Puccinia recondita*
f. sp. *secalis*) in Rye**

Gene	Origin	Chromosomal Location	Reference
Pr1 syn. *Lr1*[a]	Germany	6RL	Wehling et al. (2003)
Pr2 syn. *Lr2*[a]	Germany	7RL	Wehling et al. (2003)
Pr3 syn. *Lr3*[a]	Russia	1RS	Roux et al. (2007)
Pr4 syn. *Lr4*[a]	Canada	1RL	Roux et al. (2004)
Pr5 syn. *Lr5*[a]	Germany	1RL	Roux et al. (2004)
Pr-d syn. *Lr6*[a]	Russia	2RS	Roux et al. (2007)
Pr-e syn. *Lr7*[a]	Russia	6R	Roux et al. (2007)
Pr-f syn. *Lr8*[a]	Portugal	2RS	Roux et al. (2007)
Pr-i	Germany	1RS	Klocke (2004)
Pr-j	Germany	4R	Klocke (2004)
Pr-k	Germany	1RS	Klocke (2004)
Pr-l	Germany	4R	Klocke (2004)
Pr-n syn. *Lr9*[a]	Russia	1R	Roux et al. (2007)
Pr-p	Argentina	?	Roux et al. (2007)
Pr-r	USA	?	Roux et al. (2007)
Pr-t	Russia	?	Roux et al. (2007)

Note: ? means not yet localized.

[a] Renamed by the author in 2007 from *Lr, Lra, Pr1* to *Lr1*; the latter designation is not identical with the designation by Solodukhina and Kobylyanski (2001, 2007).

by the *Lr4* gene, in Immunnaya 1 by *Lr5*, in Tschulpan 3 and Immunnaya 4 by *Lr6*, in Novozybkovskaya 4-2 by *Lr7*, in Lovaszpatonai by *Lr8*, and in Jaroslavna 3 by *LriO*. Stem rust resistance is controlled by the dominant gene *Sri*. By pyramiding effective resistance genes, two new winter rye cultivars have been bred: Estafeta Tatarstana (1999) and Era (2001) are characterized by a high-level resistance to leaf and stem rust and to powdery mildew (Solodukhina and Kobylyanski 2003).

8.3.2.3 Stem Rust (*Puccinia graminis* f. sp. *secalis*)

Elongated, reddish brown to coffee-brown deposits of uredo spores are visible on leaves, stalks, and ears. Black-brown, elongated teleuto spores are found primarily on leaf sheaths or stalks. Remnants of the epidermis are often still visible on both sides of the bearing. Organic agriculture is especially affected by the increasing spread of stem rust in rye because there is a lack of resistant varieties. The control by fungicides is difficult. The early maturity of rye usually enables it to escape serious damage from stem rust or black rust. Common barberry is the alternate host. There are few genes with qualitative inheritance that can be used in breeding. Stem rust resistance gene *SrR* occurs in wheat lines carrying the 1RS translocation from *S. cereale* cv. Imperial. Using a mutation-based approach, recently 41 independent mutants were identified by molecular markers. The results indicated that the *SrR* region in rye is syntenic to the *Mla* region in barley or that *SrR* is possibly orthologous to the *Mla* locus (Mago et al. 2004). There is another dominant oligo-genic (Sr) stem rust resistance in var. Rossul and Kharovskaya 55 controlled by the gene *Sr1* (syn. *Rpg5*). It is assumed to have the typical R-gene domains including the nucleotide-binding site, leucine-rich repeat, and domains of serine/threonine protein kinase. The gene appears to condition compatible or incompatible interactions with the wheat stem rust races QCCJ and Ug99, because it is the only polymorphic gene correlating with resistance and susceptibility in the delimited rpg4/Rpg5 region. Sequence analysis of rpg5-susceptible alleles showed that they make up two groups. The group 1 susceptible lines contain an insertion/deletion region having a predicted functional protein phosphatase 2C gene (Arora et al. 2012).

8.3.2.4 Powdery Mildew (*Erysiphe graminis*)

In rye, powdery mildew can be controlled with fungicides. There is great genetic variability. Several resistance genes were described with high heritability. Both quantitative and qualitative inheritance can be found. Selection efficiency can be even higher by using molecular markers.

Most information about powdery mildew was gained from studies on wheat–rye translocation lines (see Section 4.8.3.4). Powdery mildew (*Pm*) in wheat is caused by *Blumeria graminis* f. sp. *tritici* (*Bgt*). It is one of the devastating diseases of wheat, which has seriously impeded wheat production. It causes white, cottony cushion in rye mainly on the top of young leaves, leaf sheaths, and blades. The cushion grows rapidly and develops into a mealy covering. Toward the end of the vegetation period, tiny black dots appear.

To date, few of the 70 known resistance genes have been successfully applied in wheat breeding. Many lost their resistance due to the rapid coevolution of the pathogen's virulence. Therefore, mining new resistance gene is an urgency for wheat improvement, and increasingly also for rye. Besides the known *Pm* genes *Pm7*, *Pm8*, *Pm1*, and *PmJZHM2RL*, recently a dominant *Pm* resistance gene was found in a Chinese wheat line "07jian126." It was designated *Pm07J126*. It originates from rye but is not associated with the 1RL.1BS translocation. Several molecular markers are tightly linked to *Pm07J126*, for example, *Xbarc183* (Yu et al. 2012). The latter can be applied in marker-assisted selection.

8.3.2.5 Scab or Head Blight [*Fusarium graminearum*
(anamorph) syn. *Gibberella zeae*]

Rye and triticale are generally less susceptible to the disease than wheat. *Fusarium* head blight is caused by the fungus *F. graminearum* (Schwabe), although the Netherlands and other areas of Central Europe report *F. culmorum* as the most prev-alent species. In Poland, *F. culmorum*, *F. graminearum*, and *F. nivale* have shown similar moderate to severe virulence levels, while *F. avenaceum* has proved to be mildly to moderately virulent. However, in several studies aimed at identifying the causal organism, as many as 18 *Fusarium* sp. were isolated and identified. Scab is prevalent in warm, humid regions where flowering coincides with rainy periods. The incidence of this disease has been increasing over the past 10 years for various reasons. Perhaps the most important reason is increased area where wheat is rotated with maize or other cereals. Other reasons are changes in the cropping system for soil protection purposes and changes in wheat cropping from traditional to more humid, nontraditional areas. *Fusarium graminearum* is a facultative parasite and is pathogenic on rye and many grasses. Disease control is effectively based on integrated management, including proper agronomic practices, utilization of resistant or tolerant cultivars, and chemical applications. Standard cultivars as well as breeding strains are denoted by various susceptibility levels to spike fusariosis. Dańkowskie Nowe and several Polish strains SMH-50, SMH-55, SMH-57, JEC 1/97/169, and LAD 2T 82R are characterized by the lowest susceptibility level. The winter hybrid variety Amando exhibited the highest level of resistance (see Table 7.6).

Miedaner et al. (1993) studied different gene pools for resistance after artificial inoculation. Single disease rating, average disease rating, and yield components (grain weight per spike, 1000-grain weight, kernel number per spike) relative to the noninoculated treatment were significantly affected by *Fusarium* head blight in all material groups. The relative grain weight per spike ranged from 26% to 88%. Significant genotypic and genotype × year interaction variances were found throughout. Heritability values were highest for homogeneous inbreds ($h^2 = 0.6$–0.8) and lowest for heterogeneous F_2 crosses ($h^2 = 0.4$–0.6). Disease rating and relative grain weight per spike were highly correlated for the inbreds and F_2 crosses ($r \approx 0.7$), but lower for the F_1 crosses ($r \simeq 0.6$). Interannual correlation coefficients for disease ratings and relative grain weight per spike ranged from $r \simeq 0.7$ (inbreds) to $r \simeq 0.5$ (F_2 crosses). The diallel analysis showed significant GCA effects only for relative 1000-grain weight in 1990, but significant specific combining ability (SCA) and SCA × year interaction variances for most traits. The resistances of 16 inbreds to *F. culmorum* and *F. graminearum* were tightly associated for all traits ($r = 0.96$–0.97). They concluded that only slow progress can be expected from selecting for *Fusarium* head blight resistance in rye due to limited amount of additive genetic variance and the great importance of environmental factors (Miedaner et al. 2001).

8.3.2.6   *Fusarium* Root Rot (*Fusarium culmorum*)

The activities of the individual cell-wall-degrading enzymes (glucanases, chitinases, xylanases, endocellulases, exocellulases, pectinases, and polygalacturonases) released by the fungus are considered to be one of the key factors responsible for the plant–fungal interactions. *In vitro* experiments showed that organic compounds (glucose, chitin, rye roots, and *Fusarium* cell walls) added to the growth medium differentially affect the activity of the individual enzymes released by the particular isolates (Jaroszuk-Sciseł et al. 2011). The activities of xylanases and endocellulases released to the plant cell wall-amended medium by the plant growth-promoting isolate are significantly lower than those of these enzymes released by the deleterious rhizosphere and the pathogenic isolates. The activity of pectinases is repressed by glucose. The activities of acidic hydrolases are greater than those of alkaline hydrolases. Principal component analysis revealed that the activities of the cell-wall-degrading enzymes found in the supernatants of the autolyzing *F. culmorum* cultures can be clustered into two distinct groups. Pectinase, exocellulase, polygalacturonase, and all the remaining hydrolases are tested, suggesting that enzymes from either group may act in synergy during cell wall degradation. Xylanase activities of the enzymatic complexes are strongly positively ($r \geq 0.9$) correlated with their efficiency in hydrolyzing plant cell wall, whereas chitinase activities were correlated with efficiency in fungal cell wall hydrolysis.

8.3.2.7 *Ramularia* Leaf Spots (*Ramularia collo-cygni*)

It causes limited oval, brown spots on the side of the leaf sheaths which are often brighter in the middle and the veins. Older spots are formed irregularly, and whose edge is often discolored dark brown. Around the spots is the yellow leaf tissue.

All cereals are attacked by smut, but each crop is host to specific species of smut fungus. Each smut species has a number of races that differ only in their ability to infect certain cultivars of that crop host. This is a very common disease on rye, particularly in Minnesota and nearby states. The symptoms appear first as lead gray, long narrow streaks on the stems, sheaths, and the blades. These streaks later turn black. Infected plants are darker green than normal and somewhat dwarfed. The stems usually are twisted or distorted, and the heads fail to emerge from the sheath. Spores can be carried both on the seed and in the soil. Disease control could be achieved by seed treatment and crop rotation where the spores are soil-borne. Resistant varieties are also available.

8.3.2.8 Anthracnose [*Colletotrichum graminicola* syn. *Glomerella graminicola* (teleomorph), *Alternaria* sp., and *Cladosporium herbarum*]

This disease is especially prevalent in the humid and subhumid part of eastern United States. Infected tissues are stained brown on the leaf sheath that surrounds the diseased stem. Head infections cause the light brown kernels to shrivel. Infected plants often ripen or die prematurely.

8.3.2.9 Black Head Molds [*Alternaria* sp., *Cladosporium herbarum* syn. *Mycosphaerella tassiana* (teleomorph), *Epicoccum* sp., *Sporobolomyces* sp., and *Stemphylium* sp.]

The natural occurrence of actively sporing colonies of *Clad. herbarum* near pollen clusters on the surfaces of rye leaves suggested that pollen might be an important source of nutrients in the phyllosphere. One month after flowering, concentrations of approximately 100 pollen grains per cm^2 are not uncommon.

8.3.2.10 Pink Snow Mold, *Fusarium* Patch, or Snow Mold [*Microdochium nivale* syn. *Fusarium nivale* syn. *Monographella nivalis* (teleomorph)]

The disease can be controlled with seed treatment. The resistance is under polygenic control with low heritability for selection.

8.3.2.11 Leaf Spots (*Rhynchosporium secalis*)

It is a dominant disease in winter rye. The severity is mainly determined by the annual weather-dependent infection pressure and cultivar susceptibility to disease. Fungicides can be dosed at two intensities: 100% (situation-dependent) and 50%

(reduced dosage), in each case, after a certain disease threshold is exceeded. Both intensity levels achieved good to very good fungicidal effectiveness, but the effectiveness of the lower dosage level is often significantly lower than that of high infestation levels. Overall (11 years' trials), the mean treatment frequency index was 1.4 in winter rye, 1.0 in winter barley, and 0.8 in winter wheat. Cultivar resistance proved to be the determining factor of the need for fungicide treatment. In winter rye, where resistance is rather low, at least one fungicide treatment is necessary in all years of the trial. The genetic variability is low and resistance genes are rare. A Polish experiment demonstrated that *Rhynchosporium* is not an agent of yield failure hazard from the agricultural perspective.

8.3.2.12 Strawbreaker, Eyespot, or Foot Rot [*Pseudocercosporella herpotrichoides* syn. *Tapesia acuformis* (teleomorph)]

It can partially be controlled with seed treatment. There is little genetic variation for this trait including low heritability. It is quantitatively inherited.

8.3.2.13 Take-All (*Gaeumannomyces graminis*)

It causes blackening of the roots and the stem base becomes softened. Plants can be easily pulled from the soil. Characteristic brown "running hypha" grows on the root surface and the stem base. The affected plants show low tillering and their growth is impaired. An experiment with hybrid rye showed that the frequency of occurrence of disease symptoms in hybrid rye at the first node stage is similar, in terms of statistics, as in the open-pollinated cultivar Amilo. Seed dressing (Tridemorph, Imazalil, Fuberidazol) reduced the incidence of take-all diseases (Szemplinski et al. 2005).

Additional fungal diseases are known in rye. Their importance is limited:

Black point [*Bipolaris sorokiniana* syn. *Cochliobolus sativus* (teleomorph), and *Fusarium* sp.]

Bunt or stinking smut (*Tilletia caries* syn. *T. tritici*, *T. laevis* syn. *T. foetida*)

Cephalosporium stripe, *Hymenula cerealis* syn. *Cephalosporium gramineum*

Common root rot and seedling blight [*Bipolaris sorokiniana* syn. *Helminthosporium sativum*, *Cochliobolus sativus* (teleomorph)]

Winter crown rot or cottony snow mold (*Coprinus psychromorbidus*)

Wist or *Dilophospora* leaf spot (*Dilophospora alopecuri*)

Dwarf bunt (*T. controversa*)

Stem or stalk smut (*U. occulta*)

Halo spot (*Pseudoseptoria donacis* syn. *Selenophoma donacis*)

Karnal bunt or partial bunt (*Neovossia indica* syn. *T. indica*)

Leaf streak (*Cercosporidium graminis* syn. *Scolicotrichum graminis*)

Leptosphaeria leaf spot (*Phaeosphaeria herpotrichoides* syn. *Leptosphaeria herpotrichoides*)

Loose smut (*Ustilago tritici*)

Pythium root rot (*Pythium aphanidermatum*, *P. arrhenomanes*, *P. debaryanum*, *P. graminicola*, *P. ultimum*)

Septoria leaf blotch (*Septoria secalis*)

Sharp eyespot and *Rhizoctonia* root rot [*Rhizoctonia cerealis* syn. *Ceratobasidium cereale* (teleomorph)]

Snow scald or *Sclerotinia* snow mold (*Myriosclerotinia borealis* syn. *Sclerotinia borealis*)

Speckled, gray snow mold, or *Typhula* blight (*Typhula idahoensis, T. incarnate, T. ishikariensis, T. ishikariensis* var. *canadensis*)

Spot blotch (*Bipolaris sorokiniana*)

Glume blotch or *Stagonospora* blotch [*Stagonospora nodorum* syn. *Septoria nodorum, Phaeosphaeria nodorum* (teleomorph) syn. *Leptosphaeria nodorum*]

Stalk or stripe smut (*U. occulta*)

Stripe or yellow rust [*P. striiformis* syn. *Uredo glumarum* (anamorph)]

Tan or yellow leaf spot [*Pyrenophora tritici-repentis* syn. *Drechslera tritici-repentis* (anamorph) syn. *Helminthosporium tritici-repentis*]

Storage molds (*Alternaria, Aspergillus, Epicoccum, Nigrospora, Penicillium, Rhizopus* sp.)

8.3.3 Viruses

8.3.3.1 Barley Yellow Dwarf Virus

Yellow dwarf of barley and wheat and red leaf of oats are collectively referred to as barley yellow dwarf virus (BYDV). It attacks wheat, barley, oats, rye, maize, and many grasses. It was first recognized as a virus disease in 1951 in California. Since then, it has been found in most agricultural areas of the world. Localized severe outbreaks of the disease are common, but the disease develops into an epidemic every 5–8 years. Losses from BYDV occur in both yield and quality of grain and depend on the incidence of infection. The rye chromosomes 1R and 2R are associated with tolerance to BYDV, at least in wheat and triticale.

8.3.3.2 Tobacco Rattle Virus

In a European area known for the occurrence of tobacco rattle virus (TRV) on potatoes, 5%–10% of rye plants of one experimental plot were severely stunted and showed distinct chlorotic mottling of leaves. In these plants, TRV was identified on the basis of particle morphology and serology [immunoelectron microscopy and enzyme-linked immunosorbent assay (ELISA)]. The virus was also detected in the weeds *Capsella bursa-pastoris* and *Stellaria media* from an area adjacent to the rye plot but not in nearby commercial rye fields (Huth and Lesemann 1984).

8.3.3.3 Soil-Borne Viruses

Soil-borne viruses such as the furoviruses SBCMV, SBWMV, and the bymovirus WSSMV, are transmitted by the fungal vector *Polymyxa graminis* Ledingham (see Table 8.7). Increasing infections are observed in different cereals (Kastirr et al. 2012). The detection of furovirus and bymovirus in field plants depends on temperature conditions during the vegetation period and the cereal species. The furoviruses tolerate a broad temperature spectrum, and once established, infection is detectable until the harvest time.

TABLE 8.7
Classification of Important Soil-Borne Viruses in Rye Transmitted by Fungal Vector *Polymyxa graminis* Ledingham

Virus	Abbreviation	Virus Group	Host
Wheat spindle streak mosaic virus	WSSMV	Bymovirus	Wheat, triticale, rye
Wheat yellow mosaic virus	WYMV		
Soil-borne wheat mosaic virus	SBWMV	Furovirus	Wheat, triticale, rye
Soil-borne cereal mosaic virus	SBCMV		
Chinese wheat mosaic virus	CWMV		

Both viruses are distinguished by the infection type in tobacco, *Nicotiana benthamiana*. Whereas SBCMV isolates spread through the whole plant causing yellowing and plant death, the SBWMV infects the inoculated leaves only. From the United States, Argentina, Brazil, China, Japan, New Zealand, and Europe, reports on soil-borne virus in cereals are common. In Germany, the first evidence of virus contamination in rye was given by Pröseler and Stanarius (1983).

8.3.3.4 Wheat Spindle Streak Mosaic Virus

The propagation of WSSMV seems to be restricted to lower temperatures. Consequently, this virus is best detected by the end of February until the middle of April. Among the cereals, rye becomes infected more early than wheat and triticale. Both furoviruses can be differentiated by variable virulence reactions on cereal hosts and indicator plants. The SBCMV infects rye, triticale, and wheat but not barley. The SBWMV is able to contaminate barley too.

The virus is small and rod-shaped, typical for bymovirus. It is transmitted through the fungus *P. graminis* that persist in the soil. It attacks the roots of the rye plant whenever the conditions for its growth are favorable. During early spring, leaves of infected plants produce chlorotic streaks, which later can change to yellow, necrotic, and dying out. Plant growth, tillering, and winter hardiness is decreased. Symptoms resemble a nutrient deficiency.

The virus is described in several European countries. During the past two decades, WSSMV is also found in different states of Germany, where rye and triticale are grown. As known for the yellow mosaic viruses of winter barley, rapid expansion of the WSSMV can be expected in rye in the near future. Yield reduction of 50%–70% is noticed after infection with WSSMV that mainly causes root damage. Since no resistant varieties are available, late sowing dates are the only measure to reduce heavy plant infections. In wheat, genetically determined resistance source can be detected which allows some hope for rye as well. The wild grass, *Haynaldia villosa*, might be a promising donor for resistance genes (Zhang et al. 2005).

8.3.3.5 Soil-Borne Cereal Mosaic Virus

SBCMV represents one of the largest disease problems of recent rye cropping in advanced agriculture. Considering the development of yellow mosaic viruses in winter barley, a fast expansion of SBCMV is expected in rye. SBCMV is a stick-like virus and member of furoviruses. It leads to a reduction of the root system and therewith of yield. Reduction of 50%–70% is possible. The virus is transmitted by a soil fungus (*P. graminis* Ledingham). Its spores may survive for decades in the ground. Under suitable soil conditions, that is, temperature, moisture, and stimulation by the host plant, the fungus starts to grow. The spores contaminated with the viruses of the vector *P. graminis* are spread in the ground by tilling devices and crop remainders as well as by wind and water so that it completely contaminates the neighboring fields in less years. The virus occurs in several European countries. In France, Italy, and England, the virus attacks only wheat; in Germany, Poland, and Denmark, it attacks only rye and triticale. In Germany, the virus could be proved as the cause of infection for the first time by the end of the 1980s in Saxony-Anhalt and Lower Saxony.

The leaves of the infected plants produce chlorotic streaks in early spring that become yellow, later necrotic, and finally die out. Plant growth gets retarded, and tillering and winter hardiness are reduced. The yellowish leaves are often misclassified as malnutrition of the plants. However, molecular identification by multiplex reverse transcription polymerase chain reaction (RT-PCR) unequivocally reveals specific virus DNA, even in mixed samples (see Figure 8.6).

Later sowing dates are the only agricultural measure to reduce virus growth on plants. The mobility of the vector is restricted with later sowing due to generally lower temperatures. Resistance breeding might be the alternative.

In the United States, SBCMV ratings indicated that resistance is present in triticale and rye varieties. Karl 92, a resistant wheat variety, received low SBCMV scores (2 in 1996 and 1 in 1997). TAM 107, a susceptible check in 1996, received an SBCMV score of 7, and Custer, a susceptible check in 1997, received an SBCMV score of 9. In 1996, the SBCMV scores of the rye and triticale varieties ranged from 2 (resistant) to 4 (moderately resistant). The triticales received scores indicating moderate resistance to the virus, whereas three of the four rye varieties (Bonel, Elbon, Oklon, and Maton) received scores of 3 and Bonel received a score of 2. The rye varieties and the triticale Pika all received scores of 2, whereas triticale Presto received a score of 3, and the remaining three triticales received scores of 5

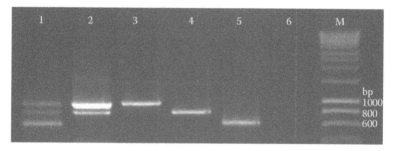

FIGURE 8.6 Detection of soil-borne cereal viruses and the fungal vector by Triplex RT-PCR: 1—mixing infection with SBCMV (978 bp), WSSMV (817 bp), and *Polymyxa graminis* (628 bp) in rye roots; 2—mixing infection with SBCMV and WSSMV in rye leaves; 3—six control lines for SBCMV; 4—WSSMV; 5—*P. graminis*; 6—negative control; M—molecular mass marker (Bioline). (Modified from Kastirr, U. et al., *J. Kulturpflanzen*, 64, 469–477, 2012.)

(Staggenborg et al. 1997). Kastirr et al. (2012, pers. comm.) identified resistant strains of *S. cereale* (R2501) and *S. anatolicum* (R797) against SBCMV, WSSMV, and BWMV among numerous genebank accessions. The SBCMV occurs usually with WSSMV. The plants are infected frequently with both pathogens. Based on the leaf symptoms they cannot be distinguished.

8.3.4 INSECTS, NEMATODES, AND OTHER PREDATORS

Rye is attacked by many insects that attack other small grains too. Serious losses on rye are not common. Early sown winter rye provides a favorable environment for the deposition of grasshopper eggs, which may promote grasshopper injury to other crops.

Parasitic nematodes are known to cause cereal cyst, *Heterodera avenae*, leaf and stem damage, *Ditylenchus dipsaci*, root gall, *Subanguina radicicola*, root knots, *Meloidogyne* sp., or seed gall, *Anguina tritici* in rye. *Ditylenchus dipsaci*, *An. tritici*, and *H. avenae* can cause serious losses. Wheat, potato, turnip, lupine, alfalfa, and white mustard are grown before rye to reduce *D. dipsaci*. Clean seed and crop rotation reduces *An. tritici*. The use of leguminous and root crops in rotation reduces *H. avenae*. Rye harbors particularly low densities of root lesion nematode (*Pratylynchus penetrans*).

There are genes that confer resistance to nematodes. Four deletion mutants of rye chromosome 6R were identified in the progeny of wheat–rye addition lines. The rye chromosome carried a resistance gene against the cereal cyst nematode (*H. avenae*). This chromosome originated in a hexaploid triticale line T701. The deletion mutants were selected on the basis of dissociation of three isozyme loci on the long arm of the rye chromosome. Three plants and their progeny showed expression of the rye genes *α-Amy-R1* and *Got-R2* but lacked the gene *Pgd-R2*. The other plant and its progeny showed expression of the *α-Amy-R1* gene, while the genes *Got-R2* and *Pgd-R2* were absent. The four deletion chromosomes displayed long-arm terminal deficiencies of different sizes, which enabled physical mapping of the rye isozyme and the nematode resistance gene. The distal to proximal order of the 6R isozyme loci was found to be *PgdR2*, *Got-R2*, and then *α-Amy-R1*. Bioassay tests demonstrated that the resistance gene (*Cre-R*) was located on an interstitial section of the long arm of 6R adjacent to *Got-R2* locus (Dundas et al. 2001).

8.3.4.1 Curl Mite (*Aceria tosichella*)

Wheat curl mite (*Ac. tosichella* Keifer) occurs mainly on wheat, but its populations can also develop on maize, sorghum, barley, rye, and a larger number of grasses. Resistance genes can be found in *Aegilops tauschii* and *S. cereale* (Malik et al. 2003). The rye-derived resistance gene, designated *Cmc3*, is present on wheat–rye translocation 1AL.1RS. Marker analysis of a segregating F_2 population revealed that the rye-specific microsatellite marker *Xscm09* can be used to select wheat lines carrying the 1RS segment and *Cmc3*. Allelism tests indicated that the *Ae. tauschii*-derived resistance gene, designated *Cmc4*, segregated independently of the *Cmc1* gene previously transferred from this species. Molecular and cytogenetic analyses located *Cmc4* distally on chromosome 6DS flanked by markers *Xgdm141* (4.1 cM) and *XksuG8* (6.4 cM).

8.3.4.2 Fall Armyworm (*Spodoptera frugiperda*)

It is probably the most serious insect pest of winter rye planted for grazing, mainly in the United States. During high population or outbreak years, fall armyworms can infest and destroy stands of newly planted rye. The insect attacks numerous grass crops such as maize, sorghum, millets, and turf and pasture grasses. Populations reach a peak in late August and September. Fall armyworms often will lay eggs in newly planted field of rye during this period. Young larvae defoliate leaves. Larger larvae can completely consume seedlings and will feed on the plant below the soil surface often killing the plant. Large larvae can be identified by four black dots on the back of the tip of the abdomen and a pale, inverted Y pattern on the head.

High populations and outbreaks occur every 2–4 years and are favored by hot, dry conditions. During these years fall armyworm infestation may destroy rye stands in all or part of a field with ~5%–10% of fields being treated and replanted. Scouting before planting and immediately after planting will help detect infestations. The treatment threshold is three larvae per square foot. However, scouting is rarely done and infestations are only noticed when stands look thin or absent. A foliar insecticide is typically applied to the area before the seed is replanted. Recommended insecticides for fall armyworm control in rye are, for instance, Methomyl or Spinosad.

8.3.4.3 True Armyworm (*Pseudaletia unipuncta*)

In the United States, true armyworm attacks rye mainly grown for grain during the stem elongation and heading stages in late winter and early spring. The larvae rest during the day under leaf litter on the soil surface. They climb the stalks and defoliate rye at night by consuming the lowest leaves first. They also sometimes eat spike glumes and cut seed heads. Damage usually occurs during cool wet springs, with rank and lodged areas of a field often being most severely damaged. Severe outbreaks of true armyworm historically occur every 10 years. Thresholds are four or more larvae per square foot before pollen shedding and eight larvae per square foot during grain fill. Because a ground sprayer would damage standing rye, insecticide applications are usually applied aerially. Rye is seldom treated, but if treatment is required, methyl parathion is used typically.

8.3.4.4 Russian Wheat Aphid (*Diuraphis noxia*)

The Russian wheat aphid is an economically significant insect pest of wheat and barley in the United States but not as severe in rye. Since 1986, only one biotype existed in the United States. In 2003, a new biotype appeared which was virulent to all known resistance genes in wheat, with the exception of *Dn7*. *Dn7* is a rye gene transferred into a wheat background via 1RS.1BL translocation (see Section 4.8.3.4). It originated from rye landrace Turkey 77 and was previously mapped. However, no molecular markers were developed for marker-assisted selection and the map was sparse. Lapitan et al. (2007) presents a higher density map of *Dn7* containing 19 markers. The markers spanning *Dn7* are *XHor2* and *Xscb241* at distances of 1.4 and 1 cM, respectively; from *Dn7* PCR markers were developed and tested on 19 wheat cultivars. Two markers, which amplified rye-specific fragments, proved to be articularly useful. *Xrems1303* amplified a 320-bp band only in cultivars with high-level resistance to biotype 2 and was effective for selection of *Dn7*. *Xib267* is linked to the susceptible locus and amplified a fragment specific for rye Petkus 1RS, and would be a good marker for selecting against the susceptible locus. The value of *Dn7* as a source of resistance has increased because of the broad spectrum resistance it provides against several biotypes.

8.4   GROWTH REGULATORS

Because of the plant height of common rye varieties, lodging resistance is reduced. Lodging significantly decreases grain yield and quality, despite more costly harvest. Even when seeding rate is low and nitrogen dressing is optimal, double growth regulators can be applied for stem reduction, considering the specific characteristics of the variety and local and climatic conditions. Before and after application of growth regulators, there should be sufficient soil moisture! As growth regulators Moddus (trinexapac-ethyl), Medax Top (mepiquat-chloride, prohexadione calcium), CCC (chlormequat chloride = 2-chloro-*N*,*N*,*N*-trimethylethanaminium), or Camposan (ethephon = 2-chloroethylphosphonic acid) are applied. The latter inhibits the activity of enforcement hormones. Therefore, best results are achieved when applied during the period of most intensive growth of culm. Mostly the middle and upper internodes are shortened. The other growth regulators are best applied before the emergence of culm.

8.5   INCORPORATION IN CROP ROTATION

Rye is very suitable for several crop rotations showing high adaptability. As an over-wintering plant, it increases soil organic matter, reduces soil erosion, and provides some degree of weed suppression through allelopathic and competitive mechanisms (see Section 8.6).

The new hybrid varieties show high yield potential, and if an optimal cropping technique is used, they outyield other cereals under comparable agronomic conditions. However, wheat is usually grown on better soils and it receives more attention within the crop rotation. Often the production is suboptimal so that the yield potential of rye is made use of in lower measure. Trials where rye is grown in monoculture or in rotation with winter wheat and winter barley showed better yield performance. Rye has been shown to be allelopathic toward other plants, but some of the suppressive effects may relate to tie-up of soil nitrogen by decomposing rye residues. It also can inhibit germination or growth of vegetable crops sown after rye is incorporated. In some countries, it is sown with legumes or other grasses. Some examples are given how rye can be incorporated in rotations depending on the soil type and utilization (see Table 8.8).

Rye may be also an excellent choice as a winter cover crop because it rapidly provides ground cover to hold the soil in place. It has deep roots that help prevent the soil from becoming compacted in annually tilled fields. Its extensive root system enables it to scavenge nutrients from the soil profile. In a no-till situation, it also can help control weeds.

TABLE 8.8
Some Examples of Crop Rotations Including Rye

Soil Type	Farming Type	Crop Rotation
Sandy soils	Cattle	Maize–rye–rye–green crop
(~18–30, soil fertility)	Grain cropping	Rye–rye as cover crop for grass seed–rye
		Lupine–rye–rye
		Rye–rye–rye–fallow
Sandy loam	Cattle	Maize–maize–rye–rapeseed–rye
(~30–40, soil fertility)	Grain cropping	Triticale/wheat–rye–rye–rapeseed
	Biogas	Fodder rye/maize–maize–triticale–rye

Source: de Vries, G., *Roggen—Anbau und Vermarktung*, Roggenforum, Germany, 2006. With permission.

8.6 ALLELOPATHIC EFFECTS

Several reports demonstrate the allelopathic ability of rye. Living rye and nonliving rye mulches have been found to inhibit both the density and growth of several weedy species. Decreased biomass was observed with common ragweed (*Ambrosia artemisiifolia*), green foxtail (*Setarria viridis*), and common purslane (*Portulca oleracea*) when grown in the presence of rye residues. Reduction in germination ranges from ~40% in common ragweed to 100% in common purslane. Vegetable crops such as cabbage and lettuce can also be affected by rye residues. The same is true for asparagus, large-seeded legumes, or cucumber.

A targeted search was carried out for allelopathic sources among accessions of *Triticum*, *Secale*, triticale, and aneuploid lines to be used in breeding programs to improve weed suppression ability of wheat. A bioassay with mustard as target plant was used for the screening. Mustard was chosen among seven tested target plants because it showed a particular high root growth inhibition when grown together with rye compared to wheat. None of the *Triticum* accessions studied showed potential allelopathic activity of interest for breeding, but rye and most triticale, that is, wheat–rye hybrids, did. Several of the wheat–rye substitution and translocation lines also showed high allelopathic activity. The highest activity was found in lines with a substitution of 1R or 2R. Some multiple substitution lines and lines with only rye chromatin also showed high allelopathic activity. It is suggested that *in vitro* selection of wheat–rye substitution lines with high allelopathic potential should be applied (Bertholdsson et al. 2011). On the other hand, there must be genes on the chromosomes that control features of allelopathy.

Rye produces several compounds that inhibit crops and weeds (Brooks et al. 2012). Analysis indicated significant variability for 2,4-dihydroxy-1,4-benzoxazin-3-one (DIBOA) content in rye harvested at the flag leaf stage.

Rye residues maintained on the soil surface release DIBOA and a breakdown product 2(3*H*)-benzoxazolone (BOA), both of which are strongly inhibitory to germination and seedling growth of several dicotyledonous and monocotyledonous plant species (Chase et al. 1991a, 1991b). Further, microbially produced transformation products of BOA demonstrate severalfold increases in phytotoxic levels. Hence, a variety of natural products contributes to the herbicidal activity of rye residues. Even soil containing rye roots influences seed germination. Willowherb and horseweed germination were inhibited up to 50% and plant growth as well (Przepiorkowski and Gorski 1994).

Numerous studies have demonstrated the allelopathic characteristics of rye residues and root exudates containing DIBOA and BOA. Experiments have shown marked reductions in germination and growth of several problematic agronomic weeds including barnyard grass (*Echinochloa crusgalli*), common lambsquarters (*Chenopodium album*), common ragweed (*Am. artemisiifolia*), green foxtail (*Setaria viridis*), and redroot pigweed (*Amaranthus retroflexus*).

This has led to interest in developing allelopathic cultivars with increased DIBOA to improve weed control in this important cover crop. The objectives of this study were to determine heritability estimates for DIBOA in rye and determine the utility of gas chromatography as a screening tool in a rye allelopathy breeding program. A synthetic population of half-sib families varying in production of DIBOA was analyzed. DIBOA concentrations ranged from 0.52 to 1.15 mg/g dwt tissue (mean = 0.70 mg/g dwt). Analysis of variance indicated significant variability for DIBOA content in rye harvested at the flag leaf stage. Year × location × genotype and block (year × location) interactions were significant. Several genotypes were consistently "high" or "low" DIBOA producers across all locations and years. Narrow sense heritability estimates were $h^2 = 0.18$ on a per plot basis and $h^2 = 0.57$ on an entry mean basis. Obviously, the allelopathic capability is inherited and can be utilized for selection (Brooks et al. 2012).

Even if weeds such as barnyard grass, horseweed, or willowherb became resistant to herbicides, for instance, atrazine, the allelopathic potential of rye is still effective! Resistance to triazine herbicides by these three weed plants has been observed in a number of countries and they are becoming a major weed control problem in both agricultural and horticultural crops.

8.7   VOLUNTEERING

Rye is often considered as a serious, introduced weed of dry land agricultural regions. For example, in the United States, it is an important weed in winter wheat production in many parts of the country. It reduces net profits by more than $27 million due to lower grain yields, increased dockage, and reduced land values.

Volunteering rye closely resembles cultivated cereal rye with the exception of having a shattering seed head. It was assumed that this so-called feral rye may have originated from hybridization of cultivated rye, *Secale cereale*, with mountain rye, *S. montanum*. It is known that the two species can hybridize, though the flowering time is not synchronous, and fertility is reduced to 30%–50% in the F_1 offspring. There are also "hybrid swarms" of the two species in Turkey, which also has a weedy rye, but the origin of weedy rye in Turkey is not known. Yet, *S. montanum* is known to be a weed species, for example, in *Triticum turgidum* crops of Turkey, a nonnatural ecosystem. An early study indicated that a weedy rye type had evolved in northern California and southern Oregon by spontaneous crossing between naturalized strands of *S. montanum* and cultivated cereal rye (Sun and Corke 1992). Common rye (*S. cereale*) is stated as a weed as well, for example, in Western Australia (Hussey et al. 2007). There is a potential for the cross-contamination of cereal rye crops by the perennial pasture type (Oram 1996).

A comprehensive study by Burger et al. (2006) disputed these results. They doubted a true hybridization, rather the feral rye shared a common ancestry including the genes for shattering spikes. The authors characterized the genetic structure of feral rye populations across a broad geographical range and reexamined evidence for hybrid origin versus direct evolution from domesticated cultivars. Eighteen feral populations were examined from three climatically distinct regions in the western United States. Seven cultivars, four mountain rye accessions, and one wild annual relative (*S. ancestrale*) were included in the analysis as possible progenitors of feral rye. Individual plants were scored for 14 allozyme and 3 microsatellite loci. Estimates of genetic diversity in feral populations were relatively high compared to those of the possible progenitors, suggesting that the weed had not undergone a genetic bottleneck. Weed populations had no geographical structure at either a broad or a local scale, suggesting idiosyncratic colonization and gene-flow histories at each site. Feral rye populations were no more closely related to mountain rye than cultivars were. They were, however, weakly clustered as a distinct lineage relative to cultivars.

Thus, rye can become a weed through volunteering or when it sometimes escapes in waste places and fields. In some areas, it grows even with only 15–20 l/m^2 of precipitation.

To date, limited research has been conducted on components that make rye a problem in various cropping systems. Herbicide-tolerant wheat technology might be used to manage rye, but current efficacy levels are not adequate for high rye densities, despite the fact that long-term effects on rye populations are unknown.

8.8 ALLERGENIC POLLEN

Rye can cause allergic sensitization in susceptible individuals. In 1978, column-chromatographic studies on allergens were carried out by purification of rye pollen extracts (Loewe et al. 1978). Several fractions were obtained from the non-dialyzable residue of aqueous extracts by discontinuously increasing ionic strength. Immunoelectrophoresis, immunodiffusion, and polyacrylamide gel electrophoresis showed that these fractions were antigen and protein mixtures of different heterogeneity. Rye pollen extracts contain proteinases, which, probably, are very hard to separate from the other pollen proteins. Even after storing the extracts for 1 year, no change in their allergenic activity was found by intracutaneous test. Exposure of plants to elevated ozone levels increases the allergen content in pollen (Eckl-Dorna et al. 2010).

Later Montero et al. (1992a) studied the proteins and allergens released by rye pollen. Dot immunobinding and immunoblotting techniques reveal that allergens leave the rye pollen, for the most part, after 5-min incubation and they are proteins with molecular weights of 28, 33, 48, and 67 kDa. Moreover, it was shown that unretained proteins and proteins eluted at 0.2 M NaCl from extract B contain the highest proportion of allergens. Sodium dodecyl sulfate polyacrylamide gel electrophoresis

(SDS-PAGE) of chromatographic peaks showed that peak 2 from extract B contained a highly purified 28 kDa band. On the skin of allergic patients, this band gave a stronger positive prick test than for the crude extract (Montero et al. 1992b).

Linkage analysis in Czech atopic families revealed suggestive linkage to plant allergens at human chromosome segment 5q33, highest linkage being observed to rye-specific immunoglobulin E (IgE) (Gusareva et al. 2009). Monoclonal antibodies recognized a major allergen, DG3, with molecular weight between 35 and 40 kDa (Walsh et al. 1990). N-terminal sequencing of DG3 purified by affinity chromatography, 2-D electrophoresis, and electroblotting to polyvinylidenedifluoride revealed significant homology with a group-V allergen (PhlpV) from timothy grass (*Phleum pratense*).

The rye pollen allergen is designated as "Secc1" with a sequence length of 239: AGC GCGCATGGCATCGCGAAGGTACCACCGGGCCCCAACATCACGGCCGAGT ACGGCGACAAGTGGCTGGACGCGAAGAGCACCTGGTACGGCAAGC CGACCGGCGCCGGTCCCAAGGACACCGGCGGCGCGTGCGGGTAC AAGGACGTCGACAAGGCGCCGTTCAACGGCATGACCGGCTGCGGCA ACACCCCCATCTTCAAGGACGGCCGTGGCTGCGGCTCCTGCTTCGAG ATCAAGTGCACCAAGCCCGAGTCCTGCTCCGGCGAGGCTG TCACCGTCACAATCACCGACGACAACGAGGAGCCCATCGCAC CCTACCACTTCGACCTCTCGGGCCACGCGTTCGGGTCCA TGGCCAAGAAGGGCGAGGAGCAGAAGCTCCGCAGCGCCG GCGAGCTGGAGCTCCAGTTCAGGCGGGTCAAGTGCA AGTACCCGGACGGCACCAAGCCAACGTTCCACGTCGAGAAGGG TTCCAATCCCAACTACCTGGCTATTCTGGTGAAGTACGTCGACGGC GACGGCGACGTGGTGGCCGTGGACATCAAGGAGAAGGGCAAGGAT AAGTGGATCGAGCTCAAGGAGTCGTGGGGAGCAGTCTGGAGGATCG ACACCCCCGATAAGCTGACGGGCCCATTCACCGTCCGCTACACCAC CGAGGGCGGCACCAAATCCGAAGTCGAGGATGTCATTCCTGAGG GCTGGAAGGCCGACAC.

Its molecular mass is 25,904 Da. The protein translation is given as follows: SAHGIAKVPPGPNITAEYGDKWLDAKSTWYGKPTGAGPKD TGGACGYKDVDKAPFNGMTGCGNTPIFKDGRGCGSCFEIKCTKPESC SGEAVTVTITDDNEEPIAPYHFDLSGHAFGSMAKKGEEQKLRSAGELE LQFRRVKCKYPDGTKPTFHVEKGSNPNYLAILVKYVDGDGDVVAV DIKEKGKDKWIELKESWGAVWRIDTPDKLTGPFTVRYTTEG GTKSEVEDVIPEGWKADT. A gene is named as *Secc1*, that is registered as "I4IY76_SECCE" in the international European Molecular Biology Laboratory-European Bioinformatics Institute (EMBL-EBI) database (Augustin et al. 2012).

9 Utilization

Rye bread makes cheeks red. (Proverb)

9.1 NUTRITIONAL VALUE

Breads baked with rye flour are rich in fiber and rye provides a lot of iron for the body. Rye breads taste particularly strong, slightly sour, and very spicy. The higher the rye flour content, the stronger the flavor. Grain products for human nutrition are full grain, flour, and flakes (see Table 9.1). Rye contains 73% insoluble dietary fiber and the proportion of soluble fibers at 27% compared to other grains is relatively high. Rye fibers are beneficial to health than dietary fibers in many ways. They positively influence bowel activity, metabolism, and the quantity, quality, and composition of the intestinal flora. Particular attention is paid to the fiber under the lignans, mainly matairesinol and secoisolariciresinol. These are converted by using microorganisms of the intestine into the only effective diphenols enterolactone and enterodiol. Altogether rye is just healthy.

9.1.1 FIBER

Although the total production of rye is slightly diminished, its utilization in human consumption is increasing. Cereals are the most important source of dietary fiber in the human diet. The fiber in cereals is located mainly in the outer layers of the kernel, particularly in the bran. Rye is of special importance in contributing dietary fiber, because it is generally consumed as whole grain products, and has a high dietary fiber content in the starchy endosperm.

Whole grains are universally recommended as an integral part of the diet. Whole grains are an important source of nutrients that are in short supply in our diet, including digestible carbohydrates, dietary fiber, resistant starch, trace minerals, certain vitamins, and other compounds of interest in disease prevention, including phytoestrogens and antioxidants.

The main components of cereal grains are the hull, pericarp, testa, aleuron, endosperm, and germ. Rye is, in contrast to wheat, a special grain because it is mostly consumed as whole grain flour in breads and other cereal products. The grains may also be fractionated into different types of flour during the milling process. Rye has also a higher nutritional value compared to other cereals. Soluble fiber that reduces heart diseases in human makes up ~17% of dietary fiber found in the whole grain food. In addition, rye seeds contain numerous important vitamins (e.g., B, E) and minerals (e.g., Ca, Mg, P, K, Fe, Zn, and folate). Commercial whole grain rye flour and rye flakes have a fructan content of 4 g/100 g. Light-refined rye flour had fructan content of 3 g/100 g, and rye bran had fructan content of 7 g/100 g. Fructan content

TABLE 9.1

Basic Grain Products for Human Nutrition and Some Important Characteristics (per 100 g) according to the Regulations of the European Community 2012

Type	Energy (kJ/kcal)	Fat (g)	Protein (N × 6.25) (g)	Carbohydrates (g)	Fiber (g)	Sugar (g)	Fatty Acids (g)	Sodium (g)	Moisture (%)
Whole grain[a]	1362/322	1.7	9.5 (>11%)	61	13.2	0.90	0.30	<0.01	<15.0
Porridge	1362/322	1.7	9.5 (>11%)	61	13.2	0.90	0.30	<0.01	<15.0
Flour	1245/294	1.7	9.5 (>11%)	61				<0.01	<15.0
Flour (extracted, type: 997)	1391/329	1.1	7.4 (>8.5%)	68	8.6	0.9	0.2	<0.01	<15.0
Flour (thermally treated)	1362/322	1.7	9.5 (>11%)	61	13.2	0.90	0.3	<0.01	<15.0
Flakes (large)	1362/322	1.7	9.5 (>11%)	61	13.2	0.90	0.30	<0.01	<13.0
Flakes (small)	1245/294	1.7	9.5 (>11%)	61					<13.0

Source: Anonymous. 2013. Requirements for the labelling of special products made of grain. http://ec.europa.eu/enterprise/policies/european-standards/harmonised-standards/index_en.htm

[a] Minimum germination capacity is 98%; min. weight of a thousand grains is 40 g; min. falling number is 200 sec.

as high as 23 g/100 g was detected in the water-extractable concentrate of rye bran (Karppinen et al. 2003). Whole grain products contain antioxidants, especially phenolic acids and polyphenols, which protect against physiologic stress. Actually, the pericarp and testa of the caryopsis (see Figure 2.7) serve as the reservoir for antioxidants, rather than the whole grain (Landberg et al. 2008).

Results from both animal and human experiments point to rye dietary fiber as a good source of butyrate generation. Whole grain rye bread also lowers total bile acid concentration in feces and reduces the concentration of free secondary bile acids, primarily because of a much higher concentration of saponifiable bile acids in feces. The concentration of fecal litocholic acid, the most toxic of the bile acids, was significantly lower in a rye bread diet than in a wheat bread diet (Grasten et al. 2000). It is believed that saponifiable bile acids are less cocarcinogenic and comutagenic than free secondary bile acids.

Fiber increases fecal volume and weight because the unfermentable residue maintains its water-holding property and reduces the intestinal transit time. This promotes proper bowel function and prevents constipation. Dietary fiber is not digested in the upper gastrointestinal tract of humans. Depending on the baking process, part of the starch in rye bread may also be resistant to digestion in the small intestine and may reach the large intestine where it is fermented by colonic bacteria. The amount of this so-called resistant starch is 1%–2% and it increases the total amount of fermentable carbohydrates (Hansen et al. 1988).

9.1.2 ACRYLAMIDE

In addition, acrylamide forms from free asparagine and sugars during cooking, and products derived from the grain of rye contribute a large proportion of total dietary intake. Compared to wheat, rye contains the highest content of asparagine, one of the precursors of acrylamide, with ~36 mg/100 g dry matter. When free amino acid and sugar concentrations were measured in the grain of a range of rye varieties grown at different locations (Hungary, France, Poland, and the United Kingdom) and harvested in 2005, 2006, and 2007, genetic and environmental (location and year) effects on the levels of acrylamide precursors were assessed (Curtis et al. 2010). The data confirmed free asparagine concentration to be the main determinant of acrylamide formation in heated rye flour. In contrast to wheat, sugar, particularly sucrose, concentration correlated with both asparagine concentration and acrylamide formed.

Rye has higher arabinoxylan content than other cereals. Arabinoxylans are considered as the major limiting factor for use of rye in animal feeding. Arabinoxylan contents and extract viscosity are influenced by the genotype. The anthocyanin-free hybrid cultivars Helltop and Hellvus (see Table 7.6) combined reduced total, soluble, and insoluble fraction levels with low extract viscosity and high crude protein contents. It is more suitable for use in animal feeding than other varieties.

9.1.3 ARABINOXYLAN

The degradation of arabinoxylans and cellulose is much lower than that of the mostly soluble β-glucan. The potential of rye arabinoxylan to selectively stimulate some groups of bacteria in the colon is not fully understood yet. However, it was shown in pig experiments that whole grain rye is a much better stimulator of fermentation and enterolactone production than white wheat bread (Glitso et al. 2000).

9.1.4 Asparagine

Free asparagine concentration was shown to be under genetic, environmental, and integrated control. The same was true for glucose, whereas maltose and fructose were affected mainly by environmental factors and sucrose was largely under genetic control. Free asparagine concentration was closely associated with bran yield, whereas sugar concentration was associated with low Hagberg falling number. In addition, rye grain was found to contain much higher concentrations of free proline than that found in wheat grain, and less acrylamide formed per unit of asparagine in rye than in wheat flour.

9.1.5 Milling

Before rye grains can be used in food production, the outer part of the grain, the hull, must be removed. After hulling, which generally occurs during threshing, the grains are used as whole, cracked, or flaked, or they are ground to make flakes or flour. The starchy endosperm constitutes ~80%–85% of the weight of the whole kernel, the germ 2%–3%, and the outer layers ~10%–15%. In the milling process, the kernel can be ground and fractionated into different types of flour and bran. Ash content is a concept used to indicate the amount of inorganic minerals in food samples. Because the majority of minerals are located in the outer layers of the kernel, the ash content of flour indicates how much of the outer layers has been included in the flour. In white wheat flour (ash content 0.7% or less), ~30% of the outer layers of the kernel have been removed. Traditional rye bread and crisp bread are made of whole grain rye flour (ash content ~2%), in which all of the components of the kernel are present.

9.1.6 Carbohydrate

It can be estimated that the amount of carbohydrates potentially available for fermentation in the large intestine is ~12 g/100 g of whole grain rye bread, of which ~80% is in the form of nonstarch polysaccharides. This level is about three times higher than in white wheat bread. Moreover, whole grain rye bread provides 2 g of lignin per 100 g of dry matter, compared to 0.3 g in white wheat bread. The large amount of carbohydrates available for bacterial fermentation is beneficial for bowel physiology (Grasten et al. 2000).

9.2 FEEDING

9.2.1 Grazing and Fodder

In the United States, rye is primarily grown as forage. It grows tall with a tendency to lodge. As an obligate outcrossing species, its breeding has been difficult. Outcrossing is enhanced by a strong self-incompatibility system. Therefore, the majority of cultivars released across the southern United States are open-pollinated varieties, also sometimes referred to as "synthetics" (see Section 7.1.2), that are both heterogeneous and heterozygous. Released cultivars have probably changed genetically over time due to unintended natural selection processes during seed increase and maintenance or failure of seed production in isolation. Outside of the United States, namely, in Germany, hybrid rye cultivars for grain production have been successfully developed (see Section 7.1.3). Whether these production levels can be transferred to forage production and whether hybrid seed production becomes economical for forage in the United States remain in question.

Forage is used in the form of green chop, pasture, haylage, or hay, and is used alone or sometimes mixed with clover or ryegrass. Although it is less palatable than other forages, it is advantageous because it grows well in lower temperatures and matures earlier than wheat. Dairy animals fed rye as forage may have milk flavor affected. Fodder rye cultivars used for green fodder and hay in spring and summer grow fast and have thick bushes and foliage. They are able to regrow after being mown and their herbage is very nourishing. Their straw is thin and does not coarsen. They resist lodging and react well to application of retardants, which increases their resistance to lodging when they are farmed for their seeds.

The major breeding programs in the United States that have contributed to improvement of rye for forage in the past have been the Samuel Roberts Noble Foundation (Ardmore, OK), Texas AgriLife Research (College Station, TX), the University of Georgia (Athens, GA), and the University of Florida (Gainesville, FL). At the Samuel Roberts Noble Foundation, recent emphasis has been placed largely on early autumn to winter forage production. With the major breeding objective, effort has been placed on forage quality traits including crude protein, dry matter percentage, acid or neutral detergent fiber, or *in vitro* dry matter digestibility (Newell and Butler 2012). Currently, only the Noble Foundation and the University of Florida maintain active rye breeding programs. All released cultivars currently grown as forage are population varieties and are derived from open-pollinated single-plant visual selections from earlier cultivars, single crosses, or polycrosses (see Section 7.1.1). Interestingly, the pedigrees of most of the current cultivars are derived from Abruzzi, originally collected from Italy, or Maton, derived from a polycross of 43 collections of Balbo made from various states and 10 cultivars including Abruzzi (see Table 7.6). Although most of the recent cultivars are derived from these few released sources, the released varieties are not static populations but most likely differ with respect to their spatial and temporal origin. For the region surrounding the Samuel Roberts Noble Foundation, much of the forage rye produced consists of Elbon and Maton gene pools. In the southeast of the United States, most of the acreage is planted with Wrens Abruzzi.

9.2.2 GRAIN

Another use of rye grain is for animal feed. Although rye has a higher feed value than oats or barley, and has a feeding value of 85%–90% of maize, its high soluble fiber content makes it better suited to ruminants than monogastrics (see Table 9.2). It is most palatable to livestock when it is 30% or less of the feed. It should not be fed to young animals, as it may cause digestive disorders.

TABLE 9.2
Starch and Protein Contents of Some Cereal Crops

Crop	Crude Protein (%)	Starch (%)	Energy Content (MJ/kg)
Rye	8.6	53.2	13.4
Wheat	11.7	59.0	14.1
Triticale	11.0	56.4	13.9
Barley	10.9	50.3	12.8

Source: de Vries, G., *Roggen: Anbau und Vermarktung*, Roggenforum, Germany, 2006. With permission.

In animal feeding, soluble dietary fibers have been considered as the main antinutritive factor. High levels of soluble pentosane content in grains decreases feed intake in chickens and piglets, hampers digestion, and makes droppings very sticky, resulting in inhibition of growth (Sodkiewicz and Makarska 2004). Since pentosanes are needed in grain rye for bread making but not for feeding, a specialized breeding is required!

Experiments were conducted with growing chicks. In one trial, the effects of soybean oil and tallow at two dietary concentrations and penicillin supplementation on the utilization of rye- and wheat-based diets were studied. Rye-fed chicks, when compared to wheat-fed chicks, had poorer performance and nutrient retention. Substitution of tallow for soybean oil in the rye-fed chicks decreased feed intake, efficiency of feed conversion, and weight gain by 8%, 17%, and 23%, respectively, while the corresponding reductions in the wheat-fed birds were only 2%, 2%, and 3%. Digestibility of dietary lipids in the rye-fed chicks was affected by the type and concentration of lipid, as inclusion of tallow at 3% and 8% decreased lipid digestibility by 18% and 34%, respectively, compared with similarly formulated wheat diets. The corresponding reduction in the digestibility of soybean oil was less (9%) and was independent of the dietary levels of soybean oil. Penicillin supplementation of rye diets improved feed intake (7%), feed conversion efficiency (9%), weight gain (19%), and lipid digestibility (9%), while it had almost no effect in the wheat-fed chicks. In a second trial, the substitution of sunflower oil, soybean oil, and coconut fat by tallow in rye diets reduced feed conversion efficiency by 18%, 11%, and 11%, respectively, while the corresponding changes for the wheat diets were only −1%, 1%, and 4%. These studies demonstrate that rye contains the antinutritional factors that interfere with the utilization of lipids, particularly when the diet contains a high concentration of long-chain saturated fats (Antoniou and Marquardt 1982).

In Poland, rye grain is used for feeding cattle, old chicken, and porkers in mixtures of 30%, 10%–20%, and 40%–60%, respectively.

9.3 BREAD MAKING

Bread was not eaten until the beginning of rye cropping. In fact, most of the high and late medieval population had to live with simple dishes, while a good number of food ingredients were reserved for the table of nobility and high clergy. Cabbage, turnips, and unspiced porridge made from rye or millet were the staple food in the general population.

Much later, the dark rye bread became a staple food. Rye flour is used in particular to make the famous bread of German origin, with its sour taste and heavy texture. It is also used in the preparation of crisp breads, spice breads, crepes, pâtés, and muffins. In the book *Physica* (1151–1158) written by the German abbess Hildegard von Bingen (1098–1179), it is already apparent that "the bread made from rye is good for healthy people, and makes them strong." "It has many powers," said the abbess. It is said "to help lose weight of obese people, without weakening them. For people with weak digestion, however, it is unfavorable!"

Various types of rye grain have come from places all over Europe, such as Finland, Denmark, Russia, Baltic countries, and Germany, and India; the type of bread varies accordingly. It can be light or dark in color, depending on the type of flour used and the addition of coloring agents, and is typically denser than bread made from wheat flour. Because of the higher fiber content of its flour, the bread shows the highest fiber content among many common types of bread. Loaves are darker in color and stronger in flavor.

In addition, bread quality depends on the ways of extracting flour. Rye flours with extraction rate of 100% (wholemeal flour), 95% (brown flour), 90% (brown flour), and 70% (light flour) show different contents of bioactive compounds, for example, total phenolic compounds, total flavonoids, inositol hexaphosphate, reduced and oxidized glutathione, tocopherols, and tocotrienols. Studies strongly support the use of rye flours with extraction rates of up to 90% in the bakery industry. It can be suggested that in the near future more and more rye products will be available in the market, especially those originated from whole meal or brown flours (Michalska and Zielinski 2006).

There are some substantial differences in the biochemistry of wheat and rye that can drastically affect the bread-making process. A key issue is amylases. While wheat amylases are generally not heat stable and have no effect on the stronger wheat gluten, rye amylase remains active at substantially higher temperatures. Since rye gluten is not particularly strong, the main structure of the bread is based on complex polysaccharides, including rye starch and pentosanes, and the amylases in the flour can break down the resulting structure, inhibiting the rise of the dough.

The starch of rye already gelatinizes at temperatures around 50°C, while the starch of wheat needs a temperature 10°C–20°C higher than rye. Therefore, the enzymatic digestion of rye starch remains longer and can weaken the crumb of the bread. Even the elasticity of the crumb may suffer from very long enzymatic activity. The analysis of the enzymatic activity and the strength features of the flour are therefore important components of routine for the examination and assessment of its baking quality. This is especially true for quality classification of rye flour (see Table 9.3). To assess the enzymatic activity of its flour, its maltose content, falling number, dextrin content, and amylograms are considered.

Amylographic investigations describe the viscosity of the flour, that is, the ability of dough swelling and gelatinization. Good rye flour may show a maximum amylogram value of 520 (a.u.) together with a maximum temperature of 67.5°C (see Figure 9.1). An amylograph with automatic data recording is shown in Figure 9.2. The loaf shape is correlated with the amylogram values. Low values result in low-quality bread whereas high values result in well-shaped bread (see Figure 9.3).

TABLE 9.3
Standard Grading of Rye Flower and Main Characteristics according to German Regulations

Grade	Falling Number (sec)	Maltose Content (%)	Characteristics
High	>180	<2.5	Low enzymatic activity; delayed fermentation; small volume; low tanning; fader taste; low shelf life; tendency to crumble
Middle	120–180	2.5–5.0	Good volume
Low	<120	>5.0	Increased enzymatic activity; fast fermentation; weaker dough; moist crumb; inelastic crumb

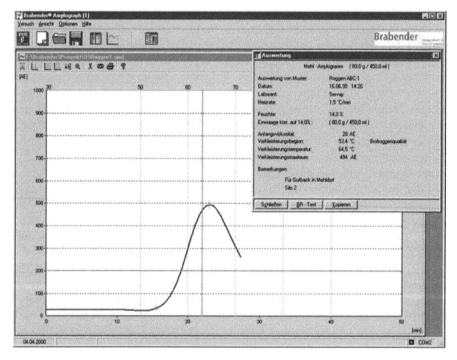

FIGURE 9.1   Amylogram of a common rye flower, prepared by an amylograph. (Courtesy of Löns, M., Brabender Ltd., Duisburg, Germany.)

Low dextrin values are associated with low enzymatic activity of the flours. High dextrin values indicate high enzymatic activity. The treatment of commercial flours with special enzymes can be detected, which is difficult to reveal just by falling number. Rye flour with low enzymatic activity shows dextrin values of 9–10, with optimal activity values of 11–14, increased activity values of 15–17, and very high activity values of >18.

When two rye cultivars, Marder (hybrid) and Motto (population), with falling numbers 314 and 309, respectively, were germinated *in vitro*, only minor changes in

FIGURE 9.2 An amylograph with automatic data recording. (Courtesy of Löns, M., Brabender Ltd., Duisburg, Germany.)

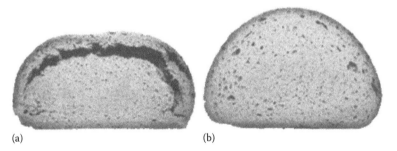

(a)   (b)

FIGURE 9.3 Loaf shape of rye breads differentiated by low (110; a) and high (430; b) amylogram values. (Courtesy of Löns, M., Brabender Ltd., Duisburg, Germany.)

the microstructure of the cell walls and proteins in the kernels could be demonstrated. Doughs made from flours of germinated grains are always softer than doughs made from flours of native grains, and Marder doughs were always more rigid than Motto doughs. The higher the water content and the longer the incubation time, the greater the rheological changes during incubation. Microstructural studies showed that germination and incubation caused changes in the cell wall structures of dough that can explain the softening of the doughs (Autio et al. 1998).

TABLE 9.4

Mean Change of Protein Content by Strong Selection within Inbred Lines of Rye

Year	Mean Protein Content (%)	Range of Variation
1935	17.2	26.1–8.3
1936	14.3	21.6–7.0
1937	19.1	24.4–13.8
1938	14.4	21.6–7.2
1939	16.8	24.0–9.6
1940	16.6	20.4–12.7

Source: Aust, S., and Ossent, H. P., *Züchter*, 13, 78–84, 1941. With permission.

Breeding for bread-making characteristics began comparably late. Neumann (1929) reported first studies from Germany in 1928. Since 1935, regular testing of protein contents was described by using self-fertile inbred lines. A strong selection for this characteristics led within 6 years to a significant increase of protein content in Petkus derivatives from ~7% to 26% (see Table 9.4). Most of those inbreds were not directly usable in breeding, but as crossing parents they have been included. In bread making, this protein requires exposure to an acid such as lactic acid so that the bread will rise. This is usually achieved by sourdough fermentation, that is, yeast—characteristic for rye bread. In addition, there is an antifungal hydroxy fatty acid produced during sourdough fermentation. *Lactobacillus hammesii* converts linoleic acid in sourdough and the resulting monohydroxy octadecenoic acid exerts antifungal activity in bread, as recent studies by Black et al. (2013) demonstrated. During the baking process, the loaf can form a crust. However, it makes rye bread hard when it is eaten after three to five days.

An old proverb says: "Hard bread is hard, no bread is even harder." The hard rye bread was once the staple food of European population.

The secalins are the typical prolamin glycoprotein found in the grain of rye and are one of the forms of gluten proteins that people with celiac disease cannot tolerate. Thus, rye should be avoided by people with this disease. It is generally recommended that such people follow a gluten-free diet.

Sixty-two DNA sequences are known for the coding regions of, for example, ω-secalin genes. Only 19 out of the 62 ω-secalin gene sequences were full-length open reading frames (ORFs), which can be expressed into functional proteins. The other 43 DNA sequences were pseudogenes, as their ORFs are interrupted by one or a few stop codons or frameshift mutations. The 19 ω-secalin genes have a typical primary structure, which is different from wheat gliadins. There is no cysteine residue in ω-secalin proteins, and the potential celiac-toxic epitope was identified to appear frequently in the repetitive domains. The ω-secalin genes from various cereal species share high homology in their gene sequences. The ω-secalin gene family has involved fewer variations after the integration of the rye chromosome or whole genome into the wheat or triticale genome. Based on the conserved sequences of ω-secalin genes, it is possible to manipulate the expression of this gene family in rye, triticale, or wheat 1BL.1RS translocation lines to reduce its negative effects on grain quality (Jiang et al. 2010).

9.3.1 BREAD FOR MILLIONS

According to the World's Healthiest Foods rating, rye ranges among the best nutrients. This estimation was published as 'Reference Values for Nutrition Labeling' by the US Food and Drug Administration. The standard considers adult women of ages 31–50, 1816 calories, and body mass index (BMI) of 18.5. In Table 9.5, a few important nutrients are given and their evaluation according to the US standard, despite rye, is a better grain choice for persons with diabetes. Rye is a good source of fiber, which is especially important in the United States and industrial countries since most of them do not get enough fiber in their diets. Rye fiber is richly endowed with noncellulose polysaccharides, which have exceptionally high water-binding capacity and quickly give a feeling of fullness and satiety, making rye bread a real help for anyone trying to lose weight.

Rye can be eaten as whole grain, steel cut kernel, crushed grain, malted and crushed grain, precooked kernel, malted kernel, whole grain flour, sifted flour, bran, flakes, four-grain flakes, toasted flakes, breakfast cereal, or sourdough rye bread mix. Known bread types are common bread, crisp bread, thin crisp bread, rolls and buns, rolls, buns and breads containing wheat, or parbaked products. Recently, novel products, such as porridge, berry pastries, pasta, snacks, and crisp bread sandwiches, have been introduced.

Various types of rye products are used daily in many countries. Today, after several decades of diminishing rye consumption in favor of wheat, the increased knowledge of the relationship between diet and well-being has again raised the consumer's interest toward rye. The cereal industry is actively developing new and innovative rye products. In 2012, the food consumption of rye in the world was ~6 million metric tons, which is ~46% of total production. The rest was used as feed (see Table 9.6). Approximately 3.14 billion people live in countries where rye is consumed. The average consumption is ~1 kg per capita per year. However, countries such as Belarus

TABLE 9.5
Important Nutrients of Rye Flour (100 g) and the Nutrient Value according to the US Food and Drug Administration, Modified

Nutrient	Amount/100 g	Daily Value (%)	Nutrient Density	Rating
Fiber	15.10 g	72.0	6.9	*
Manganese	2.58 mg	33.7	3.2	**
Phosphorus	332.00 mg	18.5	1.8	*
Magnesium	110.00 mg	15.3	1.5	*
Tryptophan	0.10 g	18.8	1.8	*
Calories	338 cal			*

Note: *Good (daily value ≥ 25%, OR density ≥ 1.5, and daily value ≥ 2.5%); **very good (daily value ≥ 50%, OR density ≥ 3.4, and daily value ≥ 5%).

TABLE 9.6
World Rye Production for Food Supply (t) according to the Food and Agriculture Organization, Modified

Country	Inhabitants (Mio)	Total Production (t)	Food Supply (t)
Albania	3	2,500	2,005
Argentina	41	44,460	?
Armenia	3	345	225
Australia	22	33,300	9,505
Austria	8	163,600	107,535
Belarus	9	734,606	319,512
Belgium	11	1,969	?
Bolivia	11	106	?
Bosnia	4	7,424	5,055
Brazil	194	3,165	3,100
Bulgaria	7	17,500	5,909
Canada	35	216,400	17,361
Chile	17	4,774	4700
China	1358	650,000	384,700
Croatia	4	2,507	2500
Czech Republic	10	118,200	104,025
Denmark	6	254,700	90,315
Ecuador	15	90	55
Egypt	82	41,000	9,428
Estonia	1	25,100	25,100
Finland	5	68,500	84,292
France	64	150,797	29,074
Germany	82	2,903,470	705,006
Greece	11	34,500	3,502
Hungary	10	79,400	11,555
Ireland	5	4,200	3,308
Italy	61	13,926	4,474
Kazakhstan	17	42,150	15,385
Kyrgyzstan	6	1,000	856
Latvia	2	86,700	56,250
Luxembourg	1	5,116	1,543
Macedonia	2	8,850	329
Mexico	116	~1,000	759
Moldova	4	~3,700	3,614
Montenegro	1	900	897
Morocco	33	3,000	2,269
The Netherlands	17	10,242	8,053
North Korea	25	70,000	46,761
Norway	5	~33,000	32,028
Peru	30	~150	140

TABLE 9.6

(Continued) World Rye Production for Food Supply (t) according to the Food and Agriculture Organization, Modified

Country	Inhabitants (Mio)	Total Production (t)	Food Supply (t)
Poland	38	3,270,300	1,240,076
Portugal	11	~33,000	32,255
Romania	21	34,281	32,462
Russia	143	1,635,630	1,282,051
Serbia	7	10,470	9,113
Slovakia	5	~44,000	43,392
Slovenia	2	~9,000	8,287
South Africa	51	~4,500	4,276
Spain	46	251,800	40,000
Sweden	9	123,400	109,229
Switzerland	8	13,708	10,378
Turkey	75	365,560	218,677
UK	63	38,000	19,718
Ukraine	46	464,900	437,179
USA	318	188,760	82,841
World, rye-growing countries	**3136**	**12,373,640**	**5,734,183**
World, all countries	**7058**		**46%**

Source: http://www.fao.org/corp/statistics/en.
Note: Bold text represents Mean. ?, no data provided.

consume ~33 kg per capita per year, followed by Poland 32 kg, Latvia 19 kg, Estonia 18 kg, Denmark and Finland 16 kg, Sweden 11 kg, or Czech Republic 10 kg.

9.4 BIOMASS AND BIOGAS PRODUCTION

Winter rye is an ideal crop for agricultural energy production because of its vigorous growth, high nutrient- and water-use efficiency, and low input production. For use in biogas plants, maximal biomass yield with dry matter contents of >30% is an essential breeding aim, for example, in Germany. Dry matter yield ranges, on an average, from 130 to 141 dt/ha. In one experiment, the best biomass donor was the American population Florida Black (173 dt/ha). In Europe, the harvest of green rye usually starts in May, so fields can be used by a second crop such as maize. If the stand is harvested during the milk ripeness, the biomass yield is higher and hybrids are superior over population varieties.

Broad genetic variation for biomass yield exists in elite hybrid materials, population varieties with forage grain use, current breeding material, tetraploid rye, and genebank accessions. In a 2-year trial, significant genetic variances in both the population per se performance and the testcross performance were demonstrated by Roux et al. (2010) for total dry matter yield at different cutting dates as well as for grain yield. At cutting dates 1 and 2, the average yields amounted to 70.1 and 131.9 dt/ha dry matter, respectively. Forage rye materials were superior with regard to dry matter yield at cutting date 1, while forage rye and a population variety achieved the highest dry matter yield at cutting date 2 in the testcross performance.

The heterotic increase averaged a substantial level of 9.3%, 11.6%, and 32.3% at cutting dates 1 and 2 and for grain yield, respectively. Traits with high relevance for biomass production were identified, such as height, date of ear emergence, and dry mass content. Correlations between the dry matter yield, the biogas production, and the content of lignin were determined.

Approximately 380 testcross progeny incorporating intrapool crosses and interpool hybrids were analyzed by Miedaner et al. (2011b) in six experiments across each of three to four locations. Mean values of biomass yield on a 100% dry matter basis ranged between 152 and 160 dt/ha, and the best entries had 175 dt/ha across locations. Genotypic variances for biomass yield were significant in all experiments, which are higher for interpool hybrids. Genotype–location interaction variances were significant and, in most instances, of similar size than genotypic variances. Estimates of entry-mean heritability were moderate. Consistently, significant genotypic correlations with biomass yield were found for plant height (0.4–0.9) and early growth (0.2–0.8). The promising results can stimulate selection for this newly adopted trait.

Another strategy is to use whole-plant silage for biogas reactors. Fast juvenile development in spring compared to other cereals and unpretentiousness in poor soils enable rye to be a universal crop in many locations. Hybrid varieties can even support this development. A superior hybrid variety is Helltop (see Table 7.6). It combines high plant growth and large seeds with unusually good lodging resistance, maximal biomass yield, and good field resistance. This particular genotype is nicknamed Stabilstroh (stable straw) and shows larger stem diameter next to thicker and more structured stem walls (Muszynska 2013).

In general, the production costs per unit (1 kWh) energy supply are 2.05 cent/kWh in maize silage, followed by sunflower (−3%) and rye and/or triticale (−5.8%) (de Vries 2006). When the supply costs are related to the yield of methane, it turns out that silages from maize and other cereals deliver the best substrates for biogas production. Thus, not just the cost for the production of the substrate is important. It is equally important how much biogas can be produced per dollar.

9.4.1 THERMAL UTILIZATION

Because of relatively low cereal prizes in the world market and the high prices of crude oil, thermal utilization of cereal crops has become increasingly important in industrial countries. Therefore, some farmers intend to use cereal crops for heating. Even the utilization of cereal grain becomes of interest. When used for heating the whole plant, the straw or the grain can be utilized. Because of the high-energy concentration of the grain (4.4 kWh/kg) and its simple handling, it becomes more and more commercially attractive. Rye has a somewhat higher energy concentration

(4.74 kWh/kg) and heating value than wheat. Based on the energy value of cereal grain, it is currently cheaper than fuel oil. Therefore, its introduction as a source of energy is useful, particularly as fuel.

9.5 CATCH CROP

This is a method of increasing agricultural or horticultural productivity by filling in the empty field spaces; for example, it is created when slow-growing vegetables are harvested with fast-growing crops. Usually, catch crop is a short duration crop grown in between two main crops in rotation to maximize cropping intensity, for example, summer green gram grown in between two main cereal crops (wheat–green gram–maize). A catch crop is often grown without any extra nutrient application and is expected to feed on the residues of the nutrients that were applied to the main crops.

Rye has a good repressive ability as other autumn-sown crops. It also shows an early closure of the stand (Percze 2006). The most typical weeds are the winter species, early spring species, and the perennials. Among the tillage variants, plowing and disking result in lowest weed covering. For example, the weediest situation in Hungary is observed after direct drilling and French cultivation. When rye is involved as catch crop, the weediness is significantly lower (see Table 9.7). When direct drilling is applied as tillage method, the effect of rye is most pronounced.

Brassicaceae and rye as cover crops alter free-living soil nematode community composition. In an experiment, the cover crop treatments included mustard blend (*Sinapis alba* and *Brassica juncea*) Caliente, rapeseed (*B. napus*) Essex/Humus, forage radish (*Raphanus sativus*) Dichon, oilseed radish (*R. sativus*) Adagio/Colonel, rye (*Secale cereale*) Wheeler, and a no cover crop (winter weeds) control. Soil samples (0–15 cm) were collected two or three times per year and the extracted nematodes were identified for genus or family. Cover crops had unique impacts on

TABLE 9.7

Average Weed Cover (%) Applying Different Tillage Methods and Rye as Catch Crop

	Tillage Method					
Vegetation	Plowing (26–30 cm)	Direct Drilling	French Cultivation (12–16 cm)	Cultivation (16–20 cm)	Disking (16–20 cm)	Loosening + Disking (40 + 16–20 cm)
Mean over six vegetative stages without rye	2.7	5.6	4.6	4.8	1.8	4.1
Mean over six vegetative stages with rye	1.5	3.3	3.4	2.9	2.6	2.5

Source: Percze, A., *Cereal Res. Comm.*, 34, 259–262, 2006. With permission.

nematode communities, but these impacts appeared to be associated more with the quality of organic matter inputs rather than biofumigation or allelopathy. Across experiments, fungi or nematode abundance was increased in either rapeseed Essex or rye compared to radishes or the control. Canonical discriminant analysis suggested that rapeseed and rye had similar effects on the nematode community composition, as did the two radish cultivars, though distinct from the effect of rye and rapeseed. Overall, the results suggest that radishes stimulated a bacterial decomposition pathway, while rapeseed and rye stimulated a proportionally greater fungus-based food web (Stocking-Gruver et al. 2010).

When used as poor cover, rye is planted in the fall, killed in the spring, and left to decompose in the same fields where soybean and other cash crops are later planted. But instead of mowing, many farmers will flatten it out by attaching a rolling, paddlewheel-like cylinder with metal slats to a tractor and barreling over the rye, tamping and crimping it into a mat. Rolling rye with a roller–crimper uses less energy than mowing, is faster, and only needs to be done once a season. Unlike mowing, it also leaves rye residue intact in the field, forming a thick mat that can provide better weed suppression. The varieties Aroostook and Wheeler (see Table 7.6) are particularly suitable for that purpose.

9.6   ETHANOL PRODUCTION

Rye—the unfinished whisky. (Bierce 2008)

9.6.1   WHISKY AND VODKA

Rye grains are used in the manufacture of alcoholic drinks and industrial alcohol. They are grown and used in the same way as wheat sprouts. In order to minimize production cost yield of the variety is the most wanted trait. Lodging resistance and resistance to diseases are needed. High starch content associated with low protein and pentosanes guarantees high ethanol output. It can be indirectly selected by bigger seeds, that is, 1000-grain weight or hector liter weight. The negative correlation between starch content and protein content favors selection work.

Since the by-products of ethanol processing often are used for feeding animals, the seed lots should be free of ergot and fusariosis. Mycotoxins can be concentrated during processing.

9.6.2   KVASS

The Russian kvass is one of the best nonalcoholic beverages. It is inferior to none in the matter of taste and alimentary properties. It was invented more than 1000 years ago. Nowadays, it has also a well-deserved popularity. In Russia, kvass was first mentioned in 989, when the Duke of Moscow, Wladimir, turned his nationals Christian. It was chronicled: "To distribute food, mead and kvass to the people."

Kvass, which is made of rye or barley malt, has not only high taste properties, but also invigorates and improves metabolism. It is similar to kefir in its health effect, as also to sour clotted milk, koumiss, and acidophilus milk. Kvass, as the product of lactic fermentation, regulates the work of the gastrointestinal tract, prevents pathogenic bacteria from reproducing, tonisizes muscles, improves metabolism, and the working of the cardiovascular system. These healing properties of kvass can be explained by the availability of lactic acid, vitamins, free amino acids, sugar, and diverse trace elements.

At home, kvass is made of rye malt, under set conditions for the rich chemical composition of this product. Rye has many healthy substances. Therefore, there is 80 g calcium, 340 g phosphorus, 13 g iron, 1.8 g copper, 8 g manganese, 5 g molybdenum, 3.5 g zinc, and 11 g cobalt in 100 g rye grain. Those elements act as an important part of metabolic control and their intake into the organism is quite desirable. Along with the trace substances, more than 10 amino acids are available, and 8 of them are essential. Considering this, the importance of this traditional malt beverage becomes greater. The number of vitamins in the kvass is not very large, but their regular intake into the organism has a positive effect. There are vitamins such as B_1 (0.2 mg), carotene (0.2 mg), pyridoxine (0.2 mg), riboflavin (0.2 mg), PP (1.2 g), and H and E (100 g) of the rye seed. There are proteins (2 g), carbohydrates (50 g), organic acids (3 g), and the above-listed vitamins in 1 l volume of the kvass.

If there is bread and kvass, then everybody is with us. (Russian proverb)

9.6.3 INDUSTRIAL ETHANOL

Ethanol made from grain has been and will continue to be an important raw material for chemical engineering, pharmaceuticals, cosmetics, food, and fuel. The carbon footprint of ethanol is important. It is drastically influenced by the consumption of energy needed for the distillation process. The use of up to 20–30 l fuel oil is the accepted standard for distillation of 100 l of ethanol (86 vol.%). The use of hammer mills is the standard procedure for bioethanol production in large distilleries. Its advantages are low cost for investment, low cleaning requirements for grain, and high performance. However, hammer mills need a much higher energy input than roller mills. They destroy the stable fiber structure of the bran largely, but it is not necessary anymore. Modern fermentation processes and enzymes are able to convert unreleased starch particles in the roller mill flours into glucose and ethanol. The analysis of the press cakes made from roller-milled grain shows dry matter contents of ~30% in rye and ~31% in wheat and triticale.

For starch hydrolysis, cereals such as rye, wheat, and barley need additional reducing steps of viscosity. They contain higher amounts of water-soluble and water-insoluble β-glucans as well as pentosanes. A new enzyme Sanferm from Novozymes®, the world leader in bioinnovation, helps by allowing distillers to achieve 2%–3% higher alcohol yields and increased fermentation efficiency, with a 10%–20% shorter fermentation time.

Rye hybrids were proved to be a very good raw material for ethanol production, as good as a regular variety. The most profound difference found in the chemical composition of hybrids and a regular variety was higher level of pentosanes, reaching from 7.6% to 9.2% of dry matter. Applying hemicellulase enzyme in the mashing process as the additional enzyme is particularly reasonable for rye hybrids. Taking into account higher productivity of hybrid varieties, the ethanol yield from a hectare of land might be up to 40% higher in comparison with a regular rye grain (Czarnecki and Nowak 2005).

9.7 OTHER USES

9.7.1 Substitution of Crude Oil Products

In Germany, there are projects to substitute common raw materials by "sustainable cereal products." Marketable products, such as Ceralith® (70% rye flour)—a granulated insulation, Rofa® (~40% rye flour)—against soil erosion and soundproofing, or Extrudat RQM (90%–100% rye flour)—used as binder for several industrial colors, are a few examples of additional and innovative utilization of rye.

9.7.2 Phytosterols

Phytosterols (including saturated forms, i.e., phytostanols) are a group of cholesterol-like molecules that have gained considerable interest during the past decade. Phytosterols are characterized by the occurrence of a methyl or ethyl group in carbon 24 of the basic cholesterol molecule. This interest in phytosterols is mainly due to the fact that dietary phytosterols inhibit cholesterol absorption, which leads to decreased plasma low-density lipoprotein (LDL) cholesterol levels, and therefore a potentially decreased risk of developing cardiovascular diseases. Furthermore, several studies report that phytosterols might protect against development of colon cancer, both in murine models and in human *in vitro* models. It has also been claimed that low doses (60 mg/day) of phytosterols have a positive impact on the human immune system, and in clinical trials, it has been shown that intake of phytosterols decreases the symptoms of benign prostata hyperplasi. Thus, intake of phytosterols has an array of potential health-promoting effects and several phytosterol-enriched functional food products are now commercially available in Europe and the United States.

As applicable to rye, a screening revealed that statistically significant differences in sterol content exist, both between varieties and between the years (Zangenberg et al. 2004). The highest phytosterol content occurred in the Russian cultivar Tschulpan 3, which contained ~967 mg/kg, and in the Finnish population cultivar Anna, which contained ~946 mg/kg. The German hybrid cultivar Amando contained the lowest amount (~761 mg/kg), while intermediate amounts of plant sterols were found in the hybrid cultivar Esprit and the population cultivars Dominator and Motto. The highest amount (~1007 mg/kg) found in the entire study occurred in Tschulpan 3 in 1997, while the concentration found for Amando in 1999 was the lowest. Thus, there was ~25% difference in phytosterol content between the highest and the lowest content.

The sterol composition was similar both between the different cultivars and between the years. It was dominated by sitosterol and campesterol, while stigmasterol, campestanol, and sitostanol constituted minor fractions. On average in the entire study, sitosterol and campesterol constituted ~59% and ~20% of the total sterol content, respectively. Sitostanol constituted ~8%, and campestanol and stigmasterol constituted ~5% and ~4%, respectively.

Thus, Tschulpan 3 would be a suitable cultivar to focus on in a breeding program aimed at increasing the phytosterol concentration in rye.

9.7.3 ANTIOXIDANTS

Phenolic compounds are the most abundant phytochemicals in rye grain, and also the most structurally diverse. They have also been widely studied as the major group of compounds with antioxidant activity, although the biological significance of this activity in human health remains to be established. They include phenolic acids, alkylresorcinols, lignans, and flavonoids. They are concentrated in the outer layers of the grain.

The most abundant are the phenolic acids, which are aromatic rings with one or more hydroxyl groups. They are divided into two groups: hydroxybenzoic acids (notably syringic and vanillic acids) and hydroxycinnamic acids (notably *p*-coumaric, ferulic, and sinapic acids). Furthermore, they occur in three forms within the grain: as free acids; as soluble conjugates with sugars, sterols, or terpenes; and bound to insoluble polymeric components (arabinoxylan, lignin, proteins). Phenolic acids exhibit antioxidant activity with strong correlations being reported between the content of the total phenolic acid and the total antioxidant activity. Hence, these have been proposed to contribute to the health benefits of whole grain cereals.

Dietary antioxidants that protect LDL from oxidation can help to prevent atherosclerosis and coronary heart disease. The antioxidant activities of purified monomeric and dimeric hydroxycinnamates and phenolic extracts from rye (whole grain, bran, and flour) were investigated using an *in vitro* copper-catalyzed human LDL oxidation assay (Andreasen et al. 2001).

The most abundant ferulic acid dehydrodimer (diFA) found in rye, 8-*O*-4-diFA, was a slightly better antioxidant than ferulic acid and *p*-coumaric acid. The antioxidant activity of the 8-5-diFA was comparable to that of ferulic acid, but neither 5-5-diFA nor 8-5-benzofuran-diFA inhibited LDL oxidation when added at 10–40 µM. The antioxidant activity of the monomeric hydroxycinnamates decreased in the following order: caffeic acid > sinapic acid > ferulic acid > *p*-coumaric acid. The antioxidant activity of rye extracts is significantly correlated with their total content of monomeric and dimeric hydroxycinnamates, and rye bran extract was the most potent. The data suggest that especially rye bran provides a source of dietary phenolic antioxidants that may have potential health effects.

9.7.4 CANCER INHIBITION

Rye bread contains a large amount of fiber and only a little fat. It does not create high spikes in blood sugar as white bread and other breads do. Moreover, rye is a cereal, which is the richest source of lignans (~1547 µg/100 g pinoresinol and 3540 µg/100 g rye bran). The lignans are a group of chemical compounds found in many plants. They are one of the major classes of phytoestrogens, which are estrogen-like chemicals and act as antioxidants, that is, cancer inhibiting. The other classes of phytoestrogens are the isoflavones and coumestans (Adlercreutz 2002). Plant lignans are polyphenolic substances derived from phenylalanine via dimerization of substituted cinnamic alcohols.

9.7.5 ALKYLRESORCINOLS

Alkylresorcinols are phenolic lipids comprising a 1,3-dihydroxylated benzene ring with an alkane chain at position 5. A range of forms occurs in wheat in which the alkane chain is usually saturated and comprises 17, 19, 21, 23, or 25 carbons. They are located in the outer layers of the grain (the nucellar epidermis, testa, and inner pericarp) and can be used as a biochemical marker for these tissues in milling fractions. Alkylresorcinols have also been found to be concentrated within the cuticles of rye leaves, with similar homolog compositions occurring on the adaxial and abaxial leaf surfaces. Ji and Jetter (2008) localized very long-chain alkylresorcinols in the intracuticular wax of rye leaves near the tissue surface.

Alkylresorcinols were proposed as biomarkers for the consumption of whole grain products in human nutrition studies and may have specific health benefits. The total alkylresorcinol content in wholemeal flours ranged from 0.1% to 0.3% of dry weight. Human subjects absorb alkylresorcinols in the small intestine via the lymph system, which circulate bound to the lipoproteins. Human urine after rye bread diets was shown to contain two metabolites, 3,5-dihydroxybenzoic acid (DHBA) and 3-(3,5-dihydroxyphenyl)-1-propanoic acid (DHPPA), but very small amounts of unchanged alkylresorcinols, supporting the hypothesis that alkylresorcinols are metabolized in humans via β-oxidation of their alkyl chains.

A quick and simple semiquantitative method of selecting single seeds of rye for their content of 5-alkylresorcinols was presented by Hoffmann and Wenzel (1977). The resorcinols are extracted from whole, intact caryopses with acetone. Under the conditions described, the germination rate is hardly decreased. The single-grain method offers especially striking advantages when self-fertile inbred lines are used. The method can be applied in breeding both for selection of high and low contents of alkylresorcinols. Later Ross et al. (2001) established a method for a fast and easy analysis of 5-n-alkylresorcinols, considering the effects of milling (intact grains vs. flour), extraction solvent (ethyl acetate, methanol, and acetone), extraction volume per gram of sample (20 and 40 ml), and extraction time (3, 6, 18, and 24 h). The ease of extraction is consistent with the presence of alkylresorcinols in the outer layers of the kernel and their absence in the endosperm. The extracts were analyzed (without further purification or derivatization) by gas chromatography using methyl behenate as the internal standard. The methodology was then applied to analyze the alkylresorcinol content in 15 rye cultivars grown at two locations in Sweden. Total alkylresorcinol contents varied within the range 549–1022 μg/g.

9.7.6 CHITINASE

Chitinases are hydrolytic enzymes that break down glycosidic bonds in chitin. As chitin is a component of the cell walls of fungi and exoskeletal elements of some animals (including worms and arthropods), chitinases are generally found in organisms that either need to reshape their own chitin or dissolve and digest the chitin of fungi or animals. Chitinases are also present in plants. Some of these are pathogenesis-related proteins that are induced as part of systemic acquired resistance (see Section 8.3). Expression is mediated by genes and the salicylic acid pathway,

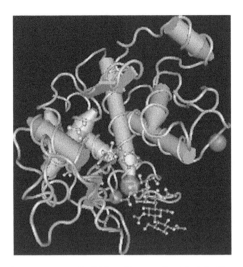

FIGURE 9.4   Crystal structure and chitin oligosaccharide-binding mode of a "loopful" family GH19 chitinase from rye seeds, *Secale cereale*, at resolution 1Å. (Courtesy of T. Fukamizo.)

both involved in resistance to fungal and insect attack. Other plant chitinases may be required for creating fungal symbioses. Rye chitinase binds the hyphal tips, lateral walls, and septa of fungi, and degrades mature chitin. It is localized in the aleuron cells of the seed endosperm (at protein level). Levels increase from 23 to 40 days after flowering and are maintained until maturation. It belongs to the glycosyl hydrolase family 19.

Three chitinases, designated RSC-a, RSC-b, and RSC-c, were purified from the seeds of rye using ammonium sulfate precipitation, carboxymethyl (CM) cellulose column chromatography, gel filtration on Sephadex G-75, and S-Sepharose column chromatography. RSC-a, RSC-b, and RSC-c are basic proteins having molecular masses of 33, 26 (GH19), and 26 kDa, and isoelectric points of 9.7, 10, and >10, respectively (Yamagami and Funatsu 1993; see Figure 9.4).

Epilogue

After early Anatolian peasants formed a new cereal from an original weedy plant, rye became one of the major food crops in medieval Europe.

It was the merit of Ferdinand von Lochow (Petkus, Germany) to develop new selection methods for rye by the end of the nineteenth century. Thus, enormous breeding progress has been achieved along with agronomic improvements that resulted in significant increases in yield. The successor of F. von Lochow, KWS Lochow Ltd., continues the famous breeding tradition in Germany with new hybrid varieties.

It took another 50 years for Professor F. Wolfgang Schnell to begin heterosis breeding in rye—as Drs. Edward Murray East (Cambridge, MA) and George Harrison Shull (Cold Spring Harbor, NY) did 70 years before in maize.

His ideas were greatly promoted as well as evaluated through one of his students, Professor Hartwig H. Geiger (Hohenheim, Germany).

The hybrid varieties resulting from these efforts have provided a new boost in grain yield with the beginning of the twenty-first century. At the end of the nineteenth century, a yield of 25 quintals per hectare was achieved, whereas 200 quintals of yield per hectare has recently been recorded, which is a 10-fold increase in yield per hectare.

This is a triumph of plant breeding, of course, accomplished by many other dedicated scientists and breeders!

References

Aastveit, K. 1967. Variation and selection for seed set in tetraploid rye. *Hereditas* 60: 294–316.

Abirached-Darmency, M., Zickler, D., and Cauderon, Y. 1983. Synaptonemal complex and recombination nodules in rye (*Secale cereale*). *Chromosoma* 88: 299–306.

Achrem, M., Kalinka, A., and Rogalska, S. M. 2013. Assessment of genetic relationships among *Secale* species and subspecies based on ISSR and IRAP markers and the distribution of SSR. *Turk. J. Bot.* doi:10.3906/bot-1207-26.

Adam, L. 2004. *Roggen, Anbau und Vermarktung*. Jahresber. Landesamt fur Verbraucherschutz, Landwirtschaft, Flurneuordnung, Brandenburg, Germany, pp. 33–34.

Adlercreutz, H. 2002. Phyto-oestrogens and cancer. *Lancet Oncol.* 3: 364–373.

Adolf, K., and Neumann, H. 1981. Probleme bei der Entwicklung und Nutzung eines Funktionssystems für die Hybridzüchtung beim Roggen. *Tag. Ber. AdL* 191: 61–72.

Adolf, K., and Winkel, A. 1985. A new source of spontaneous sterility in winter rye—Preliminary results. *Proceedings of Eucarpia Meeting of the Cereal Section on Rye*, June 11–13. Svalöv, Sweden, vol. 1, pp. 293–306.

Ahloowalia, B. S. 1967. Chromosome association and fertility in tetraploid ryegrass. *Genetica* 38: 471–484.

Ahokas, A. 2012. The ethnobotany of rye cultivation in Finland and its influence abroad. http://www.nordgen.org/

Akgün, I., and Tosun, M. 2004. Agricultural and cytological characteristics of M1 perennial rye (*Secale montanum* Guss.) as effected by the application of different doses of gamma rays. *Pak. J. Biol. Sci.* 7: 827–833.

Akgün, I., Tosun, M., Haliloglu, K., and Aydin, M. 2011. Development of autotetraploid perennial rye (*Secale montanum* Guss.) and selection for seed set. *Turk. J. Field Crops* 16: 23–28.

Altpeter, F., Popelka, J. C., and Wieser, H. 2004. Stable expression of 1Dx5 and 1Dy10 high-molecular-weight glutenin subunit genes in transgenic rye drastically increases the polymeric glutelin fraction in rye flour. *Plant Mol. Biol.* 54: 783–792.

Alves, E., Ballesteros, I., Linacero, R., and Vazquez, A. M. 2005. *RYS1*, a foldback transposon, is activated by tissue culture and shows preferential insertion points into the rye genome. *Theor. Appl. Genet.* 111: 431–436.

Amend, F. 1919. Untersuchungen über flämischen Roggen unter besonderer Berücksichtigung des veredelten flämischen Landroggens und seiner Züchtung. *Landw. Jb.* 52: 614–669.

Andreasen, M. F., Landbo, A. K., Christensen, L. P., Hansen, A., and Meyer, A. S. 2001. Antioxidant effects of phenolic rye (*Secale cereale* L.) extracts, monomeric hydroxycinnamates, and ferulic acid dehydrodimers on human low-density lipoproteins. *J. Agric. Food Chem.* 49: 4090–4096.

Anonymous. 1956–2012. DESTATIS—Statistisches Bundesamt. Wiesbaden, Germany. https://www.destatis.de

Anonymous. 2007. *Roggen—Getreide mit Zukunft*. DLG Verlag, Frankfurt, Germany.

Anonymous. 2012. http://ryeroute.eu/wp/wp-content/uploads/2012/11/Materjal-rukki-kohta.pdf

Anonymous. 2013. Requirements for the labelling of special products made of grain. http://ec.europa.eu/enterprise/policies/european-standards/harmonised-standards/index_en.htm

Antoniou, T. C., and Marquardt, R. R. 1982. Utilization of rye diets by chicks as affected by lipid type and level and penicillin supplementation. *Poult. Sci.* 61: 107–116.

Apolinarska, B. 2003. Chromosome pairing in tetraploid rye with monosomic-substitution wheat chromosomes. *J. Appl. Genet.* 44: 119–128.

Apolinarska, B. 2008. Introgression of wheat chromosomes into diploid rye by use of a hexaploid triticale with an ABRRRR genome. *Cer. Res. Comm.* 36: 33–42.

Apolinarska, B., Wioeniewska, H., and Wojciechowska, B. 2010. *Aegilops*-rye amphiploids and substitution rye used for introgression of genetic material into rye (*Secale cereale* L.). *J. Appl. Genet.* 51: 413–420.

Appels, R. 1982. The molecular cytology of wheat-rye hybrids. *Rev. Cytol.* 80: 93–132.

Appels, R., Driscoll, C., and Peacock, W. J. 1978. Heterochromatin and highly repeated DNA sequences in rye (*Secale cereale*). *Chromosoma* 70: 67–89.

Arora, D., Wang, X., Gross, P., Steffenson, B., and Brueggeman, R. 2012. Identification of a candidate barley stem rust susceptibility gene determining the recessive nature of *Rpg4*-mediated *Ug99* resistance in barley. *Ann. Wheat Newslett.* 58: 16.

Arseniuk, E., and Oleksiak, T. 2003. Rye production and breeding in Poland. *Plant Breed. Seed Sci.* 47: 7–16.

Augustin, S., Nandy, A., and Krontal, N. 2012. Primary structure of a Sec c 1 sequence, the major rye group 1 allergen. *EMBL/GenBank/DDBJ databases.* http://www.uniprot.org/uniprot/I4IY76

Augustin, C., and Schlegel, R. 1983. Chromosome manipulations. II. Production and characterization of interchromosomal translocations for genetic mapping in rye. *Arch. Züchtungsforsch.* 12: 180–184.

Aust, S. 1941. Erhöhte Saatgutgewinnung bei Roggen durch vegetative Vermehrung. *Züchter* 13: 84–87.

Aust, S., and Ossent, H. P. 1941. Qualitätszüchtung beim Roggen. *Züchter* 13: 78–84.

Autio, K., Fabritius, M., and Kinnunen, A. 1998. Effect of germination and water content on the microstructure and rheological properties of two rye doughs. *Cereal Chem.* 75: 10–14.

Ayonoadu, U. W., and Rees, H. 1968. The regulation of mitosis by B-chromosomes in rye. *Exp. Cell Res.* 52: 284–290.

Bailey, R. J., Rees, H., and Adena, M. A. 1978. Interchange heterozygosity and selection in rye. *Heredity* 41: 1–12.

Balkandschiewa, J. 1971. Morphologische und cytologische Untersuchungen an triploidem Roggen und seinen Nachkommen zum Aufbau einer Trisomenserie. PhD thesis, Martin Luther University, Halle-Wittenberg, Germany, pp. 1–147.

Banaei-Moghaddam, A. M., Schubert, V., Kumke, K., Weib, O., Klemme, S., Nagaki, K., Macas, J. et al. 2012. Nondisjunction in favor of a chromosome: The mechanism of rye B chromosome drive during pollen mitosis. *Plant Cell* 24: 1–11.

Bartos, J., Paux, P., Kofler, R., Havrankova, M., Kopecky, D., Suchanková, P., Safar, J. et al. 2008. A first survey of the rye (*Secale cereale*) genome composition through BAC end sequencing of the short arm of chromosome 1R. *BMC Plant Biol.* 8: 95–102.

Barzali, M., Lohwasser, U., Niedzielski, M., and Börner, A. 2005. Effects of different temperatures and atmospheres on seed and seedling traits in a long-term storage experiment on rye (*Secale cereale* L.). *Seed Sci. Technol.* 33: 713–721.

Bashir, A., Auger, J. A., and Rayburn, L. A. 1993. Flow cytometric DNA analysis of wheat-rye addition lines. *Cytometry* 4: 843–847.

Bates, L. S., Campos, V. A., Rodriguez, R. R., and Anderson, R. G. 1974. Progress toward novel cereal grains. *Cereal Sci. Today* 19: 283–285.

Bates, L. S., Mujeeb, K. A., Rodriguez, R. R., and Waters, R. F. 1975. Rye dwarf gene introgression into barley. *Barley Genet. Newslett.* 6: 7–8.

Batygin, N. F. 1965. Uslovija vyrascivanija M1 charakter izmencivosti rastenij. *Sbornik Rad. i. Sel. Rast.* 5: 60–64.

Bauhin, C. 1658. *Theatri Botanici*. König, Basel, Switzerland.

Bednarek, P. T., Dąbkowska, A., Kolasińska, I., and Krajewski, P. 2003. Testing cms-P-linked AFLPs for selection of rye hybrid components. *Cell. Mol. Biol. Lett.* 8: 185–193.

Benito, C., Romero, M. P., Henriques-Gil, N., Llorente, F., and Figueiras, A. M. 1996. Sex influence on recombination frequency in *Secale cereale* L. *Theor. Appl. Genet.* 93: 926–931.

Bennett, M. D. 1974. Meiotic, gametophytic and early endosperm development in *Triticale*. In *Triticale*. R. MacIntyre, and M. Campbell (eds.). International Development Research Centre, Ottawa, Canada, pp. 137–148.

Bennett, M. D. 1976. DNA amount, latitude, and crop plant distribution. *Environ. Exp. Bot.* 16: 93–108.

Bennett, M. D. 1977. The time and duration of meiosis. *Phil. Trans. R. Soc. Lond.* 277: 201–226.

Bennett, M. D., Chapman, V., and Riley, R. 1971. The duration of meiosis in pollen mother cells of wheat, rye and Triticale. *Proc. R. Soc. Lond.* 178: 259–275.

Bennett, M. D., Finch, R. A., Smith, J. B., and Rao, M. K. 1973. The time and duration of female meiosis in rye, wheat and barley. *Proc. R. Soc. Lond. B* 183: 301–319.

Bennett, M. D., Gustafson, J. P., and Smith, J. B. 1977. Variation in nuclear DNA in the genus *Secale. Chromosoma* 61: 149–176.

Bennett, M. D., and Rees, H. 1967. Natural and induced changes in chromosome size and mass in meristems. *Nature* 215: 93–94.

Bennett, M. D., and Rees, H. 1970. Induced variation in chiasma frequency in rye in response to phosphate treatments. *Genet. Res.* 16: 325–331.

Bennett, M. D., and Smith, J. B. 1976. Nuclear DNA amounts on angiosperms. *Phil. Trans. R. Soc. Lond.* 274: 227–231.

Bennett, M. D., Smith, J. B., and Barclay, I. 1975. Early seed development in Triticeae. *Phil. Trans. R. Soc. Lond.* 272: 199–227.

Bennett, M. D., Smith, J. B., and Kemble, R. 1972. The effect of temperature on meiosis and pollen development in wheat and rye. *Can. J. Genet. Cytol.* 14: 615–624.

Bensin, B. M. 1933. Autogamous Turkestan rye. *Bull. Torrey Bot. Club.* 60: 155–157.

Berkner, F., and Meyer, K. 1927. Morphologische Studien an Roggenähren aus Anatolien. *Z. Pflanzenzücht.* 12: 229–245.

Bertholdsson, N.-O., Andersson, S. C., and Merker, A. 2011. Allelopathic potential of *Triticum* spp., *Secale* spp. and *Triticosecale* spp. and use of chromosome substitutions and translocations to improve weed suppression ability in winter wheat. *Plant Breed.* 131: 75–80.

Bhattacharyya, N. K., and Jenkins, B. C. 1960. Karyotype analysis and chromosome designation for *Secale cereale* L. "Dakold." *Can. J. Genet. Cytol.* 2: 168–277.

Bierce, A. 2008. *The Devil's Dictionary*. Bloomsbury Publishing PLC, London.

Black, B. A., Zannini, E., Curtis, J. M., and Gänzle, M. G. 2013. Antifungal hydroxy fatty acids produced during sourdough fermentation: Microbial and enzymatic pathways, and antifungal activity in bread. *Appl. Environ. Microbiol.* 79: 1866–73.

Blackman, E., and Parry, D. W. 1968. Opaline silica deposition in rye (*Secale cereale* L.). *Ann. Bot.* 32: 199–206.

Blunden, R., Wilkes, T. J., Forster, J. W., Jimenez, M. M., Sandery, M. J., Karp, A., and Jones, R. N. 1993. Identification of the E3900 family, a 2nd family of rye chromosome-B specific repeated sequences. *Genome* 36: 706–711.

Boczkowska, M. K., and Puchalski, J. 2012. SSR studies of genetic changes in relation to long-term storage and field regeneration of rye (*Secale cereale*) seeds. *Seed Sci. Technol.* 40: 63–72.

Bogomolov, A. M., and Cyganok, E. K. 1971. Reakcija sortov ozimoj rzi na obrabotku semjan chimiceskimi mutagenami. *Vsb. Prakt. Chim. Mut.* 11: 59–63.

Bolibok, H., Rakoczy-Trojanowska, M., Hromada, A., and Pietrzykowski, R. 2005. Efficiency of different PCR-based marker systems in assessing genetic diversity among winter rye (*Secale cereale* L.) inbred lines. *Euphytica* 126: 109–116.

Bolibok-Bragoszewska, H., Heller-Uszynska, K., Wenzl, P., Uszynski, G., Kilian, A., and Rakoczy-Trojanowska, M. 2009. DArT markers for the rye genome—Genetic diversity and mapping. *BMC Genomics* 10: 578–586.

Bonnett, O. T. 1940. Inflorescences of maize, wheat, rye, barley, and oats: Their initiation and development. *J. Agric. Res.* 60: 25–38.

Börner, A., Korzun, V., Polley, A., Malyshev, S., and Melz, G. 1998b. Genetics and molecular mapping of a male fertility restoration locus (*Rfg1*) in rye. (*Secale cereale* L.). *Theor. Appl. Genet.* 97: 99–102.

Börner, A., Korzun, V., Voylokov, A. V., Worland, A. J., and Weber, W. E. 2000. Genetic mapping of quantitative trait loci in rye (*Secale cereale* L.). *Euphytica* 116: 203–209.

Börner, A., Korzun, V., and Worland, A. J. 1998a. Comparative genetic mapping of mutant loci affecting plant height and development in cereals. *Euphytica* 100: 245–248.

Börner, A., and Melz, G. 1988. Response of rye genotypes differing in plant height to exogenous gibberellic acid application. *Arch. Züchtungsforsch.* 18: 71–74.

Börner, A., Melz, G., and Lenton, J. R. 1992. Genetical and physiological studies of gibberellic acid insensitivity in semidwarf rye. *Hereditas* 116: 199–201.

Böttcher, H., Garz, J., and Weipert, D. 2000. Long-term effects of different fertilization on yield and processing performance of rye grown continuously or in crop rotation—Results of the long-term experiment "Eternal Rye Cropping" Halle. *Pflanzenbauwiss.* 4: 1–8.

Bredemann, C., and Heuser, W. 1931. Beitäge zur Heterosis bei Roggen. *Z. Züchtg.* 16: 1–56.

Bremer, G., and Bremer-Reinders, D. E. 1954. Breeding of tetraploid rye in the Netherlands. I. Methods and cytological investigations. *Euphytica* 3: 49–63.

Bremer-Reinders, D. E. 1958. The early stages of the development of the rye spike. *Acta Bot. Nederlandica* 7, 223–232.

Brewbaker, H. E. 1926. Studies of self-fertilization in rye. *Univ. Minn. Agric. Exp. State Tech. Bull.* 40: 1–24.

Brink, R. A., Cooper, C. D., and Ausherman, E. 1944. A new hybrid between *Hordeum jubatum* and *Secale cereale* reared from an artificially cultivated embryo. *J. Hered.* 35: 67–75.

Brooks, A. M., Danehower, D. A., Murphy, J. P., Reberg-Horton, S. C., and Burton, J. D. 2012. Estimation of heritability of benzoxazinoid production in rye (*Secale cereale*) using gas chromatographic analysis. *Plant Breed.* 131: 104–109.

Brule-Babel, A. L., and Fowler, D. B. 1989. Genetic control of cold hardiness and vernalization requirement in rye. *Genome* 32: 19–23.

Bühring, J. 1960. Observations and investigations on winter rye on delaying earing and maintaining clones by photoperiodic treatment. *Z. Pflanzenzücht.* 43: 266–296.

Bukharov, A. A., Kolosov, V. I., Klezovich, O. N., and Zolotarev, A. S. 1989. Nucleotide sequence of rye chloroplast DNA fragment, comprising *psbD*, *psbC* and *trnS* genes. *Nucleic Acids Res.* 17: 798.

Bukharov, A. A., Kolosov, V. I., and Zolotarev, A. S. 1988. Nucleotide sequence of rye chloroplast DNA fragment encoding *psbB* and *psbH* genes. *Nucleic Acids Res.* 16: 8737–8738.

Burger, J. C., Lee, S., and Ellstrand, N. C. 2006. Origin and genetic structure of feral rye in the western United States. *Mol. Ecol.* 15: 2527–2539.

Bushuk, W. (ed.). 1976. *Rye: Production, Chemistry, and Technology*. American Association of Cereal Chemists, St. Paul, MN, pp. 1–11.

Cakmak, I., Derici, R., Torun, B., Tolay, I., Braun, H. J., and Schlegel, R. 1997. Role of rye chromosomes in improvement of zinc efficiency in wheat and triticale. *Plant Soil* 196: 249–253.

Camacho, M. V., Matos, M., Gonzales, C., Perez-Flores, V., Pernauta, B., Pinto-Carnida, O., and Benito, C. 2005. *Secale cereale* inter-microsatellites (SCIMs): Chromosomal location and genetic inheritance. *Genetica* 123: 303–311.

Campos-Dias, S., Mattar da Silva, M. C., Teixeira, F. R., Figueira, E. L. Z., Brilhante de Oliveira-Neto, O., Alves de Lima, L., Franco, O. L., and Grossi-de-Sa, M. F. 2010. Investigation of insecticidal activity of rye α-amylase inhibitor gene expressed in transgenic tobacco (*Nicotiana tabacum*) toward cotton boll weevil (*Anthonomus grandis*). *Pestic. Biochem. Physiol.* 98: 39–44.

Candela, M., Figueiras, A. M., and Lacadena, J. R. 1979. Maintenance of interchange heterozygosity in cultivated rye, *Secale cereale* L. *Heredity* 42: 283–289.

Caperta, A., Neves, N., Morais-Cecilio, L., Malho, R., and Viegas, W. 2002. Genome restructuring in rye affects the expression, organization and disposition of homologous rDNA loci. *J. Cell Sci.* 115: 2839–2846.

Carchilan, M., Delgado, M., Ribeiro, T., Costa-Nunes, P., Caperta, A., Morais-Cecilio, L., Jones, R. N., Viegas, W., and Houben, A. 2007. Transcriptionally active heterochromatin in rye B chromosomes. *Plant Cell* 19: 1738–1749.

Carlton, P. M., Cowan, C. R., and Cande, W. Z. 2003. Directed motion of telomeres in the formation of the meiotic bouquet revealed by time course and simulation analysis. *Mol. Biol. Cell* 14: 2832–2843.

Castillo, A. M., Vasil, V., and Vasil, I. K. 1994. Rapid production of fertile transgenic plants of rye (*Secale cereale* L.). *Biotechnology* 12: 1366–1371.

Chao, S., Sharp, P. J., and Gale, M. D. 1988. A linkage map of wheat homoeologous group 7 chromosomes using RFLP markers. *Proceedings of the 7th International Wheat Genetics Symposium*. T. E. Miller, and R. M. D. Koebner (eds.). July 13–19. IPSR, Cambridge Laboratory, Cambridge, pp. 493–498.

Chapman, C. G. D. 1984. Does preferential pairing occur in *Secale cereale* × *Secale montanum* tetraploids hybrids? *Heredity* 52: 317–322.

Chase, W. R., Nair, M. G., and Putnam, A. R. 1991a. 2,2′-oxo-1,1′-azobenzene: Selective toxicity of rye (*Secale cereale* L.) allelochemicals to weed and crop species. II. *J. Chem. Ecol.* 17: 9–19.

Chase, W. R., Nair, M. G., Putnam, A. R., and Mishra, S. K. 1991b. 2,2′-oxo-1,1′-azobenzene: Microbial transformation of rye (*Secale cereale* L.) allelochemical in field soils by *Acinetobacter calcoaceticus*. III. *J. Chem. Ecol.* 17: 1575–1584.

Chaudhary, H. K. 2013. New frontiers in chromosome elimination-mediated DH breeding. *Proceedings of the 21st International Plant & Animal Genome Conference*, January 13–18, 2012. San Diego, CA, Abstract P219.

Chebotar, S., Röder, M. S., Korzun, V., Saal, B., Weber, W. E., and Börner, A. 2003. Molecular studies on genetic integrity of open-pollinating species rye (*Secale cereale* L.) after long-term genebank maintenance. *Theor. Appl. Genet.* 107: 1468–1476.

Chikmawati, T., Skovmand, B., and Gustafson, J. P. 2005. Phylogenetic relationships among *Secale* species revealed by amplified fragment length polymorphisms. *Genome* 48: 792–801.

Clarke, B. C., and Appels, R. 1999. Sequence variation at the *Sec-1* locus of rye, *Secale cereale* (Poaceae). *Plant Syst. Evol.* 214: 1–14.

Clarke, B. C., Mukai, Y., and Appels, R. 1996. The *Sec-1* locus on the short arm of chromosome 1R of rye (*Secale cereale*). *Chromosoma* 105: 269–275.

Collins, F. C., Lertmongkol, V., and Jones, J. P. 1973. Pollen storage of certain agronomic species in liquid air. *Crop Sci.* 13: 493–494.

Corriveau, J. L., and Coleman, A. W. 1988. Rapid screening method to detect potential biparental inheritance of plastid DNA and results for over 200 angiosperm species. *Am. J. Bot.* 75: 1443–1458.

Coulthart, M. B., Spencer, D. F., Huh, G. S., and Gray, M. W. 1994. Polymorphism for ribosomal RNA gene arrangement in the mitochondrial genome of fall rye (*Secale cereale* L.). *Curr. Genet.* 26: 269–275.

Crecelius, F., Streb, P., and Feierabend, J. 2003. Malate metabolism and reactions of oxidoreduction in cold-hardened winter rye (*Secale cereale* L.) leaves. *J. Exp. Bot.* 54: 1075–1083.

Cuadrado, A., Ceoloni, C., and Jouve, N. 1995. Variation in highly repetitive DNA composition of heterochromatin in rye studied by fluorescence in situ hybridization. *Genome* 38: 1061–1069.

Cuadrado, A., and Jouve, N. 2002. Evolutionary trends of different repetitive DNA sequences during speciation in the genus *Secale. J. Hered.* 93: 339–345.

Cunfer, B. M., and Youmans, J. 1979. Effect of irradiance upon the population of *Pseudomonas coronafaciens* in leaves and symptom expression of halo blight of rye. *Can. J. Microbiol.* 25: 163–166.

Curtis, T. Y., Powers, S. J., Balagiannis, D., Elmore, J. S., Mottram, D. S., Parry, M. A. J., Rakszegi, M., Bedö, Z., Shewry, P. R., and Halford, N. G. 2010. Free amino acids and sugars in rye grain: Implications for acrylamide formation. *J. Agric. Food Chem.* 58: 1959–1969.

Czarnecki, Z., and Nowak, J. 2005. Hemicellulase supplementation of regular and hybrid rye grain mashes for ethanol fermentation. *EJPAU* 8, http://www.ejpau.media.pl/volume8/issue4/art-56.html

Dainel, G. 1993. Anther culture in rye: Improved plant regeneration using modified MS-media. *Plant Breed.* 110: 259–261.

Davies, E. D. G., and Jones, G. H. 1974. Chiasma variation and control in pollen mother cells and embryo-sac mother cells of rye. *Genet. Res.* 23: 185–190.

De Bustos, A., and Jouve, N. 2002. Phylogenetic relationships of the genus *Secale* based on the characterisation of rDNA ITS sequences. *Plant. Syst. Evol.* 235: 147–154.

de la Pena, A., Lörz, H., and Schell, J. 1987. Transgenic rye plants obtained by injecting DNA into young floral tillers. *Nature* 325: 274.

de la Puente, R., González, A. I., Ruiz, M. L., and Polanco, C. 2008. Somaclonal variation in rye (*Secale cereale* L.) analyzed using polymorphic and sequenced AFLP markers *in vitro. Cell. Dev. Biol. Plant* 44: 419–426.

de Vries, G. 2006. *Roggen—Anbau und Vermarktung.* Roggenforum, Germany, pp. 1–47.

de Vries, J. N., and Sybenga, J. 1976. Identification of rye chromosomes: The Giemsa banding pattern and the translocation tester set. *Theor. Appl. Genet.* 48: 35–43.

de Vries, J. N., and Sybenga, J. 1984. Chromosomal location of 17 monogenically inherited morphological markers in rye (*Secale cereale* L.) using the translocation tester set. *Z. Pflanzenzücht.* 92: 117–139.

de Vries, J. N., and Sybenga, J. 1989. Meiotic behaviour of telo-tertiary compensating trisomics of rye: Evaluation for use in hybrid varieties. *Theor. Appl. Genet.* 78: 889–896.

Deimling, S., Flehinghaus-Roux, T., Röber, F., Schechert, A., Roux, S. R., and Geiger, H. H. 1994. Doubled haploid production—Now reproducible in rye. *Abstracts VIIIth International Congress of Plant Tissue and Cell Culture*, June 12–17. Florence, Italy, p. 95.

Dennert, J., and Fischbeck, G. 1999. Anbaumanagement von Winterroggen. *Pflanzenbauwiss* 2: 2–8.

Derr, H. B. 1892. *Origin of Cultivated Plants.* D. Appleton and Co., New York, pp. 370–382.

Derzhavin, A. I. 1938. Resultaty robot po vyvedenije mnogoletnych sortov psenicy i rzi (Russian). *Bull. Acad. Nauk* 3: 663–665.

Derzhavin, A. I. 1960. The theory and practice of producing perennial rye varieties. In *Wide Hybridization in Plants.* N. V. Tsitsin (ed.). Israel Program for Scientific Translations, Jerusalem, Israel, pp. 143–152.

Devos, K. M., Atkinson, M. D., Chinoy, C. N., Francis, H. A., Harcourt, R. L., Koebner, R. M. D., Liu, C. J., Masojć, P., Xie, D. X., and Gale, M. D. 1993. Chromosomal rearrangements in the rye genome relative to that of wheat. *Theor. Appl. Genet.* 85: 673–680.

Devos, K. M., and Gale, M. D. 1997. Comparative genetics in the grasses. *Plant Mol. Biol.* 35: 3–15.

Dierks, W., and Reimann-Philipp, R. 1966. Die Züchtung eines perennierenden Roggens als Möglichkeit zur Verbesserung der Zuchtmethodik und zur Schaffung eines mehrfach nutzbaren Grünfutter- und Körnerroggens. *Z. Pflanzenücht.* 56: 343–368.

Dittmer, H. J. 1937. A quantitative study of the roots and the root hairs of a winter rye plant (*Secale cereale*). *Am. J. Bot.* 24: 417–420.

Dobrovolskaya, O., Martinek, P., Voylokov, A. V., Korzun, V., and Börner, A. 2009. Microsatellite mapping of genes that determine supernumerary spikelets in wheat (*T. aestivum*) and rye (*S. cereale*). *Theor. Appl. Genet.* 119: 867–874.

Dohmen, G., Hessberg, H., Geiger, H. H., and Tudzynski, P. 1994. CMS in rye: Comparative RFLP and transcript analyses of mitochondria from fertile and male-sterile plants. *Theor. Appl. Genet.* 89: 1014–1018.

Dohmen, G., and Tudzynski, P. 1994. A DNA-polymerase-related reading frame (pol-r) in the mtDNA of *Secale cereale*. *Curr. Genet.* 25: 59–65.

Dorn, J., and Metz, R. 1996. Wirkung von organischen Schadstoffen (PAK; PCB) und Schwermetallen in Rieselfeldböden auf Biomasseertrag und Schwermetalltransfer bei Roggen (*Secale cereale*). *J. Plant Nutr. Soil Sci.* 159: 87–91.

Dowrick, G. J. 1957. The influence of temperature on meiosis. *Heredity* 11: 37–49.

Dreyer, F., Miedaner, T., and Geiger, H. H. 1996. Chromosomale Lokalisation von Restorergenen aus argentinischen und iranischem Roggen (*Secale cereale* L.). *Vortr. Pflanzenzücht.* 32: 49–51.

Driscoll, C. J., and Anderson, L. M. 1967. Cytogenetic studies in Transec—A wheat-rye translocation line. *Can. J. Genet. Cytol.* 9: 375–380.

Dubovets, N. I., Silkova, O. G., Shchapova, A. I., Solovey, L. A., and Shtyk, T. I. 2005. Transmission of univalent chromosome 5R via gametes of di-monosomic 5D-5R (Russian). *Vestnik VOGS* 9: 495–498.

Duckart, J. 1927. Ergebnisse neunjähriger Inzestzuchtversuche bei Roggen. *Verh. Int. Kongr. Vererbungswiss.* 1: 603–608.

Dundas, I. S., Frappell, D. E., Crack, D. M., and Fisher, J. M. 2001. Deletion mapping of a nematode resistance gene on rye chromosome 6R in wheat. *Crop Sci.* 4: 1771–1778.

Eckl-Dorna, J., Klein, B., Reichenauer, T. G., Niederberger, V., and Valenta, R. 2010. Exposure of rye (*Secale cereale*) cultivars to elevated ozone levels increases the allergen content in pollen. *J. Allergy Clin. Immun.* 126: 1315–1317.

Ehdaie, B., Whitkus, R. W., and Waines, J. G. 2003. Root biomass, water-use efficiency, and performance of wheat-rye translocations of chromosomes 1 and 2 in spring bread wheat "Pavon." *Crop Sci.* 43: 710–717.

Endo, T. R., Nasuda, S., Jones, N., Dou, Q., Akahori, A., Wkimoto, M., Tanaka, H., Niwa, K., and Tsujimoto, H. 2008. Dissecction of B chromosomes, and nondisjunction properties of the dissected segments in a common wheat background. *Genes Genet. Syst.* 83: 23–30.

England, F. J. W. 1974. The use of incompatibility for the production of F1 hybrids in forage grasses. *Heredity* 32: 183–188.

Erenoglu, B., Cakmak, I., Römheld, V., Derici, R., and Rengel, Z. 1999. Uptake of zinc by rye, bread wheat and durum wheat cultivars differing in zinc efficiency. *Plant Soil* 209: 245–252.

Ermishev, V. Y., Naroditsky, B. S., and Khavkin, E. E. 2002. A fragment of the rpoC2 chloroplast RNA polymerase gene in perennial and annual rye. *Russ. J. Plant Physiol.* 49: 387–392.

Evans, G. M. 1976. Rye, *Secale cereale*. In *Evolution of Crop Plants*. N. W. Simmonds (ed.). Longman, New York, pp. 108–111.

Evans, P. K., Keates, A. O., and Cocking, E. C. 1972. Isolation of protoplasts from cereal leaves. *Planta* 104: 178–181.

Falke, K. C., Susic, Z., Hackauf, B., Korzun, V., Schondelmaier, J., Wilde, P., Wehling, P. et al. 2008. Establishment of introgression libraries in hybrid rye (*Secale cereale* L.) from an Iranian primitive accession as a new tool for rye breeding and genomics. *Theor. Appl. Genet.* 117: 641–652.

Falke, K. C., Susic, Z., Wilde, P., Wortmann, H., Möhring, J., Piepho, H.-P., Geiger, H. H., and Miedaner, T. 2009. Testcross performance of rye introgression lines developed by marker-assisted backcrossing using an Iranian accession as donor. *Theor. Appl. Genet.* 118: 1225–1238.

Fallahi, M., Crosthwait, J., Calixte, S., and Bonen, L. 2005. Fate of mitochondrially located S19 ribosomal protein genes after transfer of a functional copy to the nucleus in cereals. *Mol. Genet. Genom.* 273: 76–83.

Fedak, G. 1978. A viable hybrid between *Hordeum vulgare* and *Secale cereale. Cer. Res. Comm.* 6: 353–358.

Fedak, G. 1985. Cytogenetics of a hybrid and amphiploid between *Hordeum pubiflorum* and *Secale africanum. Can. J. Genet. Cytol.* 27: 1–5.

Fedak, G., and Armstrong, K. C. 1986. Intergeneric hybrids between *Secale cereale* (2×) and *Thinopyrum intermedium* (6×). *Can. J. Genet. Cytol.* 28: 426–429.

Fedak, G., and Nakamura, C. 1982. Chromosomal instability in a hybrid between *Hordeum vulgare* and *Secale vavilovii. Can. J. Genet. Cytol.* 24: 207–212.

Fedorov, V. S., Smirnov, V. G., and Soshichina, S. P. 1967. The genetics of rye (*Secale cereale* L.). VI. On the genetics of wax bloom. *Res. Genet.* 3: 104–111.

Fedorov, V. S., Smirnov, V. G., and Soshichina, S. P. 1970a. The genetics of rye (*Secale cereale* L.). X. The inheritance of dwarfness. *Genetika* 6: 273–282.

Fedorov, V. S., Smirnov, V. G., and Sosnihina, S. P. 1970b. Genetics of rye (*Secale cereale* L.). XI. Inheritance of the liguleless condition. *Genetika* 6: 5–14.

Fedorov, V. S., Smirnov, V. G., and Sosnihina, S. P. 1970c. Genetics of rye (*Secale cereale* L.). XII. Inheritance of pubescence of outer glumes. *Genetika* 6: 5–16.

Feil, B., and Schmid, E. J. 2002. *Dispersal of Maize, Wheat and Rye Pollen—A Contribution to Determining the Necessary Isolation Distances for the Cultivation of Transgenic Crops*. Shaker Verl., Aachen, Germany.

Ferwerda, F. P. 1956. Recurrent selection as a breeding procedure for rye and other cross-pollinated plants. *Euphytica* 5: 175–184.

Figueiras, A. M., de la Peña, A., and Benito, M. C. 1991. High mutability in rye (*Secale cereale* L.). *Mutat. Res.* 264: 171–176.

Flavell, R. B., Bennett, M. D., Smith, J. B., and Smith, D. B. 1974. Genome size and proportion of repeated sequence DNA in plants. *Biochem. Genet.* 12: 257–269.

Flehinghaus, T., Deimling, S., and Geiger, H. H. 1991. Methodical improvements in rye anther culture. *Plant Cell Rep.* 10: 397–400.

Flehinghaus-Roux, T., Deimling, S., and Geiger, H. H. 1995. Anther culture ability in *Secale cereale* L. *Plant Breed.* 114: 259–261.

Fluch, S., Kopecky, D., Burg, K., Simkova, H., Taudien, S., Petzold, A., Kubalakova, M. et al. 2012. Sequence composition and gene content of the short arm of rye (*Secale cereale*) chromosome 1. *PLoS ONE* 7: e30784. doi:10.1371/journal.pone.0030784.

Focke, O. W. 1881. *Die Pflanzenmischlinge*. Bornträger, Berlin, Germany.

Fra-Mon, P., Salcedo, G., Aragoncillo, C., and Garcia-Olmedo, G. 1984. Chromosomal assignment of genes controlling salt-soluble proteins (albumins and globulins) in wheat and related species. *Theor. Appl. Genet.* 69: 167–172.

Frederiksen, S., and Peterson, G. 1998. A taxonomic revision of *Secale* (Triticeae, Poaceae). *Nordic J. Bot.* 18: 399–420.

Freer, M., Donnelly, J. R., Axelson, A., Myers, L. F., Davidson, J. L., and Dymock, J. 1997. Comparison of secale with other perennial grasses under grazing at a cool site in the high rainfall zone of New South Wales. *Aust. J. Exp. Agric.* 37: 19–25.

Friebe, B., Hatchett, J. H., Sears, R. G., and Gill, B. S. 1990. Transfer of Hessian fly resistance from "Chaupon" rye to hexaploid wheat via a 2BS/2RL wheat-rye chromosome translocation. *Theor. Appl. Genet.* 79: 385–389.

Friedt, W. 1979. The use of *Secale vavilovii* in rye breeding. *Proceedings Conference on Broad Genetic Base of Crops.* A. C. Zeven (ed.). July 3–7, 1978. Wageningen, the Netherlands, pp. 221–224.

Friedt, W., Lind, V., Walther, H., Foroughi-Wehr, B., Züchner, S., and Wenzel, G. 1983. The value of inbred lines derived from *Secale cereale* × *Secale vavilovii* via classical inbreeding and adrogenetic haploids. *Z. Pflanzenzücht.* 91: 89–103.

Frimmel, F., and Baranek, F. 1935. Beitrag zur Methodik der Roggenzüchtung und des Saatgutbaues. *Z. Zücht.* 20: 1–20.

Fröst, S., Vaivars, L., and Carlbom, C. 1970. Reciprocal extrachromosomal inheritance in rye (*Secale cereale* L.). *Hereditas* 65: 251–260.

Fruwirth, C. 1907. Untersuchungen über den Erfolg von Veredelungsauslese. *Z. Versuchsw.* 10: 476–535.

Fruwirth, C., and Roemer, T. 1923. *Einführung in die landwirtschaftliche Pflanzenzüchtung.* 2. Auflage. P. Parey Verlag, Berlin, Germany.

Füle, L., Hodos-Kotvics, G., Galli, Z., Acs, E., and Heszky, L. 2005. Grain quality and baking value of perennial rye (cv. "Perenne") of interspecific origin (*Secale cereale × S. montanum*). *Cer. Res. Comm.* 33: 809–816.

Gacek, E. 1972. Badania cytoemriologiczne nad karlowym mutantem zyta vydrebnicnym z materialow napromienionych szybkimi neutronami. *Rostlinna* 18: 253–255.

Galenius, C. 1965. De alimentorum facultatibus. In *Claudii Galeni Opera Omnia.* C. G. Kühn (ed.). T. VI, 2. Lipsiae 1823 (reprinted Hildesheim 1965), Verlag G. Olms, Hildesheim, Germany, pp. 453–748.

Gallego, F. J., Calles, B., and Benito, C. 1998. Molecular markers linked to the aluminium tolerance gene *Alt1* in rye. *Theor. Appl. Genet.* 97: 1104–1109.

Gawrońska, H., and Nalborczyk, E. 1989. Photosynthetic productivity of winter rye (*Secale cereale* L.). II. Biomass accumulation and distribution by six cultivar in winter rye (*Secale cereale* L.). *Acta Physiol. Plant.* 11: 265–277.

Geiger, H. H. 1972. Wiederherstellung der Pollenfertilität in cytoplasmatisch männlich sterilem Roggen. *Theor. Appl. Genet.* 42: 32–33.

Geiger, H. H. 1982. Breeding methods in diploid rye (*Secale cereale* L.). *Tag. Ber. AdL* 198: 305–332.

Geiger, H. H. 1990. Wege, Fortschritte und Aussichten der Hybridzüchtung. In *Pflanzen-produktion im Wandel.* G. Haug et al. (eds.). VCH Verlag, Weinheim, Germany, pp. 41–47.

Geiger, H. H., Diener, C., and Singh, R. K. 1981. Influence of self-fertility on the performance of synthetic populations in rye (*Secale cereale* L.). In *Quantitative Genetics and Plant Breeding.* A. Gallais (ed.). INRA Service des Publications, Versailles, France, pp. 169–177.

Geiger, H. H., and Miedaner, T. 1996. Genetic basis and phenotypic stability of male fertility restoration in rye. *Vortr. Pflanzenzücht.* 35: 27–38.

Geiger, H. H., and Miedaner, T. 1999. Hybrid rye and heterosis. In *Genetics and Exploitation of Heterosis in Crops.* J. G. Coors, and S. Pandey (eds.). Crop Science Society of America, Madison, WI, pp. 439–450.

Geiger, H. H., and Miedaner, T. 2009. Rye breeding. In *Handbook of Plant Breeding,* vol. 3, Cereals, M. J. Carena (ed.). Springer Publishing, New York, pp. 157–181.

Geiger, H. H., and Schnell, F. W. 1970. Cytoplasmic male sterility in rye (*Secale cereale* L.). *Crop Sci.* 10: 590–593.

Geiger, H. H., Yuan, Y., Miedaner, T., and Wilde, P. 1995. Environmental sensitivity of cytoplasmic genic male sterility (CMS) in *Secale cereale* L. *Adv. Plant. Breed.* 18: 7–17.

Geisler, G. 1980. *Pflanzenbau: Biologische Grundlagen und Technik der Pflanzenproduktion.* P. Parey Verlag, Berlin, Germany.

Gertz, A., and Wricke, G. 1991. Inheritance of temperature-induced pseudocompatibility in rye. *Plant Breed.* 107: 89–96.

Gielo, G., and Starzycki, S. 1979. Indukowanie zmiennosci mutacyjnej zyta tetraploidalnego promienieami gamma 60Co. *Hod. Ros. Akkl. i Nas.* 22: 191–201.

Gillies, C. B. 1985. An electron microscope study of synaptonemal complex formation at zygotene in rye. *Chromosoma* 92: 165–175.

Glass, C., Miedaner, T., and Geiger, H. H. 1995. Lokalisation von Restorergenen der Winterroggenlinie L18 mit RFLP-Markern. *Vortr. Pflanzenzücht.* 31: 52–55.

Glitso, L. V., Mazur, W. M., Adlercreutz, H., Wähälä, K., Mäkelä, T., Sandström, B., and Bach-Knudsen, K. E. 2000. Intestinal metabolism of rye lignans in pigs. *Br. J. Nutr.* 84: 429–437.

Gonzalez-Garcia, M., Cuacos, M., Gomez-Revilla, D., Puertas, M. J., and Vega, J. M. 2012. A ty3/gypsy retrotransposon specific of rye centromeres. *Abstracts Proceedings of the 11th Gatersleben Research Conference Chromosome Biology, Genome Evolution, and Speciation*, April 23–25. Gatersleben, Germany, p. 59.

Gordon-Kamm, W. J., and Steponkus, P. L. 1984. Lamellar-to-hexagonalII phase transitions in the plasma membrane of isolated protoplasts after freeze-induced dehydration. *Proc. Natl. Acad. Sci. U.S.A.* 81: 6373–6377.

Gordon-Werner, E., and Dörffling, K. 1988. Morphological and pyhsiological studies concerning the drought tolerance of the *Secale cereale* × *Secale montanum* cross Permontra. *Agron. Crop. Sci.* 160: 277–285.

Graham, R. D. 1984. Breeding for nutritional characteristics in cereals. *Adv. Plant Nutr.* 1: 57–102.

Graham, R. D. 1988a. Development of wheats with enhanced nutrient efficiency: Progress and potential. In *Wheat Production Constraints in Tropical Environments*. A. R. Klatt (ed.). CIMMYT, Mexico, DF, pp. 305–320.

Graham, R. D. 1988b. Genotypic differences in tolerance to manganese deficiency. Chapter 17. In *Manganese in Soils and Plants*. R. D. Graham, R. J. Hannam, and N. C. Uren (eds.). Kluwer Academic Publishers, Dordrecht, the Netherlands, pp. 261–276.

Graham, R. D., Ascher, J. S., Ellis, P. A. E., and Shepherd, K. W. 1987. Transfer to wheat of the copper efficiency factor carried on rye chromosome arm 5RL. *Plant Soil* 99: 107–114.

Grasten, S. M., Juntunen, K. S., Poutanen, K. S., Gylling, H. K., Miettinen, T. A., and Mykkänen, H. M. 2000. Rye bread improves bowel function and decreases the concentrations of some compounds that are putative colon cancer risk markers in middle-aged women and men. *J. Nutr.* 130: 2215–2221.

Grochowski, L., Kaczmarek, J., Kadlubiec, W., and Bujak, H. 1995. Using xenia in the breeding of rye hybrids. *Acta Soc. Bot. Pol.* 65: 175–179.

Grosse, B. A., Deimling, S., and Geiger, H. H. 1996. Mapping of gene loci for anther culturability of rye using molecular markers. *Vortr. Pflanzenzücht.* 32: 46–48.

Growley, J. G., and Rees, H. 1968. Fertility and selection in tetraploid *Lolium*. *Chromosoma* 24: 300–308.

Gruszecka, D., and Piettrusiak, A. 2001. Characterization of translocation rye strains created using *Dasypyrum villosum* (Crimea, USSR). *Proceedings of the 7th EUCARPIA Conference*, July 4–7. Radzikow, Poland, pp. 303–309.

Grzesik, H., and Nalepa, S. 1974. Wplyw N-nitroaometylomocznika (NMH) i predkich neutronow na zmiennosc morfologiczn. *Biull. IHAR* 5–6: 49–51.

Guo, Y. D., and Pulli, S. 2000. Isolated microspore culture and plant regeneration in rye (*Secale cereale* L.). *Plant Cell Rep.* 19: 875–880.

Gupta, P. K. 1971. Homoeologous relationship between wheat and rye chromosomes. *Genetica* 42: 199–213.

Gusareva, E. S., Havelkova, H., Blazkova, H., Kosarova, M., Kucera, P., Kral, V., Salyakina, D., Müller-Myhsok, B., and Lipoldova, M. 2009. Mouse to human comparative genetics reveals a novel immunoglobulin E-controlling locus on Hsa8q12. *Immunogenetics* 61: 15–25.

Gustafson, J. P., Butler, E., and McIntyre, C. L. 1990. Physical mapping of a low-copy DNA sequence in rye (*Secale cereale* L.). *Proc. Natl. Acad. Sci. U.S.A.* 87: 1899–1902.

Gustafson, J. P., and Ross, K. 2010. DNA sequence characterization of the malate transporter gene complex. *Crop Sci.* 294: 4.

Hackauf, B., Korzun, V., Wortmann, H., Wilde, P., and Wehling, P. 2012. Development of conserved ortholog set markers linked to the restorer gene *Rfp1* in rye. *Mol. Breed.* 30: 1507–1518.

Hackauf, B., Rudd, S., van der Voort, J. R., Miedaner, T., and Wehling, P. 2009a. Comparative mapping of DNA sequences in rye (*Secale cereale* L.) in relation to the rice genome. *Theor. Appl. Genet.* 118: 371–384.

Hackauf, B., Stojalowski, S., Wortmann, H., Wilde, P., Fromme, F. J., Menzel, J., Korzun, V., and Wehling, P. 2009b. Minimierung des Mutterkornbefalls im Hybridroggen durch Ansätze der Präzisionszüchtung. *J. Kulturpflanz.* 61: 15–20.

Hackauf, B., and Wehling, P. 2003. Development of microsatellite markers in rye: Map construction. *Plant Breed. Seed Sci.* 48: 143–151.

Hagberg, S. 1960. A rapid method for determining alpha-amylase activity. *Cereal Chem.* 37: 218–222.

Halasz, E., and Sipos, T. 2007. Perennial rye (*Secale cereanum*) breeding in Hungary, at University of Debrecen. *Anal. Univ. Oradea* 12: 57–62.

Hallauer, A. R., and Miranda, J. B. 1981. Quantitative genetics in maize breeding. Iowa State University Press, Ames, IA, pp. 1–58.

Hammer, K. 1990. Breeding system and phylogenetic relationships in *Secale* L. *Biol. Zbl.* 109: 45–50.

Hammer, K., Skolimowska, E., and Knüpffer, H. 1987. Vorarbeiten zur monographischen Darstellung von Wildplanzenzsortimenten *Secale* L. *Kulturpflanze* 33: 135–177.

Hanelt, P. 2001. *Mansfeld's Encyclopedia of Agricultural and Horticultural Crops*, vols. 1–6. Springer, Berlin, Germany, pp. 1–3716.

Hansen, H. B., Ostergaard, K., and Bach-Knudsen, K. E. 1988. Effect of baking and staling on carbohydrate composition in rye bread and on digestibility of starch and dietary fibre *in vivo*. *J. Cer. Sci.* 7: 135–144.

Hart, G., and Gale, M. D. 1988. Guidelines for nomenclature biochemical/molecular markers. *Proceedings of the 7th International Wheat Genetics Symposium*, July 13–19. Cambridge, pp. 1215–1218.

Haseneyer, G., Schmutzer, T., Seidel, M., Zhou, R., Mascher, M., Schön, C. C., Taudien, S. et al. 2011. From RNA-seq to large-scale genotyping—Genomics resources for rye (*Secale cereale* L.). *BMC Plant Biol.* 11: 131. doi:10.1186/1471-2229-11-131.

Hayes, H. K., and Garber, R. J. 1919. Synthetic production of high protein corn in relation to breeding. *Agron. J.* 11: 309–318.

Hazarika, M. H., and Rees, H. 1967. Genotypical control of chromosome behaviour in rye. X. Chromosome pairing and fertility in autotetraploids. *Heredity* 22: 317–321.

Heneen, W. K. 1962. Chromosome morphology in inbred rye. *Hereditas*, 48: 182–200.

Heneen, W. K. 1965. On the meiosis of haploid rye. *Hereditas* 52: 421–424.

Hepting, L. 1978. Analyse eines 7 × 7-Sortendialles zur Ermittlung geeigneten Ausgangsmaterials für die Hybridzüchtung bei Roggen. *Z. Pflanzenzücht.* 80: 188–197.

Heribert-Nilsson, N. 1916a. Populationsanalyse und Erblichkeitsversuche. *Z. Pflanzenzücht.* 4: 1–44.

Heribert-Nilsson, N. 1916b. Morphologische Studien an Roggenähren aus Anatolien. *Z. Pflanzenzücht.* 5: 89–113.

Heribert-Nilsson, N. 1921. Selektive Verschiebung der Gametenfrequenz in einer Kreuzungspopulation von Roggen. *Hereditas* 2: 364–368.

Heribert-Nilsson, N. 1937. Eine Prüfung der Wege und Theorien der Inzucht. *Hereditas* 23: 236–256.

Herzfeld, J. C. P. 2002. Development of a genetic transformation protocol for rye (*Secale cereale* L.) and characterisation of transgene expression after biolistic or *Agrobacterium*-mediated gene transfer. PhD thesis, Martin Luther University, Halle-Wittenberg, Germany, pp. 1–98.

Heslop-Harrison, J. 1979. Aspects of the structure, cytochemistry and germination of the pollen of rye (*Secale cereale* L.). *Ann. Bot.* 44: 1–7.

Hillman, G. 2001. New evidence of late glacial cereal cultivation at Abu Hureyra on the Euphrates. *Holocene* 11: 383–393.

Hillman, G. C. 1978. On the origins of domestic rye—*Secale cereale*: The finds from a ceramic Can Hasan III in Turkey. *Anatolian Stud.* 28: 157–174.

Hodosne-Kotvics, G., Krisztian, H., and Dornbach, L. 1999. Perennial rye: A novel variety released from a new interspecific hybrid rye. *Gyakorlati Agroforum* 10: 63–64.

Hoffmann, B. 2008. Alteration of drought tolerance of winter wheat caused by translocation of rye chromosome segment 1RS. *Cer. Res. Comm.* 36: 269–278.

Hoffmann, F., and Wenzel, G. 1977. A single grain screening technique for breeding alkylresorcinol-poor rye. *Theor. Appl. Genet.* 50: 1–2.

Hoffmann, W., Mudra, A., and Plarre, W. 1970. *Lehrbuch der Züchtung landwirtschaftlicher Kulturpflanzen*. P. Parey Verlag, Berlin, Germany.

Holloway, R. E. 1996. Zn as a subsoil nutrient for cereal. PhD thesis, University of Adelaide, Adelaide, Australia, pp. 1–95.

Hossain, M. G. 1978. Effects of external environmental factors on chromosome pairing in autotetraploid rye. *Cytologia* 43: 21–34.

Houben, A., Kumke, K., Nagaki, K., and Hause, G. 2011. CENH3 distribution and differential chromatin modifications during pollen development in rye (*Secale cereale* L.). *Chromos. Res.* 19: 471–480.

Houben, A., Kynast, R. G., Heim, U., Hermann, H., Jones, R. N., and Forster, J. W. 1996. Molecular cytogenetic characterisation of the terminal heterochromatic segment of the B-chromosome of rye (*Secale cereale*). *Chromosoma* 105: 97–103.

Hsam, S. L. K., and Zeller, F. J. 1996. Allelism of mildew resistance genes *Pm8* and *Pm17* in wheat-rye translocated chromosome T1BL.1RS. *Vortr. Pflanzenzücht.* 35: 186–187.

Huang, T., Böhlenius, H., Eriksson, S., Parcy, F., and Nilsson, O. 2005. The mRNA of the *Arabidopsis* gene *FT* moves from leaf to shoot apex and induces flowering. *Science* 309: 1694–1696.

Hübner, M., Wilde, P., Schmiedchen, B., Dopierala, P., Gowda, M., Reif, J. C., and Miedaner, T. 2013. Hybrid rye performance under natural drought stress in Europe. *Theor. Appl. Genet.* 126: 475–482.

Hull, G. A., Halford, N. G., Kreis, M., and Shewry, P. R. 1991. Isolation and characterisation of genes encoding rye prolamins containing a highly repetitive sequence motif. *Plant Mol. Biol.* 17: 1111–1115.

Hull, G. A., Sabelli, P. A., and Shewry, P. R. 1992. Restriction fragment analysis of the secalin loci of rye. *Biochem. Genet.* 30: 85–97.

Hussey, B. M. J., Keighery, G. J., Dodd, J., Lloyd, S. J., and Cousens, R. D. 2007. *A Guide to the Weeds of Western Australia*. The Weeds Society of Western Australia (Inc.), Victoria Park, Australia.

Huth, W., and Lesemann, D.-E. 1984. Vorkommen von Tobacco Rattle-Virus in Roggen (*Secale cereale*) in der Bundesrepublik Deutschland. *J. Phytopathol.* 111: 1–4.

Immonen, S., and Anntila, H. 1996. Success in anther culture of rye. *Vortr. Plfanzenzüchtg.* 35: 237–244.

In, O., Berberich, T., Romdhane, S., and Feierabend, J. 2005. Changes in gene expression during dehardening of cold-hardened winter rye (*Secale cereale* L.) leaves and potential role of a peptide methionine sulfoxide reductase in cold-acclimation. *Planta* 220: 941–950.

Iqbal, M. J., and Rayburn, A. L. 1995. Identification of the 1RS rye chromosomal segment in wheat by RAPD analysis. *Theor. Appl. Genet.* 91: 1048–1053.

Isik, Z., Parmaksiz, I., Coruh, C., Geylan-Su, Y. S., Cebeci, O., Beecher, B., and Budak, H. 2007. Organellar genome analysis of rye (*Secale cereale*) representing diverse geographic regions. *Genome* 50: 724–734.

Jain, S. K. 1960. Cytogenetics of rye (*Secale* ssp.). *Bibliogr. Genet.* 19: 1–86.

Janse, J. 1987. Certation between euploid and aneuploid pollen grains from a tertiary trisomic of rye, *Secale cereale* L. *Genome* 29: 353–356.

Jaroszuk-Scisel, J., Kurek, E., Slomka, A., Janczarek, M., and Rodzik, B. 2011. Activities of cell wall degrading enzymes in autolyzing cultures of three *Fusarium culmorum* isolates: Growth-promoting, deleterious and pathogenic to rye (*Secale cereale*). *Mycologia* 103: 929–945.

Ji, X., and Jetter, R. 2008. Very long chain alkylresorcinols accumulate in the intracuticular wax of rye (*Secale cereale* L.) leaves near the tissue surface. *Phytochemistry* 69: 1197–1207.

Jiang, Q. T., Wei, Y. M., Andre, L., Lu, Z. X., Pu, Z. E., Peng, Y. Y., and Zheng, Y. L. 2010. Characterization of omega-secalin genes from rye, triticale, and a wheat 1BL/1RS translocation line. *J. Appl. Genet.* 51: 403–411.

Jones, G. H., and Rees, H. 1964. Genotypic control of chromosome behaviour in rye. VIII. The distribution of chiasmata within pollen mother cells. *Heredity* 22: 333–347.

Jones, J. D., and Flavell, R. B. 1982. The structure, amount and chromosomal localization of defined repeated DNA sequences in species of the genus *Secale*. *Chromosoma* 86: 613–641.

Joshi, B. C., Singh, D., Kumar, S., Bansal, H. C., Varmani, S., Kamanoi, M., and Jenkins, B. C. 1962. Induction of chlorophyll deficient mutations in *Secale cereale*. *Wheat Inf. Serv.* 33–34: 30–31.

Kagawa, N., Nagaki, K., and Tsujimoto, H. 2002. Tetrad-FISH analysis reveals recombination suppression by interstitial heterochromatin sequences in rye (*Secale cereale*). *Mol. Genet. Genomics* 267: 10–15.

Kamanoi, M., and Jenkins, B. C. 1962. Trisomics in common rye, *Secale cereale* L. *Seiken Ziho* 13: 118–123.

Karppinen, S., Myllymäki, O., Forssell, P., and Poutanen, K. 2003. Fructans in rye. *Cereal Chem.* 80: 168–171.

Kaspar, T. C., Jaynes, D. B., Parkin, T. B., and Moorman, T. B. 2007. Rye cover crop and gamagrass strip effects on NO_3 concentration and load in tile drainage. *J. Environ. Qual.* 36: 1503–1511.

Kastirr, U., Ziegler, A., Perovic, D., and Golecki, B. 2012. Occurrence of *soil-borne wheat mosaic virus* (SBWMV-NY) in Schleswig-Holstein and its importance for wheat cultivation. *J. Kulturpflanz.* 64: 469–477.

Kattermann, G. 1937. Zur Cytologie halbbehaarter Stämme aus Weizen-Roggen-Bastardierungen. *Züchter* 9: 196–199.

Kattermann, G. 1939. Ein neuer Karyotyp beim Roggen. *Chromosoma* 1: 284–299.

Khabaz-Saberi, H., Graham, R. D., and Rathjen, A. J. 1998. Inheritance of Mn efficiency in durum wheat. *J. Plant Nutr.* 22: 11–21.

Khalil, H. B., Xu, Y., Frick, M., Elzein, S., Ehdaeiv, M.-R., Laroche, A., and Gulick, P. 2013. Rye subgenome suppression in triticale: An insight into genome restructuring in recent polyploids. *Proceedings of the 21st International Plant & Animal Genome Conference*, January 13–18, 2012. San Diego, CA, Abstract P98.

Khlestkina, E. K., Salina, E. A., Matthies, I. E., Leonova, I. N., Börner, A., and Röder, M. S. 2011. Comparative molecular marker-based genetic mapping of flavanone 3-hydroxylase genes in wheat, rye and barley. *Euphytica* 179: 333–341.

Khlestkina, E. K., Than, M. H., Pestsova, E. G., Röder, M. S., Malyshev, S. V., Korzun, V., and Börner, A. 2004. Mapping of 99 new microsatellite-derived loci in rye (*Secale cereale* L.) including 39 expressed sequence tags. *Theor. Appl. Genet.* 109: 725–732.

Khush, G. S., and Stebbins, G. L. 1961. Cytogenetic and evolutionary studies in *Secale*. I. Some new data on the ancestry of *S. cereale*. *Am. J. Bot.* 48: 723–730.

Kleijer, G., Häner, R., and Knüpffer, H. 2007. Triticale and rye genetic resources in Europe. *Ad hoc* Meeting, September 28, 2006, Nyon, Switzerland. Bioversity International, Rome, Italy, pp. 1–15.

Klocke, B. 2004. Virulenzstruktur und -dynamik des Roggenbraunrostes (*Puccinia recondita* f. sp. *secalis*) in der Bundesrepublik Deutschland. PhD thesis, Martin Luther University, Halle-Wittenberg, Germany, pp. 1–141.

Kobyljanski, V. D. 1969. Cytoplasmic male sterility in diploid rye (Russian). *Vestn. Selskokhooz. Nauk.* 6: 18–22.

Kobyljanski, V. D. 1975. The systematics and phylogeny of the genus *Secale* L. (Russian). *Bull. Vest. Rast.* 48: 64–71.

Kobyljanski, V. D., and Katerova, A. G. 1973. Nasledovanije citoplasmiceskoj i jaderno-muzskoj sterilnosti u osimoj diploidnij rzi. *Genetika* 9: 5–11.

Koebner, R. M. D., and Sing, N. K. 1984. Amelioration of the quality of a wheat-rye translocation lines. *Proceedings of the 4th Wheat Breeding Society*, August, 27–30. Toowoomba, Australia, pp. 86–88.

Koller, D. L., and Zeller, F. J. 1976. The homoeologous relationships of rye chromosome 4R and 7R with wheat chromosomes. *Genet. Res.* 28: 177–188.

Kolosov, V. I., Bukharov, A. A., and Zolotarev, A. S. 1989. Nucleotide sequence of the rye chloroplast *psbA* gene, encoding D1 protein of photosystem II. *Nucleic Acids Res.* 17: 1759–1760.

Kolosov, V. I., Bukharov, A. A., and Zolotarev, A. S. 1991. Rye photosystem II. Nucleotide sequence of *psbD* and *psbI* genes coding for polypeptides of the reaction centre (Russian). *Bioorg. Khim.* 17: 448–455.

Korzun, V., Malyshev, S., Kartel, N., Westermann, T., Weber, W. E., and Börner, A. 1998. A genetic linkage map of rye (*Secale cereale* L.). *Theor. Appl. Genet.* 97: 99–102.

Korzun, V., Malyshev, S., Voylokov, A., and Börner, A. 2001. A genetic map of rye (*Secale cereale* L.) combining RFLP, isozyme, protein, microsatellite and gene loci. *Theor. Appl. Genet.* 102: 709–717.

Korzun, V., Melz, G., and Börner, A. 1996. RFLP mapping of the dwarfing (*Ddw1*) and hairy peduncle (*Hp*) genes on chromosome 5 of rye (*Secale cereale* L.). *Theor. Appl. Genet.* 92: 1073–1077.

Kostoff, D. 1939. Frequency of polyembryony and chlorophyll deficiencies in rye. *Dokl. Akad. Sci. USSR* 24: 479–482.

Kota, R. S., Gill, B. S., and Hulbert, S. H. 1994. Presence of various rye-specific repeated DNA sequences on the midget chromosome of rye. *Genome* 37: 619–624.

Kotvics, G. 1970. Investigations os increasing the protein content of *Secale cereale* L. In *Protein Growth by Plant Breeding*. A. Balint (ed.). Akademiai Kiado, Budapest, Hungary, pp. 89–98.

Kousaka, R., and Endo, T. R. 2012. Nondisjunction of rye B chromosomes is gametophytically controlled by the transacting control element. *iWIS* 113: 17–19.

Kovarsky, A. E. 1939. Formative process of interspecies hybrids under different conditions. *Breed. Seed* 10/11: 39–47.

Kramer, S., Piotrowski, M., Kuhnemann, F., and Edelmann, H. G. 2003. Physiological and biochemical characterization of ethylene-generated gravicompetence in primary shoots of coleoptile-less gravi-incompetent rye seedlings. *J. Exp. Bot.* 54: 2723–2732.

Kranz, A. R. 1963. Die anatomischen, ökologischen und genetischen Grundlagen der Ährenbrüchigkeit des Roggens. *Beitr. Biol. Pflanzen* 38: 445–471.

Krasnyuk, A. A. 1936. *Secale cereale* × *Agropyron cristatum* hybrids. *Soc. Grain Farm.* 1: 106–115.

Kratovalieva, S., Srbinoska, M., Popsimonova, G., Selamovska, A., Meglic, V., and Andjelkovic, V. 2012. Ultrasound influence on coleoptile length at *Poaceae* seedlings as valuable criteria in prebreeding and breeding processes. *Genetika* 44: 561–570.

Kreis, M., Ford, B. G., Rahman, S., Miflin, B. J., and Shewry, P. R. 1985. Molecular evolution of the seed storage proteins of barley, rye and wheat. *J. Mol. Biol.* 183: 499–502.

Kubalakova, M., Valarik, M., Bartos, J., Vrana, J., Cíhalíkova, J., Molnar-Lang, M., and Dolezel, J. 2003. Analysis and sorting of rye (*Secale cereale* L.) chromosomes using flow cytometry. *Genome* 46: 893–905.

Kubicka, H., Gabara, B., and Wolska, A. 2007. The trait of transversal, yellow stripes on leaves of rye (*Secale cereale* L.). *Caryologia* 60: 325–330.

Kubo, N., Salomon, B., Komatsuda, T., von Bothmer, R., and Kadowaki, K. 2004. Structural and distributional variation of mitochondrial *rps2* genes in the tribe Triticeae (Poaceae). *Theor. Appl. Genet.* 110: 995–1002.

Kuckuck, H., 1973. Utilization of Iranian genetic material. *Wheat Inf. Serv.* 36: 21–26.

Kuckuck, H., and Peters, R. 1970. Weitere genetisch-züchterische Untersuchungen an *Secale vavilovii* Grossh. *Z. Pflanzenzüchtg.* 64: 182–200.

Kuckuck, H., and Peters, R. 1977. Die Verwendung von *Secale vavilovii* in der Züchtung selbststfertiler tetraploider Roggenformen. *Z. Pflanzenzüchtg.* 79: 265–286.

Kwiatek, M., Blaszczyk, L., Wisniewska, H., and Apolinarska, B. 2012. *Aegilops-Secale* amphiploids: Chromosome categorisation, pollen viability and identification of fungal disease resistance genes. *J. Appl. Genet.* 53: 37–40.

Kyzlasov, V. G. 2005. Apomictic development of seed in embryos of rye, soft wheat, and triticale. *Ann. Wheat Newslett.* 51: 99–100.

Kyzlasov, V. G. 2008. Amphimixis in soft wheat and apomixis in rye after mutual pollination of flowers. *Ann. Wheat Newslett.* 54: 107–110.

Lacadena, J. R. 1969. Microsporogenesis in alloplasmic rye. *Wheat Inf. Serv.* 29: 21–22.

Lamm, R. 1936. Cytological studies on inbred rye. *Hereditas* 22: 217–240.

Lamm, R. 1944. Chromosome behaviour in a triploid plant. *Hereditas* 30: 137–144.

Lancashire, P. D., Bleiholder, H., Langelüddecke, P., Stauss, R., van den Boom, T., Weber, E., and Witzenberger, A. 1991. A uniform decimal code for growth stages of crops and weeds. *Ann. Appl. Biol.* 119: 561–601.

Landberg, R., Kamal-Eldin, A., Salmenkallio-Marttila, M., Rouau, X., and Aman, P. 2008. Localization of alkylresorcinols in wheat, rye and barley kernels. *J. Cereal Sci.* 48: 401–406.

Langdon, T., Seago, C., Jones, R. N., Ougham, H., Thomas, H., Forster, J. W., and Jenkins, G. 2000. *De novo* evolution of satellite DNA on the rye B chromosome. *Genetics* 154: 869–884.

Langis, R., and Steponkus, P. L. 1990. Cryopreservation of rye protoplasts by vitrification. *Plant Physiol.* 92: 666–671.

Łapiński, M. 1972. Cytoplasmic-genic type of male sterility in *Secale montanum* Guss. *Wheat Inf. Serv.* 35: 25–28.

Lapitan, N. L. V., Peng, J., and Sharma, V. 2007. A high-density map and PCR markers for Russian wheat aphid resistance gene *Dn7* on chromosome 1RS/1BL. *Crop Sci.* 47: 811–818.

Laube, W. 1959. Roggen. Blütenbiologie und Samenbildung. In *Züchtung der Getreidearten. Handbuch der Pflanzenzüchtung.* II. H. Kappert et al. (eds.). P. Parey Verlag, Berlin, Germany, pp. 2–55.

Laube, W., and Quadt, F. 1959. Roggen (*Secale cereale* L.). In *Züchtung der Getreidearten. Handbuch der Pflanzenzüchtung.* H. Kappert et al. (eds.). P. Parey Verlag, Berlin, Germany, pp. 35–102.

Law, C. N. 1963. An effect of potassium on chiasma frequency and recombination. *Genetica* 33: 313–329.

Lawrence, C. W. 1958. Genotypic control of chromosome behaviour in rye. VI. Selection for disjunction frequency. *Heredity* 12: 127–131.

Lei, M. P., Li, G. R., Liu, C., and Yang, Z. J. 2012. Characterization of wheat: *Secale africanum* introgression lines reveals evolutionary aspects of chromosome 1R in rye. *Genome* 55: 765–774.

Lein, A. 1949. Über alloplasmatische Roggen (Roggen mit Weizenplasma). *Züchter* 19: 101–108.

Leith, B. D., and Shands, H. L. 1935. The development of improved rye by inbreeding and hybridization. *Wisconsin Agric. Exp. Stat. Bull.* 430: 38.

Lelley, T., Machmoud, A. S., and Lein, V. 1987. Genetics and cytology of unreduced gametes in rye (*Secale cereale* L.). *Genome* 29: 635–638.

Levan, A. 1942. Studies on the meiotic mechanism of haploid rye. *Hereditas* 28: 177–211.

Lewitzky, G. A. 1931. The morphology of chromosomes. *Bull. Appl. Bot. Gent. Plant Breed.* 27: 19–174.

Li, X. F., Ma, J. F., and Matsumoto, H. 2000. Pattern of aluminum-induced secretion of organic acids differs between rye and wheat. *Plant Physiol.* 123: 1537–1544.

Li, Y., Böck. A., Haseneyer, G., Korzun, V., Wilde, P., Schön, C.-C., Ankerst, D. P., and Bauer, E. 2011a. Association analysis of frost tolerance in rye using candidate genes and phenotypic data from controlled, semi-controlled, and field phenotyping platforms. *BMC Plant Biol.* 11. doi:10.1186/1471-2229-11-146.

Li, Y., Haseneyer, G., Schon, C.-C., Ankerst, D., Korzun, V., Wilde, P., and Bauer, E. 2011b. High levels of nucleotide diversity and fast decline of linkage disequilibrium in rye (*Secale cereale* L.) genes involved in frost response. *BMC Plant Biol.* 11. doi:10.1186/1471-2229-11-6.

Lili, Q., Friebe, B., and Gill, B. S. 2006. Complex genome rearrangements reveal evolutionary dynamics of pericentromeric regions in the Triticeae. *Genome* 49: 1628–1639.

Lima de Faria, A. 1952. Chrommere analysis of the chromosome complement of rye. *Chromosoma* 5: 1–68.

Lima de Faria, A. 1962. Genetic interaction in rye expressed at the chromosome phenotype. *Genetics* 47: 1455–1462.

Limin, A. E., and Fowler, D. B. 1984. The effect on cytoplasm on cold hardiness in alloplasmic rye (*Secale cereale* L.) and triticale. *Can. J. Genet. Cytol.* 26: 405–408.

Linacero, R., Freitas-Alves, E., and Vazquez, A. M. 2000. Hot spots of DNA instability revealed through the study of somaclonal variation in rye. *Theor. Appl. Genet* 100: 506–511.

Lindström, J. 1965. Transfer to wheat of accessory chromosomes from rye. *Hereditas* 54: 149–155.

Liu, C., Yang, C.-J., Li, G.-R., Zeng, Z.-X., Zhang, Y., Zhou, J.-P., Liu, Z.-H., and Ren, Z.-L. 2008. Isolation of a new repetitive DNA sequence from *Secale africanum* enables targeting of *Secale* chromatin in wheat background. *Euphytica* 159: 249–258.

Loarce, Y., Gallego, R., and Ferrer, E. 1996. A comparative analysis of the genetic relationship between rye cultivars using RFLP and RAPD markers. *Euphytica* 88: 107–115.

Loewe, G., Scheibe, F., Ludwig, C., and Diezel, W. 1978. Zur säulenchromatographischen Fraktionierung von Roggenpollenallergenen (*Secale cereale*). *Eur. Arch. Oto-rhino-laryngol.* 220: 277–286.

Lonnquist, J. H., and McGill, D. P. 1956. Performance of corn synthetics in advanced generations of synthesis and after two cycles of recurrent selection. *Agron. J.* 48: 249–253.

Lukaszewski, A. J., Brzezinski, W., and Klockiewicz-Kaminska, E. 2000. Transfer of the *Glu-D1* locus encoding high molecular weight glutenin subunits 5+10 from bread wheat to diploid rye. *Euphytica* 115: 49–57.

Lukaszewski, A. J., and Kopecky, D. 2010. The *Ph1* locus from wheat controls meiotic chromosome pairing in autotetraploid rye (*Secale cereale* L.). *Cytogen. Genome Res.* 129: 117–123.

Lukyanenko, P. P. 1973. *Izabranie trudi. Selekcia i semenovodstvo ozimoj psenici* (Russian). Kolos Isd., Moscow, Russia, pp. 1–48.

Lundqvist, A. 1954. Studies on self-sterility in rye, *Secale cereale* L. *Hereditas* 40: 278–294.

Lundqvist, A. 1956. Self-incompatibility in rye. I. Genetic control in the diploid. *Hereditas* 42: 293–395.

Lundqvist, A. 1957. Self-incompatibility in rye. II. Genetic control in the autotetraploid. *Hereditas* 43: 467–511.

Lundqvist, A. 1958. Self-incompatibility in rye. IV. Factors relating to self-seeding. *Hereditas* 44: 193–256.

Luo, M. C., Deal, K. R., Akhunov, E. D., Akhunova, A. R., Anderson, O. D., Anderson, J. A., Blake, N. et al. 2009. Genome comparisons reveal a dominant mechanism of chromosome number reduction in grasses and accelerated genome evolution in *Triticeae*. *Proc. Nat. Acad. U.S.A.* 106: 15780–15785.

Ma, J. F., Taketa, S., Chang, Y.-C., Takeda, K., and Matsumoto, H. 1999. Biosynthesis of phytosiderophores in several Triticeae species with different genomes. *J. Exp. Botany* 50: 723–726.

Ma, R., Guo, Y. D., and Pulli, S. 2003. Somatic embryogenesis and fertile green plant regeneration from suspension cell-derived protoplasts of rye (*Secale cereale* L.) *Plant Cell Rep.* 22: 320–327.

Ma, R., Guo, Y.-D., and Pulli, S. 2004a. Comparison of anther and microspore culture in the embryogenesis and regeneration of rye (*Secale cereale*). *Plant Cell Tissue Organ Cult.* 76: 147–157.

Ma, R., Yli-Mattila, T., and Pulli, S. 2004b. Phylogenetic relationships among genotypes of worldwide collection of spring and winter ryes (*Secale cereale* L.) determined by RAPD-PCR markers. *Hereditas* 140: 210–221.

Ma, X.-F., Wanous, M. K., Houchins, K., Milla, M. A. R., Goicoechea, P. G., Wang, Z., Xie, M., and Gustafson, J. P. 2001. Molecular linkage mapping in rye (*Secale cereale* L.). *Theor. Appl. Genet.* 102: 517–523.

Mago, R., Spielmeyer, W., Lawrence, G. J., Ellis, J. G., and Pryor, A. J. 2004. Resistance genes for rye stem rust (*SrR*) and barley powdery mildew (*Mla*) are located in syntenic regions on short arm of chromosome. *Genome* 47: 112–121.

Mahone, G. S., Frisch, M., Miedaner, T., Wilde, P., Wortmann, H., and Falke, K. C. 2013. Identification of quantitative trait loci in rye introgression lines carrying multiple donor chromosome segments. *Theor. Appl. Genet.* 126: 49–58.

Malik, R., Brown-Guedira, G. L., Smith, C. M., Harvey, T. L., and Gill, B. S. 2003. Genetic mapping of wheat curl mite resistance genes *Cmc3* and *Cmc4* in common wheat. *Crop Sci.* 43: 644–650.

Malyshev, S., Korzun, V., Voylokov, V., Smirnov, V., and Börner, A. 2001. Linkage mapping of mutant loci in rye (*Secale cereale* L.). *Theor. Appl. Genet.* 103: 70–74.

Manzanero, S., Puertas, M., Jimenez, G., and Vega, J. M. 2000. Neocentric activity of rye 5RL chromosome in wheat. *Chromosome Res.* 8: 543–554.

Marais, G. F., Marais, A. S., and Badenhorst, P. C. 1998a. Hybridization of *Thinopyrum distichum* and *Secale cereale*. *Plant Breed.* 117: 107–111.

Marais, G. F., Wessels, W. G., Horn, M., and du Toit, F. 1998b. Association of stem rust resistance gene (*Sr45*) and two Russian wheat aphid resistance genes (*Dn5* and *Dn7*) with mapped structural loci in common wheat. *S. Afr. J. Plant Soil*. 15: 67–71.

Marschner, H., Treeby, M., and Römheld, V. 1989. Role of root-induced changes in the rhizosphere for iron acquisition in higher plants. *Z. Pflanzenern. Bodenk.* 152: 197–204.

Martincic, J., Guberac, V., and Maric, S. 1997. Influence of winter rye seed size *Secale cereale* L. on shoot and rootlet length, and grain yield. *Rostl. Vyroba* 43: 95–100.

Martiny, B., 1870. Der mehrblütige Roggen—Eine Pflanzenkulturstudie. *Schriften Naturforsch. Gesell.*, Verlag A. W. Kasemann, Danzig, Poland, pp. 1–13.

Martis, M., Klemme, S., Banaei-Moghaddam, A. M., Blattner, F. R., Macas, J., Schmutzer, T. et al. 2012. A selfish supernumerary chromosome reveals its origin as a mosaic of host genome and organellar sequences. *Proc. Nat. Acad. Sci. U.S.A.* www.pnas.org/cgi/content/short/1204237109

Martis, M. M., Zhou, R., Haseneyer, G., Hackauf, B., Schmutzer, T., Scholz, U., Kubalakova, M. et al. 2013. The genome structure of rye (*Secale cereale* L.) and its multiple parents. *Proceedings of the 21st International Plant & Animal Genome Conference*, January 12–16. San Diego, CA, Abstract W400.

Masojć, P., Banek-Tabor, A., Milczarski, P., and Twardowska, M. 2007. QTLs for resistance to preharvest sprouting in rye (*Secale cereale* L.). *J. Appl. Genet.* 48: 211–217.

Masojć, P., and Kosmala, A. 2012. Proteomic analysis of developing rye grain with contrasting resistance to preharvest sprouting. *Mol. Breed.* 30: 1355–1361.

Masojć, P., Lebiecka, K., Milczarski, P., Wisniewska, M., Lan, A., and Owsianicki, R. 2009. Three classes of loci controlling preharvest sprouting in rye (*Secale cereale* L.) discerned by means of bidirectional selective genotyping (BSG). *Euphytica* 170: 123–129.

Masojć, P., and Milczarski, P. 2004. The identification of QTL for water extract viscosity of rye grain. *Folia Univ. Agric. Steninensis. Ser Agric.* 234: 145–147.

Masojć, P., and Milczarski, P. 2005. Mapping QTLs for α-amylase activity in rye grain. *J. Appl. Genet.* 46: 115–123.

Masojć, P., and Milczarski, P. 2009. Relationship between QTLs for preharvest sprouting and alpha-amylase activity in rye grain. *Mol. Breed.* 23: 75–84.

Masojć, P., Milczarski, P., and Mysków, B. 1999. Identification of genes underlying sprouting resistance in rye. *Proceedings of the 8th International Symposium on Pre-Harvest Sprouting in Cereals*. D. Weipert (ed.). Association of Cereal Research, Federal Centre for Cereal, Potato and Lipid Research, Detmold, Germany, pp. 131–136.

Masojć, P., Mysków, B., and Milczarski, P. 2001. Extending a RFLP-based genetic map of rye using random amplified polymorphic DNA (RAPD) and isozyme markers. *Theor. Appl. Genet.* 102: 1273–1279.

Matossian, M. K. 1989. *Poisons of the Past: Molds, Epidemics, and History*. Yale University Press, New Haven, CT, pp. 1–190.

McIntosh, R. A. 1988. Catalogue of gene symbols for wheat. *Proceedings of the 7th International Wheat Genetics Symposium*. Miller, T. E., and Koebner, R. M. D. (eds.). Cambridge, pp. 1225–1323.

McIntosh, R. A., Yamazaki, Y., Dubcovsky, J., Rogers, J., Morris, C., Somers, D. J., Appels, R., and Devos, K. M. 2008. Catalogue of gene symbols for wheat. *Proceedings of the 11th International Wheat Genetics Symposium*. August 24–29. Brisbane, Australia, pp. 1–59.

Meister, N. G., and Tyumyakov, N. A. 1927. The first generation of rye-wheat hybrids of direct and reciprocal hybridization (Russian). *J. Exp. Agron. South-East Russia* 4: 87–96.

Melz, G. 1989. Beiträge zur Genetik des Roggens (*Secale cereale* L.). DSc thesis, AdL, Berlin, Germany, pp. 1–173.

Melz, G., Melz, G., and Hartmann, F. 2001. Genetics of a male-sterile rye of "G-type" with results of the first F1-hybrids. *Proceedings of the 7th EUCARPIA Conference*, July 4–7. Radzikow, Poland, pp. 43–50.

Melz, G., Melz, G., and Hartmann, F. 2003. Genetics of a male-sterile rye of "G-type" with results of the first F1-hybrids. *Plant Breed. Seed Sci.* 47: 47–55.

Melz, G., Neumann, H., Müller, H. W., and Sturm, W. 1984. Genetic analysis of rye (*Secale cereale* L.). I. Results of gene location on rye chromosomes using primary trisomics. *Genet. Polon.* 25: 111–115.

Melz, G., and Schlegel, R. 1985. Identification of seven telotrisomics of rye (*Secale cereale* L.). *Euphytica* 34: 361–366.

Melz, G., and Sybenga, J. 1994. The 3rd workshop of rye genetics and cytogenetics—Revision and completion of the genetic map of rye. (October 25–28, 1993, Gross Lüsewitz, Germany.) *Genet. Polon.* 35: 131–132.

Mettin, D., Balkandschiewa, J., and Müller, F. 1972a. [Studies on trisomics of rye, *Secale cereale* L.] German. *Tag. Ber. Akad. Landwirtsch. Wiss.* 119: 153–161.

Mettin, D., Blüthner, W.-D., and Schlegel, R. 1973. Additional evidence on spontaneous 1B/1R wheat-rye substitutions and translocations. *Proceedings of the 4th International Wheat Genetics Symposium*, August 6–11. Columbia, MO, pp. 179–184.

Mettin, D., Müller, F., and Schlegel, R. 1972b. Untersuchungen an Valenzkreuzungen beim Roggen (*Secale cereale* L.) 3. Über die Häufigkeit von 3x- und 4x-Bastarden unter verschiedenen Bestäubungsbedingungen. [Studies on ploidy crosses in rye, *Secale cereale* L. 3. About the frequency of triploid and tetraploid hybrids under different crossing conditions]. *Tag. Ber. AdL (DDR)* 119: 135–152.

Michalska, A., and Zielinski, H. 2006. Effect of flour extraction rate on bioactive compounds content of two rye varieties. *Pol. J. Food Nutr. Sci.* 15: 297–303.

Miedaner, T., Borchardt, D. C., and Geiger, H. H. 1993. Genetic analysis of inbred lines and their crosses for resistance to head blight (*Fusarium culmorum, F. graminearum*) in winter rye. *Euphytica* 65: 123–133.

Miedaner, T., Daume, C., Mirdita, V., Schmiedchen, B., Wilde, P., und Geiger, H. H. 2007. Verminderung von Alkaloiden in der Nahrungskette durch die züchterische Verbesserung der Mutterkorn-Resistenz von Winterroggen. In *Zwischen Tradition und Globalisierung 1*, Zikeli, S. et al. (eds.). Köster Verlag, Berlin, Germany, pp. 225–228.

Miedaner, T., Fromme, F. J., and Geiger, H. H. 1995. Genetic variation for foot-rot and *Fusarium* head-blight resistance among full-sib families of a self-incompatible winter rye (*Secale cereale* L.) population. *Theor. Appl. Genet.* 91: 862–878.

Miedaner, T., and Geiger, H. H. 1999. Vererbung quantitativer Resistenzen gegen Pilzkrankheiten bei Roggen. *Vortr. Pflanzenzücht.* 46: 157–168.

Miedaner, T., Glass, C., Dreyer, F., Wilde, P., Wortmann, H., and Geiger, H. H. 2000. Mapping of genes for male-fertility restoration in "Pampa A" CMS winter rye (*Secale cereale* L.). *Theor. Appl. Genet.* 101: 1226–1233.

Miedaner, T., Koch, S., Seggl, A., Schmiedchen, B., and Wilde, P. 2011b. Quantitative genetic parameters for selection of biomass yield in hybrid rye. *Plant Breed.* 131: 100–103.

Miedaner, T., Müller, B. U., Piepho, H.-P., and Falke, K. C. 2011a. Genetic architecture of plant height in winter rye introgression libraries. *Plant Breed.* 130: 209–216.

Miedaner, T., Reinbrecht, C., Lauber, U., Schollenberger, M., and Geiger, H. H. 2001. Effects of genotype and genotype-environment interaction on deoxynivalenol accumulation and resistance to *Fusarium* head blight in rye, triticale, and wheat. *Plant Breed.* 120: 97–105.

Miedaner, T., and Sperling, U. 2004. Effect of leaf rust on yield components of winter rye hybrids and assessment of quantitative resistance. *J. Phytopathol.* 143: 725–730.

Milczarski, P. 2008. Identification of QTLs of morphological traits related to lodging of rye (*Secale cereale* L.). *Biul. Inst. Hod. Aklimat. Ros.* 250: 211–216.

Milczarski, P., Banek-Tabor, A., Lebiecka, K., Stojaowski, S., Mysków, B., and Masojć, P. 2007. New genetic map of rye composed of PCR-based molecular markers and its alignment with the reference map of the DS2RXL10 intercross. *J. Appl. Genet.* 48: 11–24.

Milczarski, P., and Masojć, P. 1999. Application of molecular map of rye genome and QTL analysis for the identification of genes that control earliness of heading. *Bull. Inst. Hod. Aklim. Rosl.* 211: 205–209.

Milczarski, P., and Masojć, P. 2002. The mapping of QTLs for chlorophyll content and responsiveness to gibberellic (*GA3*) and abscisic (ABA) acids in rye. *Cell Mol. Biol. Lett.* 7: 449–455.

Milczarski, P., and Masojć, P. 2003. Interval mapping of genes controlling growth of rye plants. *Plant Breed. Seed Sci.* 48: 135–142.

Milczarski, P., and Masojć, P. 2007. Mapping of QTLs for agronomic traits in rye. Proc. Int. Symposium on Rye Breeding and Genetics, Rostock (Germany), June 28–30, 2006. *Vortr. Pflanzenzücht.* 71: 263–265.

Mirdita, V., Dhillon, B. S., Geiger, H. H., and Miedaner, T. 2008. Genetic variation for resistance to ergot (*Claviceps purpurea* [Fr.] Tul.) among full-sib families of five populations of winter rye (*Secale cereale* L.). *Theor. Appl. Genet.* 118: 85–90.

Mogensen, H. L., and Rusche, M. L. 2000. Occurrence of plastids in rye (Poaceae) sperm cells. *Am. J. Bot.* 87: 1189–1192.

Mohammadi, R., Farshadfar, E., Aghaee-Sarbarzeh, M., and Sutka, J. 2003. Locating QTLs controlling drought tolerance criteria in rye using disomic addition lines. *Cer. Res. Comm.* 31: 257–262.

Mohan, A., Goyal, A., Singh, R., Balyan, H. S., and Gupta, P. K. 2007. Physical mapping of wheat and rye expressed sequence tag–simple sequence repeats on wheat chromosomes. *Crop Sci.* 47: 3–13.

Montero, M. T., Alonso, E., and Sainz, T. 1992a. Allergens from rye pollen (*Secale cereale*). I. Study of protein release by rye pollen during a 19-hour extraction process. Allergen identification. *Allergy* 47: 22–25.

Montero, M. T., Alonso, E., and Sainz, T. 1992b. Allergens from rye pollen (*Secale cereale*). II. Characterization and partial purification. *Allergy* 47: 26–29.

Moore, G., Devos, K., Wang, Z., and Gale, M. 1995. Cereal genome evolution: Grasses, line up and form a circle. *Curr. Biol.* 5: 737–739.

Morgenstern, K., and Geiger, H. H. 1982. Plasmotype/genotype interaction with regard to male sterility in rye (*Secale cereale* L.). *Tag. Ber. Akad. Landwirtsch. Wiss.* 198: 381–388.

Mori, S., Kishi-Nishizawa, N., and Fujigaki, J. 1990. Identification of rye chromosome 5R as carrier of the genes for mugineic acid synthetase and 3-hydroxymugineic acid synthetase using wheat-rye addition lines. *Jpn. J. Genet.* 65: 343–350.

Moseman, J. G., Martin, J. H., and Adair, C. R. 1993. Origin and history of research on wheat, rye, corn, sorghum, barley, oats, rice, and weeds by the U.S. Department of Agriculture from 1836 to 1972. *Ann. Wheat Newslett.* 39: 27–70.

Müller, F. 1969. Ergebnisse bei der Züchtung von kurzstrohigem Roggen. *Ber. Pflanzentüchtg.* 3: 97–101.

Müller, F. 1975. Untersuchungen zur Mutationszüchtung bei di- und tetraploidem Roggen. *Tag. Ber. Martin Luther Univ.* 14: 3–7.

Müller, G., Vahl, U., and Wiberg, A. 1989. Die Nutzung der Antherenkulturmethode im Zuchtprozess von Winterweizen. II. Die Bereitstellung neuer Winterweizen-dh-Linien mit 1AL-1RS-Translokationen. *Plant Breed.* 103: 81–87.

Munoz-Florez, C., Campoli, C., Matuz-Cadiz, M., Pozniak, C., and Fowler, B. 2009. Effect of photoperiod on the regulation of rye (*Secale cereale*) *ScVrn-1* vernalization gene and its repressors *ScVrn-2* and *ScVrt-2*. *Proceedings of the 17th Plant & Animal Genomes Conference*, January 10–14. San Diego, CA, p. 281.

Müntzing, A. 1937. Note on a haploid rye plant. *Hereditas* 23: 401–404.

Müntzing, A., and Akdik, S. 1948. Cytological disturbances in the first inbred generation of rye. *Hereditas* 34: 485–509.

Müntzing, A., and Bose, S. 1969. EMS treatments of inbred lines of rye. *Wheat Inf. Serv.* 28: 25–26.

Murai, K., Naiyu, X. U., and Tsunewaki, K. 1989. Studies on the origin of crop species by restriction endonuclease analysis of organellar DNA. III. Chloroplast DNA variation and interspecific relationships in the genus *Secale*. *Jpn. J. Genet.* 64: 35–47.

Murashige, T., and Skoog, F. 1962. A revised medium for rapid growth and bioassays with tobacco tissue cultures. *Physiol. Plant.* 15: 473–497.

Musmann, D., Lornsen, E., Jansen, G., Enders, M., Kusterer, B., Jürgens, H.-U., Ordon, F. et al. 2012. Association mapping in a breeding population of hybrid rye. *62th Tagung der Vereinigung der Pflanzenzüchter und Saatgutkaufleute*, November 22–23. Gumpenstein, Austria, 2011. pp. 9–18.

Muszynska, A. 2013. Molecular genetic and histological characterization of the trait "Stabilstroh." http://www.ipk-gatersleben.de/en/dept-physiology-and-cell-biology/structural-cell-biology/projects-and-cooperations/molecular-genetic-and-histological-characterization-of-the-trait-stabilstroh/

Mysków, B. 2010. *Identification of QTL for Earliness and Pre-Harvest Sprouting Resistance in Rye with Using High Density Genetic Maps of RIL Populations* (Polish). West-Pomeranian University Print, Szczecin, Poland, pp. 1–86.

Mysków, B., Stojalowski, S., Lan, A., Bolibok-Bragoszewska, H., Rakoczy-Trojanowska, M., and Kilian, A. 2012. Detection of the quantitative trait loci for α-amylase activity on a high-density genetic map of rye and comparison of their localization to loci controlling preharvest sprouting and earliness. *Mol. Breed.* 30: 367–376.

Mysków, B., Stojalowski, S., Milczarski, P., and Masojć, P. 2010. Mapping of sequence-specific markers and loci controlling preharvest sprouting and alpha-amylase activity in rye (*Secale cereale* L.) on the genetic map of an F2 (S120×S76) population. *J. Appl. Genet.* 51: 283–287.

Nagy, E. D., and Lelley, T. 2003. Genetic and physical mapping of sequence-specific amplified polymorphic (SSAP) markers on the 1RS chromosome arm of rye in a wheat background. *Theor. Appl. Genet.* 107: 1271–1277.

Naranjo, T., Valenzuela, N. T., and Perera, E. 2010. Chiasma frequency is region specific and chromosome conformation dependent in a rye chromosome added to wheat. *Cytogenet. Genome Res.* 129: 133–142.

Natrova, Z., and Hlavac, M. 1975. The effect of gametocides on microsporogenesis of winter rye. *Biol. Plant.* 17: 256–262.

Navashin, M. 1934. Chromosomal alterations caused by hybridization and their bearing upon certain general genetic problems. *Cytologia* 5: 169–203.

Neijzing, M. G. 1982. Chiasma formation in duplicated segments of the haploid rye genome. *Chromosoma* 85: 287–298.

Neumann, M. P. 1929. *Brotgetreide und Brot.* P. Parey Verlag, Berlin, Germany.

Newell, M. A., and Butler, T. J. 2012. Forage rye improvement in the Southern United States: A review. *Crop. Sci.* doi:10.2135/cropsci2012.05.0319.

Nicolaisen, W. 1931. Über quantitative Xenien bei Roggen und Erbsen. *Z. Pflanzenzüchtg.* 17: 265–276.

Niedziela, A., Bednarek, P. T., Cichy, H., Budzianowski, G., Kilian, A., and Aniol, A. 2012. Aluminum tolerance association mapping in triticale. *BMC Genomics* 13: 67. doi:10.1186/1471-2164-13-67.

Nilsson-Ehle, H. 1913. Einige Beobachtungen über erbliche Variationen der Chlorophylleigenschaft bei den Getreidearten. *Z. Ind. Abst. Vererbungsl.* 9: 289–300.

Nilsson-Ehle, H. 1928. Inzucht als Züchtungsmethode. *Züchtungskunde* 3: 257–271.

Nobbe, F. 1869. Die Controle landwirtschaftlicher Handels-Sämereien betreffend. *Landw. Vers. Station* 12: 48–51.

Nordenskiöld, H. 1939. Studies of a haploid rye plant. *Hereditas* 25: 204–210.

Nürnberg-Krüger, U. 1951. Über die Auswirkung des Plasmas auf Leistungsmerkmale beim Roggen. *Theor. Appl. Genet.* 21: 232–240.

Nürnberg-Krüger, U. 1960. Cytogenetische Untersuchungen an *Secale silvestre* Host. *Züchter* 30: 147–150.

Nürnberg-Krüger, U. 1974. Möglichkeiten der Einführung von Wildroggen-Genmaterial zur Erzielung des Merkmals Kurzstrohigkeit in den Kulturroggen. *Tag. Ber. AdL* 127: 103–110.

O'Mara, J. G. 1947. The substitution of a specific *Secale cereale* chromosome for a specific *Triticum aestivum* chromosome. *Genetics* 32: 99–100.

Ohnuma, T., Numata, T., Osawa, T., Inanaga, H., Okazaki, Y., Shinya, S., Kondo. K., Fukuda. T., and Fukamizo, T. 2012. Crystal structure and chitin oligosaccharide-binding mode of a "loopful" family GH19 chitinase from rye, *Secale cereale*, seeds. *FEBS J.* 279: 3639–3651.

Ohnuma, T., Taira, T., Yamagami, T., Aso, Y., and Ishiguro, M. 2004. Molecular cloning, functional expression, and mutagenesis of cDNA encoding class I chitinase from rye (*Secale cereale*) seeds. *Biosci. Biotech. Biochem.* 68: 324–332.

Oinuma, T. 1953. Karyomorphology of cereals. Karyotype alteration of rye, *Secale cereale* L. *Jpn. J. Genet.* 28: 28–34.

Oram, R. N. 1996. *Secale montanum*—A wider role in Australia? *N. Z. J. Agric. Res.* 39: 629–633.

Ossent, H. P. 1934. Roggenzüchtung. *Z. Naturwiss.* 22: 271–274.

Pabst, G. 1887. *Köhler's Medizinal-Pflanzen.* Verlag F. E. Köhler, Gera, Germany.

Pammer, G. 1905. Über Veredelungszüchtungen mit einigen Landsorten des Roggens in Niederösterreich. *Z. Landw. Versuchswesen* 8: 1015–1053.

Pan, W. H., Houben, A., and Schlegel, R. 1993. Highly effective cell synchronization in plant roots by hydroxyurea and amiprophos-methyl or colchicine. *Genome* 36: 387–390.

Pathak, G. N. 1940. Studies on the cytology of cereals. *J. Genet.* 39: 437–467.

Percze, A. 2006. Role of rye (*Secale cereale*) as a catch crop in elemental transport of integrated crop production. *Cereal Res. Comm.* 34: 259–262.

Pershina, L. A., Rakovtseva, T. S., Belova, L. I., Devyatkina, E. P., Silkova, O. G., Kravisova, L. A., and Shchapova, A. I. 2007. Effect of rye *Secale cereale* L. chromosomes 1R and 3R on polyembryony expression in hybrid combinations between (*Hordeum* L.)-*Triticum aestivum* L. alloplasmic recombinant lines and wheat-rye substitution lines *T. aestivum* L.-*S. cereale* L. *Genetika* 43: 955–962.

Persson, K., and von Bothmer, R. 2002. Genetic diversity amongst landraces of rye (*Secale cereale* L.) from northern Europe. *Hereditas* 136: 29–38.

Perten, H. 1964. Application of the falling number method for evaluating alpha-amylase activity. *Cereal Chem.* 41: 127–140.

Petersen, G. 1991. Intergeneric hybridization between *Hordeum* and *Secale*. II. Analysis of meiosis in hybrids. *Hereditas* 114: 141–159.

Petersen, G., and Doebley, J. F. 1993. Chloroplast DNA variation in the genus *Secale* (Poaceae). *Plant System. Evol.* 187: 115–125.

Peterson, R. F. 1934. Improvement of rye through in-breeding. *Sci. Agric.* 14: 651–668.

Pfahler, P. L. 1965. *In vitro* germination of rye pollen. *Crop Sci.* 5: 597–598.

Philipp, U., Wehling, P., and Wricke, G. 1994. A linkage map of rye. *Theor. Appl. Genet.* 88: 243–248.

Phillips, D., Mikhailova, E. I., Timofejeva, L., Mitchell, J. L., Osina, O., Sosnikhina, S. P., Jones, R. N., and Jenkins, G. 2008. Dissecting meiosis of rye using translational proteomics. *Ann. Bot.* 101: 873–880.

Pillipcuk, B. V. 1970. Izucenie induciirovannnych mutacij u tetraploidoj rzi. *Tezisy Dokl.* 16: 1–41.

Piper, J. K. 1993. A grain agriculture fashioned in nature's image: The work of the land institute. *Great Plains Res.* 3: 249–272.

Pitt, R. E., Parks, J. E., Huber, S. C., and Sangree, J. A. 1997. Glycerol permeability of rye leaf protoplasts as affected by temperature and electroporation. *Plant Cell Tissue Organ Cult.* 50: 215–219.

Plaha, P., and Sethi, G. S. 2000. Long anther trait of rye (*Secale cereale* L.)—Its chromosomal location and expression in bread wheat (*Triticum aestivum* L.). *Wheat Inf. Serv.* 90: 47–48.

Plarre, W. 1954. Vergleichende Untersuchungen an diploidem und tetraploidem Roggen (*Secale cereale* L.) unter besonderer Berüksichtigung von Inzuchterscheinungen und Fertiltätsstörungen. *Z. Pflanzenzüchtg.* 33: 303–353.

Plaschke, J., Boerner, A., Xie, D. X., Koebner, R. M. D., Schlegel, R., and Gale, M. D. 1993. RFLP mapping of genes affecting plant height and growth habit in rye. *Theor. Appl. Genet.* 85: 1049–1054.

Plinius Secundus Maior, G. 1669. *Naturalis Historiae* (anno 77), *ex officina* A. Hackios (ed.). Vol. XVIII, Chapters 39–40. Leiden; Rotterdam, the Netherlands.

Podlesak, W., Werner, T., Grün, M., Schlegel, R., and Hülgenhof, E. 1990. Genetic differences in the copper efficiency of cereals. In *Plant Nutrition—Physiology and Applications.* M. L. van Beusichem (ed.). Kluwer Academic Publishers, Amsterdam, the Netherlands, pp. 297–301.

Podyma, W. 2003. Rye genetic resources in Europe. *Plant Breed. Seed Sci.* 48: 37–44.

Pohler, W., and Schlegel, R. 1990. A rye plant with frequent A-B chromosome pairing *Hereditas* 112: 217–220.

Popelka, J. C., and Altpeter, F. 2003. *Agrobacterium tumefaciens*-mediated genetic transformation of rye (*Secale cereale* L.). *Mol. Breed.* 11: 203–211.

Popelka, J. C., Xu, J., and Altpeter, F. 2003. Generation of rye (*Secale cereale* L.) plants with low transgene copy number after biolistic gene transfer and production of instantly marker-free transgenic rye. *Transgenic Res.* 12: 587–596.

Price, H. J. 1975. Unilateral seed failure after reciprocal crosses of *Secale cereale* with *Secale africanum. Can. J. Bot.* 59: 929–932.

Price, S. A. M. 1955. Irradiation and interspecific hybridization in *Secale. Genetics* 40: 651–667.

Pröseler, G., and Stanarius, A. 1983. Nachweis des Weizenspindelstrichelmosaic-Virus (wheat spindle streak mosaic virus) in der DDR. *Arch. Phytopathol. Pflanzenschutz* 19: 345–349.

Przepiorkowski, T., and Gorski, S. F. 1994. Influence of rye (*Secale cereale*) plant residues on germination and growth of three triazine-resistant and susceptible weeds. *Weed Tech.* 8: 744–747.

Puertas, M. J., and Giraldez, R. 1979. Meiotic pairing in haploid rye. *Genet. Iber.* 30–31: 39–47.

Puertas, M. J., Romera, F., and de la Pena, A. 1985. Comparison of B chromosome effects on *Secale cereale* and *Secale vavilovii. Heredity* 55: 229–234.

Qifu, M. A., Rengel, Z., and Kuo, J. 2002. Aluminium toxicity in rye (*Secale cereale*): Root growth and dynamics of cytoplasmic Ca2+ in intact root tips. *Ann. Bot.* 89: 241–244.

Quarrie, S. A., Steed, A., Semikhodsky, A., Calestani, C., Fish, L., Galiba, G., Sutka, J., and Snape, J. W. 1994. Identifying quantitative trait loci (QTL) for stress response in cereals. Annual Report on John Innes Centre & Sainsbury Laboratory, Norwich, pp. 25–26.

Rakoczy-Trojanowska, M. 2002. The effects of growth regulators on somaclonal variation in rye (*Secale cereale* L.) and selection of somaclonal variants with increased agronomic traits. *Cell Mol. Biol. Lett.* 7: 1111–1120.

Rakoczy-Trojanowska, M., and Malepszy, S. 1993. Genetic factors influencing regeneration ability in rye (*Secale cereale* L.). I. Immature inflorescences. *Theor. Appl. Genet.* 86: 406–410.

Rakoczy-Trojanowska, M., and Malepszy, S. 1995. Genetic factors influencing regeneration ability in rye (*Secale cereale* L.). II. Immature embryos. *Theor. Appl. Genet.* 83: 233–239.

Rakoczy-Trojanowska, M., Smiech, M., and Malepszy, S. 1997. The influence of genotype and medium on rye (*Secale cereale* L.) anther culture. *Plant Cell Tiss. Org. Cult.* 48: 15–21.

Ranjekar, P. K., Lafontaine, J. G., and Palotta, D. 1974. Characterization of repetitive DNA in rye (*Secale cereale*). *Chromosoma* 48: 427–440.

Rees, H. 1955a. Genotypic control of chromosome behaviour in rye. I. Inbred lines. *Heredity* 9: 93–116.

Rees, H. 1955b. Genotypic control of chromosome behaviour in rye. III. Chiasma frequency in homozygotes and heterozygotes. *Heredity* 10: 409–424.

Rees, H., and Naylor, B. 1960. Developmental variation in chromosome behavior. *Heredity* 15: 17–27.

Rees, H., and Thompson, J. B. 1955. Localisation of chromosome breakage at meiosis. *Heredity* 9: 399–407.

Rees, H., and Thompson, J. B. 1956. Genotypic control of chromosome behaviour in rye. III. Chiasma frequency in homozygotes and heterozygotes. *Heredity* 10: 409–424.

Rehse, E. 1960. On the relationships between length of straw and yield in populations from crossing two rye varieties. *Z. Pflanzenzüchtg.* 43: 157–204.

Reimann-Philipp, R. 1986. Perennial spring rye as a crop alternative. *J. Agron. Crop Sci.* 157: 281–285.

Reimann-Philipp, R. 1995. Breeding perennial rye. *Plant Breed. Rev.* 13: 265–292.

Reimann-Philipp, R., and Eichhorn-Rhode, H. 1970. Über die Beziehungen von Fertilität und Modus der Chromosomenpaarung bei Tetraploiden, für drei Chromosomen struktur-beterozygoten Bastarden aus der Kreuzung *Secale cereale* × *Secale montanum. Theor. Appl. Genet.* 40: 991–1005.

Reimann-Philipp, R., and Rhode, H. 1968. Die cytologische Identifizierung der gene-tisch unterschiedlichen Gruppen von Artbastarden in den späteren Generationen der Kreuzung von *Secale cereale* × *S. montanum* in ihrer Bedeutung für die Züchtung eines perennierenden Kulturroggens. *Z. Pflanzenzücht.* 60: 212–218.

Ren, T. H., Che, F., Zou, Y. T., Jia, Y. H., Zhang, H. Q., Yan, B. J., and Ren, Z. L. 2011. Evolutionary trends of microsatellites during the speciation process and phylogenetic relationships within the genus *Secale. Genome* 54: 316–326.

Resch, V. 2009. SSR based genetic and physical mapping of the short arm of rye chromosome 1 (1RS). Master thesis, Universität für Bodenkultur, Vienna, Austria, pp. 1–53.

Richard, A. 1824. *Medizinische Botanik.* T. C. F. Enslin Verlag, Berlin, Germany, pp. 1–934.

Riebesel, G. 1937. Vegetative Vermehrung von Getreide-Bastarden. *Züchter* 9: 24–27.

Riley, R., and Chapman, V. 1958. The production and phenotypes of wheat-rye chromosome addition lines. *Heredity* 12: 301–315.

Rimpau, J., Smith, D. B., and Flavell, R. B. 1978. Sequence organization analysis of the wheat and rye genomes by interspecies DNA/DNA hybridization. *J. Mol. Biol.* 123: 327–358.

Rimpau, W. 1877. Die Selbst-Sterilität des Roggens. *Landw. Jahrb.* 6: 1073–1076.

Rimpau, W. 1883. Züchtung auf dem Gebiete der landwirtschaftlichen Kulturpflanzen. In *Mentzel & von Lengerke's Landwirtschaftl. Kalender II*, pp. 33–92.

Rimpau, W. 1891. *Kreuzungsprodukte landwirtschaftlicher Kulturpflanzen.* P. Parey Verlag, Berlin, Germany, pp. 1–12.

Rimpau, W. 1899. Monstrositäten am Roggen. *Dtsch. Landw.* 26: 878–901.

Römer, T., and Rudorf, W. 1950. *Handbuch der Pflanzenzüchtung*, Bd. 2, Grundlagen der Pflanzenzüchtung. P. Parey Verlag, Berlin/Hamburg, Germany, pp. 1–529.

Rogowsky, P. M., Shepherd, K. W., and Langridge, P. 1992. Polymerase chain reaction based mapping of rye involving repeated DNA sequences. *Genome* 35: 621–626.

Roshevitz, R. 1947. Monographic of the genus *Secale* L. *Act. Inst. Bot. Nom. Komarov Acad. Sci.* 1: 105–163.

Ross, A. B., Kamal-Eldin, A., Jung, C., Shepherd, M. J., and Aman, P. 2001. Gas chromato-graphic analysis of alkylresorcinols in rye (*Secale cereale* L.) grains. *J. Sci. Food Agric.* 81: 1405–1411.

Roux, S. R., Hackauf, B., Linz, A., Ruge, B., Klocke, B., and Wehling, P. 2004. Leaf-rust resis-tance in rye (*Secale cereale* L.). 2. Genetic analysis and mapping of resistance genes *Pr3*, *Pr4*, and *Pr5. Theor. Appl. Genet.* 110: 192–201.

Roux, S. R., Hackauf, B., Ruge-Wehling, B., Linz, A., and Wehling, P. 2007. Exploitation and comprehensive characterization of leaf-rust resistance in rye. Proc. Int. Symposium on Rye Breeding and Genetics, Rostock (Germany), June 28–30, 2006. *Vortr. Pflanzenzücht.* 71: 144–150.

Roux, S. R., Wortmann, H., and Schlathölter, M. 2010. Züchterisches Potenzial von Roggen (*Secale cereale* L.) für die Biogaserzeugung. *J. Kulturpflanzen* 62: 173–182.

Ruebenbauer, T., Kubara-Szpunar, L., and Pajak, K. 1983. Inheritance of violet coloration in rye kernels (*Secale cereale* L.). *Genet. Polon.* 24: 313–317.

Rui, M., Yli-Mattila, T., and Pulli, S. 2004. Phylogenetic relationships among genotypes of worldwide collection of spring and winter ryes (*Secale cereale* L.) determined by RAPD-PCR markers. *Hereditas* 140: 210–221.

Russell, R. S. 1977. *Plant Root Systems.* McGraw-Hill, New York.

Saal, B., and Wricke, G. 1999. Development of simple sequence repeat markers in rye (*Secale cereale* L.). *Genome* 42: 964–972.

Salse, J., Bolot, S., Throude, M., Jouffe, V., Piegu, B., Quraishi, U. M., Calcagno, T., Cooke, R., Delseny, M., and Feuillet, C. 2008. Identification and characterization of shared duplications between rice and wheat provide new insight into grass genome evolution. *Plant Cell* 20: 11–24.

Sandery, M. J., Forster, J. W., Blunden, R., and Jones, R. N. 1990. Identification of a family of repeated sequences on the rye B chromosome. *Genome* 33: 908–913.

Sasaki, M. 1956. Pistillody in a rye strain with wheat plasms. *Jpn. J. Genet.* 31: 310–311.

Schiemann, E., and Nürnberg-Krüger, U. 1952. New studies on *S. africanum*. II. *S. africanum* and its hybrids with *S. montanum* and *S. cereale*. *Naturwiss.* 39: 136–137.

Schlegel, R. 1973. Experimentelle Untersuchungen zur chromosomalen Stabilität und deren Bedeutung für die Fertiltät des autotetraploiden Roggens (*Secale cereale* L.). PhD thesis, Martin Luther University, Halle-Wittenberg, Germany, pp. 1–172.

Schlegel, R. 1976. The relationship between meiosis and fertility in autotetraploid rye, *Secale cereale* L. *Tag. Ber. AdL* 143: 31–37.

Schlegel, R. 1982a. Die Differentialfärbung von somatischen und meiotischen Chromosomen— Ihre Nutzung in der Züchtungsforschung und angewandten Züchtung. DSc thesis, Martin Luther University, Halle-Wittenberg, Germany, pp. 1–279.

Schlegel, R. 1982b. First evidence for rye-wheat additions. *Biol. Zbl.* 101: 641–646.

Schlegel, R. 1987. Neocentric activity in chromosome 5R of rye revealed by haploidy. *Hereditas* 107: 1–6.

Schlegel, R. 1990. Efficiency and stability of interspecific chromosome and gene transfer in hexaploid wheat, *Triticum aestivum* L. *Kulturpflanze* 38: 67–78.

Schlegel, R. 1996. Triticale—Today and tomorrow. In *Triticale: Today and Tomorrow*, Developments in Plant Breeding, vol. 5. H. Guedes-Pinto et al. (eds.). Kluwer Academic Publishers, Amsterdam, the Netherlands, pp. 21–32.

Schlegel, R. 1997. Current list of wheats with rye introgressions of homoeologous group 1. *Wheat Inf. Serv.* 84: 64–69.

Schlegel, R. 2007. *Concise Encyclopedia of Crop Improvement: Institutions, Persons, Theories, Methods, and Histories*. Haworth Press, New York.

Schlegel, R. 2009. *Dictionary of Plant Breeding*, 2nd ed. Taylor & Francis, Inc., New York.

Schlegel, R. 2013. Current list of wheats with rye and alien introgression. V01-13, 1–18, http://www.rye-gene-map.de/rye-introgression

Schlegel, R., Belchev, I., Kostov, K., and Atanasova, M. 2000. Inheritance of high anther culture response in hexaploid wheat, *Triticum aestivum* L. var. "Svilena." *Bulg. J. Agric. Sci.* 6: 261–270.

Schlegel, R., Börner, A., Thiele, V., and Melz, G. 1991a. The effect of the *Ph1* gene in diploid rye, *Secale cereale* L. *Genome* 34: 913–917.

Schlegel, R., and Cakmak, I. 1997. Physical mapping of rye genes determining micronutritional efficiency in wheat. In *Plant Nutrition—For Sustainable Food Production and Environment*. T. Ando (ed.). Kluwer Academic Publishers, Tokyo, Japan, pp. 287–288.

Schlegel, R., Cakmak, I., Ekiz, H., Kalayci, M., and Braun, H. J. 1998a. Screening for zinc efficiency among wheat relatives and their utilisation for an alien gene transfer. *Euphytica* 100: 281–286.

Schlegel, R., Cakmak, I., Torun, B., Eker, S., and Köleli, N. 1997. The effect of rye genetic information on zinc, copper, manganese and iron concentration of wheat shoots in zinc deficient soil. *Cer. Res. Comm.* 25: 177–184.

Schlegel, R., and Gill, B. S. 1984. N-banding analysis of rye chromosomes and the relationship between N-banded and C-banded heterochromatin. *Can. J. Genet. Cytol.* 2: 765–769.

Schlegel, R., and Korzun, V. 1997. About the origin of 1RS.1BL wheat-rye chromosome translocations from Germany. *Plant Breed.* 116: 537–540.

Schlegel, R., and Korzun, V. 2007. Genes, marker and linkage data of Rye (*Secale cereale* L.), 6th updated inventory. *Vortr. Pflanzenzücht.* 71: 225–245.

Schlegel, R., and Korzun, V. 2013. Genes, markers and linkage data of rye (*Secale cereale* L.), 9th uptaded inventory, V. 03.13, pp. 1–115. http://www.rye-gene-map.de

Schlegel, R., and Kynast, R. 1988. Wheat chromosome 6B compensates genetic information of diploid rye, *Secale cereale* L. *Proceedings of the 7th International Wheat Genetics Symposium*. Cambridge, July 13–19, pp. 421–426.

Schlegel, R., Kynast, R., and Schmidt, J. C. 1986a. Alien chromosome transfer from wheat into rye. In *Genetic Manipulation in Plant Breeding*. W. Horn et al. (eds.). Gruyter Publishers, Berlin, Germany, pp. 129–136.

Schlegel, R., Kynast, R., Schwarzacher, T., Römheld, V., and Walter, A. 1993. Mapping of genes for copper efficiency in rye and the relationship between copper and iron efficiency. *Plant Soil* 154: 61–65.

Schlegel, R., and Meinel, A. 1994. A quantitative trait locus (QTL) on chromosome arm 1RS of rye and its effect on yield performance of hexaploid wheat. *Cer. Res. Comm.* 22: 7–13.

Schlegel, R., and Melz, G. 1996. A comprehensive gene map of rye, *Secale cereale* L. *Vortr. Pflanzenzüchtg.* 35: 311–321.

Schlegel, R., Melz, G., and Korzun, V. 1998b. Genes marker and linkage data of rye (*Secale cereale* L.), 5th updated inventory. *Euphytica* 101: 23–67.

Schlegel, R., Melz, G., and Mettin, D. 1986b. Rye cytology and cytogenetics—Current status. *Theor. Appl. Genet.* 72: 721–734.

Schlegel, R., Melz, G., and Nestrowicz, R. 1987. A universal reference karyotype in rye, *Secale cereale* L. *Theor. Appl. Genet.* 74: 820–826.

Schlegel, R., and Mettin, D. 1975a. Untersuchungen zur intra- und interindividuellen Variabilität der Chromosomenpaarung bei di- und tetraploidem Roggen aus Populationen (*Secale cereale* L). I. Intraindividuelle Variabilität. *Biol. Zbl.* 94: 539–555.

Schlegel, R., and Mettin, D. 1975b. Untersuchungen zur intra- und interindividuellen Variabilität der Chromosomenpaarung bei di- und tetraploidem Roggen (*Secale cereale* L.). II. Interindividuelle Variabilität. *Biol. Zbl.* 94: 703–715.

Schlegel, R., and Mettin, D. 1975c. Studies of valence crosses in rye (*Secale cereale* L.). IV. The relationship between meiosis and fertility in tetraploid hybrids. *Biol. Zbl.* 94: 295–315.

Schlegel, R., and Mettin, D. 1982. The present status of chromosome recognition and gene localization in rye, *Secale cereale* L. *Tag. Ber. Akad. Landwirtsch. Wiss.* 198: 131–152.

Schlegel, R., Özdemir, A., Tolay, I., Cakmak, I., Saberi, H., and Atanasova, M. 1999. Localisation of genes for zinc and manganese efficiency in wheat and rye. In *Plant Nutrition—Molecular Biology and Genetics*. G. Gissel-Nielsen, and A. Jensen (eds.). Kluwer Academic Publishers, Amsterdam, the Netherlands, pp. 417–424.

Schlegel, R., and Pohler, W. 1994. Identification of an A-B translocation in diploid rye. *Breed. Sci.* 44: 279–283.

Schlegel, R., Scholz, I., and Fischer, K. 1985. Aneusomy as source of meiotic disturbances in tetraploid rye? *Biol. Zbl.* 104: 375–384.

Schlegel, R., and Schrader, O. 1991. Pairing restriction of homologous rye chromosomes in amphidiploid wheat-rye hybrids determined by genome dosage and temperature. *Proceedings of the 2nd International Triticale Symposium*, October 1–5. Passo Fundo, Brazil, pp. 359–367.

Schlegel, R., and Sturm, W. 1982. Meiotic chromosome pairing of primary trisomics of rye, *Secale cereale* L. *Tag. Ber. Akad. Landwirtsch. Wiss.* 198: 225–243.

Schlegel, R., Werner, T., and Hülgenhof, E. 1991b. Confirmation of a 4BL/5RL wheat-rye chromosome translocation in the wheat variety "Viking" showing high copper efficiency. *Plant Breed.* 107: 226–234.

Schlegel, R., and Weryszko, E. 1979. Intergeneric chromosome pairing in different wheat-rye hybrids revealed by the Giemsa banding technique and some implications on karyotype evolution in the genus *Secale*. *Biol. Zbl.* 98: 399–407.

Schmidt, K.-D. 1980. Untersuchungen zur Mutationszüchtung bei di- und tetraploidem Roggen (*Secale cereale* L.). PhD thesis, Martin Luther University, Halle-Wittenberg, Germany, pp. 1–332.

Schmitz, G., Schmidt, M., and Feierabend, J. 1996. Characterization of a plastid-specific HSP90 homologue: Identification of a cDNA sequence, phylogenetic descendence and analysis of its mRNA and protein expression. *Plant Mol. Biol.* 30: 479–492.

Schnell, F. W. 1959. Roggen. In *Dreißig Jahre Züchtungsforschung*. W. Rudorf (ed.). G. Fischer Verlag, Stuttgart, Germany, pp. 135–141.

Schnell, F. W. 1982. A synoptic study of the methods and categories of plant breeding. *Z. Pflanzenzüchtg.* 89: 1–18.

Schooler, A. B. 1967. Intergeneric hybrids between allohexaploid (*Hordeum depressum* × *H. compressum*) and autotetraploid Svalöf rye (*Secale cereale*). *Bot. Gaz.* 128: 113–116.

Senft, P., and Wricke, G. 1996. An extended genetic map of rye (*Secale cereale* L.). *Plant Breed.* 115: 508–510.

Shakir, H. S., and Waines, J. G. 1994. Drought resistance in bread wheat, rye, barley, and dasypyrum. *Barley Genet. Newslett.* 22: 65.

Shang, H.-Y., Wei, Y.-M., Long, H., Yan, Z.-H., and Zheng, Y.-L. 2005. Identification of LMW glutenin-like genes from *Secale sylvestre* host. *Plant Genet.* 41: 1372–1380.

Shang, H.-Y., Wei, Y.-M., Wang, X.-R., and Zheng, Y.-L. 2006. Genetic diversity and phylogenetic relationships in the rye genus *Secale* L. (rye) based on *Secale cereale* microsatellite markers. *Genet. Mol. Biol.* 29: 685–691.

Shang, W., Schmidt, M., and Feierabend, J. 2003. Increased capacity for synthesis of the D1 protein and of catalase at low temperature in leaves of cold-hardened winter rye (*Secale cereale* L.). *Planta* 216: 865–873.

Sharma, S., Bhat, P., Ehdaie, B., Close, T., Lukaszewski, A. J., and Waines, J. G. 2009. Integrated genetic map and genetic analysis of a region associated with root traits on the short arm of rye chromosome 1 in bread wheat. *Theor. Appl. Genet.* 119: 783–993.

Sharma, S., Xu, S., Ehdaie, B., Hops, A., Close, T. J., Lukaszewski, A. J., and Waines, J. G. 2011. Dissection of QTL effects for root traits using a chromosome arm-specific mapping population in bread wheat. *Theor. Appl. Genet.* 122: 759–769.

Shepherd, K. W., and Islam, A. K. M. R. 1988. Fourth compendium of wheat-alien chromosome additions. *Proceedings of the 7th International Wheat Genetics Symposium.* Cambridge, July 13–19, pp. 1373–1398.

Shi, B. J., Gustafson, J. P., Button, J., Miyazaki, J., Pallotta, M., Gustafson, N., Zhou, H., Langridge, P., and Collins, N. C. 2009. Physical analysis of the complex rye (*Secale cereale* L.) *Alt4* aluminium (aluminum) tolerance locus using a whole-genome BAC library of rye cv. Blanco. *Theor. Appl. Genet.* 119: 695–704.

Silva-Navas, J., Benito, C., Tellez-Robledo, B., Abd El-Moneim, D., and Gallego, F. J. 2012. The *ScAACT1* gene at the Qalt5 locus as a candidate for increased aluminum tolerance in rye (*Secale cereale* L.). *Mol. Breed.* 30: 845–856.

Siminovitch, D., Singh, J., and de la Roche, I. A. 1978. Freezing behavior of free protoplasts of winter rye. *Cryobiology* 15: 205–213.

Simkova, H., Safar, J., Suchankova, P., Kovarova, P., Bartos, J., Kubalakova, M., Janda, J., Cíhalíkova, J., Mago, R., Lelley, T., and Dolezel, J. 2008. A novel resource for genomics of Triticeae: BAC library specific for the short arm of rye (*Secale cereale* L.) chromosome 1R (1RS). *BMC Genomics* 9: 237–246.

Singh, B. P., Joshi, L. K., Kaul, B. L., and Zutshi, U. 1972. On the cytogenetic activity of butane sultone. A new mutagen. *Sci. Cult.* 38: 284–285.

Singh, R. J., and Röbbelen, G. 1977. Identification by Giemsa technique of the translocations separating cultivated rye from three wild species of *Secale*. *Chromosoma* 59: 217–229.

Sirks, M. J. 1929. Über einen Fall vererbbarer Lichtempfindlichkeit des Chlorophylls beim Roggen (*Secale cereale*). *Genetica* 11: 375–386.

Skuza, L., Rogalska, S. M., and Bocianowski, J. 2007. RFLP analysis of mitochondrial DNA in the genus *Secale*. *Acta Biol. Cracoviensi Bot.* 49: 77–87.

Slotta, M. 1992. Mittlere Nährstoffaufnahme (Korn + Stroh) von Roggen. In *Roggen Anbau und Vermarktung*. Verlag Roggenforum, Bergen, Germany, pp. 1–47.

Smith, D. B., and Flavell, R. B. 1977. Nucleotide sequence organization in rye genome. *Biochim. Biophys. Acta* 474: 82–97.

Smolik, M. 2012. R-ISSR—Tool for generation of a new type of products, applied for the identification of putative molecular marker linked to QTL determined tolerance to nutrient deprivation stress in rye (*Secale cereale* L.). *Not. Bot. Hort. Agrobot. Cluj-Napoca* 40: 238–246.

Smolik, M., Cieluch, P., and Mazurkiewicz-Zapalowicz, K. 2012. Morphological and cytological response of selected recombinant inbred lines of rye (*Secale cereale* L.) to nutrient deprivation stress assessed at the seedling stage. *Not. Bot. Horti Agrobot. Cluj-Napoca* 40: 282–289.

Sodkiewicz, W., and Makarska, E. 2004. Recombinative variability of soluble pentosan content among introgressive triticale lines derived from hybridization of hexaploid triticale with the synthetic amphitetraploid *T. monococcum/S. cereale* (A^mA^mRR). *Acta Sci. Polon. Technol. Aliment.* 3: 119–126.

Solodukhina, O. V., and Kobylyanski, V. D. 2001. Problems of winter rye breeding for resistance of brown and stem rust. *Proceedings of the 7th EUCARPIA Conference*. Radzikow, Poland, July 4–7, pp. 217–225.

Solodukhina, O. V., and Kobylyanski, V. D. 2003. Problems of winter rye breeding for resistance to leaf and stem rusts. *Plant Breed. Seed Sci.* 48: 87–97.

Solodukhina, O. V., and Kobylyanski, V. D. 2007. Possibility of rye breeding for long-term resistance to leaf and stem rust. *Vortr. Pflanzenzücht.* 71: 151–157.

Sosnikhina, S. P., Mikhailova, E. I., Tikholiz, O. A., Priyatkina, S. N., Smirnov, V. G., and Voilokov, A. V. 2005. Genetic collection of meiotic mutants of rye, *Secale cereale* L. *Russ. J. Genet.* 41: 1071–1080.

Specht, C.-E., and Börner, A. 1998. An interim report on a long term storage experiment on rye (*Secale cereale* L.) under a range of temperatures and atmospheres. *Genet. Resour. Crop Evol.* 45: 483–488.

Sprague, G. F., and Jenkins, M. T. 1943. A comparison of synthetic varieties, multiple crosses and double crosses in corn. *J. Am. Soc. Agron.* 35: 137–147.

Sprengel, K. P. J. 1807. *Historia rei herbaria?* vol. 2, Sumtibus Tabernae Librariae et Artium, Amsteldami, p. 9.

Staggenborg, S., Bowden, R., Marsh, B., and Martin, V. 1997. Forage yield and soilborne mosaic virus resistance of several varieties of rye, triticale, and wheat. Kansas State University, Manhattan, KS. http://www.ksre.ksu.edu/library/crpsl2/srl118.pdf

Stahl, M., and Steiner, A. M. 1998. Seed storage potential of population varieties and hybrid varieties and their breeding components in rye (*Secale cereale* L.). *Plant Breed.* 117: 179–181.

Starzycki, S. 1976. Disease, Pests and Physiology of Rye. In *Rye: Production, Chemistry and Technology*. W. Bushuk (ed.). American Association of Cereal Chemists, St. Paul, MN, pp. 27–61.

Steglich, R., and Pieper, H. 1922. Vererbungs- und Züchtungsversuche mit Roggen. *Fühlings Landw. Ztg.* 71: 201–221.

Steinborn, R., Schwabe, W., Weihe, A., Adolf, K., Melz, G., and Börner, T. 1993. A new type of cytoplasmic male sterility in rye (*Secale cereale* L.): Analysis of mitochondrial DNA. *Theor. Appl. Genet.* 85: 822–824.

Stocking-Gruver, L., Weil, R. R., Zasada, I. A., Sardanelli, S., and Momen, B. 2010. Brassicaceous and rye cover crops altered free-living soil nematode community composition. *Appl. Soil Ecol.* 45: 1–12.

Stojalowski, S., Łapiński, M., and Masojć, P. 2004. RAPD markers linked with restorer genes for the C-source of cytoplasmic male sterility in rye (*Secale cereale* L.). *Plant Breed.* 123: 428–433.

Stojalowski, S., Łapiński, M., and Szklarczyk, M. 2006. Identification of sterility-inducing cytoplasms in rye using the plasmotype–genotype interaction test and newly developed SCAR markers. *Theor. Appl. Genet.* 112: 627–633.

Stojałowski, S. A., Milczarski, P., Anek, M., Bolibok-Bragoszewska, H., Mysków, B., Kilian, A., and Rakoczy-Trojanowska, M. 2011. DArT markers tightly linked with the *Rfc1* gene controlling restoration of male fertility in the CMS-C system in cultivated rye (*Secale cereale* L.). *J. Appl. Genet.* 52: 313–318.

Stoskopf, N. C. 1985. *Cereal grain crops*. Reston Publishing Company Inc., Reston, VA pp. 403–414.

Stracke, S., Schilling, A. G., Förster, J., Weiss, C., Glass, C., Miedaner, T., and Geiger, H. H. 2003. Development of PCR-based markers linked to dominant genes for male-fertility restoration in Pampa CMS of rye (*Secale cereale* L.). *Theor. Appl. Genet.* 106: 1184–1190.

Stressmann, M., Kitao, S., Griffith, M., Moresoli, C., Bravo, L. A., and Marangoni, A. G., 2004. Calcium interacts with antifreeze proteins and chitinase from cold-acclimated winter rye. *Plant Physiol.* 135: 364–376.

Strum, W. 1974. Untersuchungen zur morphologischen und cytologischen Variabilität bei Winterroggen-Mutanten unter besonderer Berücksichtigung der bei trisomen Formen und Inzuchtlinien gewonnenen Erkenntnisse. MSc thesis, Martin Luther University, Halle-Wittenberg, Germany, pp. 1–48.

Sturm, W. 1978. [Identification of trisomics in the rye variety "Esto" and trisomic analysis of short straw gene *Hl.*] (German). PhD thesis, Akad. Landwirtsch. Wiss., Berlin, Germany, pp. 1–164.

Sturm, W., and Melz, G. 1982. Telotrisomics. *Tag. Ber. Akad. Landwirtsch. Wiss.* 198: 217–225.

Stutz, H. C. 1957. A cytogenetic analysis of the hybrid *Secale cereale* L. × *Secale montanum* Guss. and its progeny. *Genetics* 42: 199–221.

Stutz, H. C. 1972. The origin of cultivated rye. *Am. J. Bot.* 59: 59–70.

Sun, M., and Corke, H. 1992. Population genetics of colonizing success of weedy rye in Northern California. *Theor. Appl. Genet.* 83: 321–329.

Swami, U. B. S., and Thomas, H. 1968. Chromosome pairing in autotetraploid *Avena*. *Z. Pflanzenzüchtung* 59: 163–170.

Sybenga, J. 1958. Inbreeding effects in inbred rye. *Z. Vererbungslehre* 89: 338–354.

Sybenga, J. 1964. The use of chromosomal aberrations in the autopolyploidization of autopolyploids. *Proceedings of the Symposium IAEA/FAO*. Rome, Italy, May 25–June 1, pp. 741–768.

Sybenga, J. 1965. The quantitative analysis of chromosome pairing and chiasma formation based on the relative frequencies of MI configurations. II. Primary trisomics. *Genetica* 36: 339–350.

Sybenga, J. 1983. Rye chromosome nomenclature and homoeology relationships. 1st workshop on rye. *Z. Pflanzenzücht.* 90: 297–304.

Sybenga, J. 1985. Critical appraisal of balanced chromosomal ms systems in hybrid rye breeding. *EUCARPIA Meeting of the Cereal Section on Rye*. Svalöv, Sweden, June 11–13, pp. 307–316.

Sybenga, J. 2012. Can epigenetic control explain pronounced within-plant heterogeneity of meiosis in a translocation trisome of *Secale* L.? *Genome* 55: 257–264.

Sybenga, J., van Eden, J., van der Meijs, Q. G., and Roeterding, B. W. 1985. Identification of the chromosomes of the rye translocation tester set. *Theor. Appl. Genet.* 69: 313–316.

Sybenga, J., Verhaar, H., and Botje, D. G. A. 2008. Epigenetic control may explain large within-plant heterogeneity of meiotic behavior in telocentric trisomics of rye. *Genetics* 178: 1915–1926.

Sybenga, J., Verhaar, H., and Botje, D. G. A. 2012. Trisomy greatly enhances interstitial crossing over in a translocation heterozygote of *Secale*. *Genome* 55: 15–25.

Sybenga, J., and Wolters, Z. H. G. 1972. The classification of the chromosomes of rye (*Secale cereale* L.): A translocation tester set. *Genetica* 43: 453–464.

Szakacs, E., and Barnabas, B. 2004. Development of diploid pollen in spikelet cultures of barley (*Hordeum vulgare* L.) and rye (*Secale cereale* L.). *Acta Agron. Hungarica* 52: 1–8.

Szatanik-Kloc, A., Warchulska, P., and Jozefaciuk, G. 2009. Changes in variable charge and acidity of rye (*Secale cereale* L.) roots surface under Zn-stress. *Acta Physiol. Plant.* 6: 59–64.

Szemplinski, W., Dubis, B., and Pszczolkowska, A. 2005. Rye infection by fungal pathogens under conditions of various sowing rates and disease control methods. *Polish J. Nat. Sci.* 2: 243–254.

Takagi, F. 1935. Karyogenetical studies on rye. I. A trisomic plant. *Cytologia* 6: 496–501.

Takahashi, M., Yamaguchi, H., Nakanishi, H., Shioiri, T., Nishizawa, N.-K., and Mori, M. 1999. Cloning two genes for nicotianamine aminotransferase, a critical enzyme in iron acquisition (strategy II) in graminaceous plants. *Plant Physiol.* 121: 947–956.

Taketa, S. Nakazaki, T., Schwarzacher, T., and Heslop-Harrison, J. S. 1997. Detection of a 4DL chromosome segment translocated to rye chromosome 5R in an advanced hexaploid triticale line Bronco 90. *Euphytica* 97: 91–96.

Targonska, M., Hromada-Judycka, A., Bolibok-Bragoszewska, H., and Rakoczy-Trojanowska, M. 2013. The specificity and genetic background of the rye (*Secale cereale* L.) tissue culture response. *Plant Cell Rep.* 32: 1–9.

Tenhola-Roininen, T. 2009. Rye doubled haploids—Production and use in mapping studies. PhD thesis, Universit of Jyväskkylä, Finland, pp. 1–93.

Tenhola-Roininen, T., Immonen, S., and Tanhuanpää, P. 2006. Rye doubled haploids as a research and breeding tool—A practical point of view. *Plant Breed.* 125: 584–590.

Tenhola-Roininen, T., Kalendar, R., Schulman, A. H., and Tanhuanpää, P. 2011. A doubled haploid rye linkage map with a QTL affecting α-amylase activity. *J. Appl. Genet.* 52: 299–304.

Theodorus, J. 1590. *Eicones plantarum, seu stirpium, arborum nempe, fructicum, herbarum, fructuum, lignorum, radicum, omnis generis; tam inquilinorum, quam exoticorum: quae partim Germania sponte producit: partim ab exteris regionibus allata, in Germania plantantur*. Verlag N. Basse (ed.). Frankfurt/M., Germany, p. 1128.

Theophrastus. 1916. *Enquiry into Plants*, vol. 1, lib. 1–5, cap. 9, Treatise on Odours. Concerning Weather Signs. A. Hort (trans.). Harvard University Press, Cambridge, MA.

Thomas, E., Hoffmann, F., and Wenzel, G. 1975. Haploid plantlets from microspores of rye. *Z. Pflanzenzüchtg.* 75: 106–113.

Thome, O. W. 1885. *Flora von Deutschland, Österreich und der Schweiz*. Köhler Verlag, Gera, Germany, pp. 1–53.

Tijo, J. N., and Levan, A. 1950. The use of oxyquinoline in chromosome analysis. *Anal. Est. Exp. Aula Dei* 2: 21–64.

Tikhenko, N., Tsvetkova, N., Priyatkina, S., Voylokov, A., and Boerner, A. 2011. Gene mutations in rye causing embryo lethality in hybrids with wheat: Allelism and chromosomal localization. *Biol. Plant.* 55: 448–452.

Timar, I., Ponya, Z., Szabo, L., and Barnabas, B. 2002. Development of female and male gametophytes in cereal species. *Acta Agron. Hung.* 50: 321–335.

Tiwari, F., and Sisodia, N. S. 1978. Amber-white seed mutant through chemical mutagenesis in rye (*Secale cereale* L.). *Mut. Breed. Newslett.* 11: 2–3.

Tomita, M., and Seno, A. 2012. Rye chromosome-specific polymerase chain reaction products developed by primers designed from the Eco0109I recognition site. *Genome* 55: 370–382.

Tomita, M., Shinohara, K., and Morimoto, M. 2008. Revolver is a new class of transposon-like gene composing the triticeae genome. *DNA Res.* 29: 49–62.

Trang, Q. S., Wricke, G., and Weber, W. E. 1982. Number of alleles at the incompatibility locus in *Secale cereale* L. *Theor. Appl. Genet.* 63: 245–248.

Treboux, O. 1925. Beobachtungen über Vererbung yon Kornfarbe und Anthocyan beim Roggen. *Z. Pflanzenzücht.* 10: 288–291.

Treeby, M., Marschner, H., and Römheld, V. 1989. Mobilization of iron and other micronutrient cations from a calcareous soil by soil-borne, microbial and synthetic metal chelators. *Plant Soil* 114: 217–226.

Tschermak, E. 1906. Die Züchtung neuer Getreiderassen mittels künstlicher Kreuzung. II. Kreuzungsstudien an Roggen. *Z. landw. Versuchsw.* 9: 699–743.

Tschermak, E. 1915. Der veredelte Marchfelder Roggen. *Wiener landw. Zeitschr.* 65: 744–745.

Tschermak, E. 1932. Bemerkungen über echte und falsche Größen-Xenien. *Z. Züchtg.* 17: 447–450.

Tsuchida, M., Fukushima, T., Nasuda, S., Masoudi-Nejad, A., Ishikawa, G., Nakamura, T., and Endo, T. R. 2008. Dissection of rye chromosome 1R in common wheat. *Genes Genet. Syst.* 83: 43–53.

Twardowska, M., Masojć, P., and Milczarski, P. 2005. Pyramiding genes affecting sprouting resistance in rye by means of marker assisted selection. *Euphytica* 143: 257–260.

Tyrka, M., and Chekowski, J. 2004. Enhancing the resistance of triticale by using genes from wheat and rye. *J. Appl. Genet.* 42: 283–295.

Valenzuela, N. T., Perera, E., and Naranjo, T. 2012. Dynamics of rye chromosome 1R regions with high or low crossover frequency in homology search and synapsis development. *PLoS ONE* 7: e36385.

Varshney, R. K., Beier, U., Khlestkina, E. K., Kota, R., Korzun, V., Graner, A., and Börner, A. 2007. Single nucleotide polymorphisms in rye (*Secale cereale* L.): Discovery, frequency, and applications for genome mapping and diversity studies. *Theor. Appl. Genet.* 114: 1105–1116.

Vasil, I. K. 1987. Developing cell and tissue culture systems for the improvement of cereal and grass crops. *J. Plant Physiol.* 128: 193–218.

Vavilov, N. I. 1927. Geographical regularities in relation to the distribution of the genes of cultivated plants. *Appl. Bot. Genet. Plant Breed.* 17: 411–428.

Vavilov, N. J. 1917. On the origin of cultivated rye. *Appl. Bot.* 10: 561–590.

Verma, S. G., and Rees, H. 1974. Giemsa staining and the distribution of heterochromatin in rye chromosomes. *Heredity* 32: 118–122.

Vettel, F., and Plarre, W. 1955. Mehrjährige Heterosisversuche mit Winterroggen. *Z. Pflanzenzücht.* 42: 233–239.

Viinikka, Y. 1985. Identification of the chromosomes showing neocentric activity in rye. *Theor. Appl. Genet.* 70: 66–71.

von Bingen, H. 1151–1158. *Physica (Liber subtilitatum diversarum naturanim creaturarum)*, 8 vols.; The classic work on health and healing was translated from the Latin by Priscilla Throop; illustrations by Mary Elder Jacobsen. Healing Arts Press, Rochester, 1998, pp. 1–250.

von Lochow, F. III., 1894. Der Petkuser Roggen. *Frühlings Landw. Zeitg.* 43: 548–552.

von Lochow, F. III., 1900. Die Züchtung des Petkuser Roggens. *Deutsche Landw. Presse* 27: 117–118.

von Rümker, K. 1909. *Methoden der Pflanzenzüchtung in experimenteller Prüfung.* P. Parey Verlag, Berlin, Germany, pp. 1–322.

von Rümker, K., and Leidner, R. 1914. Ein Beitrag zur Frage der Inzucht bei Roggen. *Z. Pflanzenzücht.* 2: 427–444.

von Sengbusch, R. 1940a. Pärchenzüchtung unter Ausschaltung von Inzuchtschäden. *Forschungsdienst* 10: 545–549.

von Sengbusch, R. 1940b. Polyploid rye. *Züchter* 12: 185–189.

von Sengbusch, R. 1968. Meine Gedanken und Vorschläge zur Sicherung der Fortsetzung der Arbeiten des Max-Planck-Instituts für Kulturpflanzenzüchtung nach seiner Schließung, pp. 1–46. http://coreservice.mpdl.mpg.de:8080/ir/item/escidoc:37201/components/component/escidoc:49261/content

Vosa, C. G. 1974. The basic karyotype of rye (*Secale cereale*) analysed with Giemsa and fluorescence method. *Heredity* 33: 403–408.

Voylokov, A. V., Korzun, V., and Börner, A. 1998. Mapping of three self-fertility mutations in rye (*Secale cereale* L.) using RFLP, isozyme and morphological markers. *Theor. Appl. Genet.* 97: 147–153.

Wagenaar, E. B. 1959. Intergeneric hybrids between *Hordeum jubatum* L. and *Secale cereale* L. *Heredity* 50: 195–202.

Wagoner, P. 1990. Perennial grain development: Past efforts and potential for the future. *Crit. Rev. Plant Sci.* 9: 381–407.

Walsh, D. J., Matthews, J. A., Denmeade, R., Maxwell, P., Davidson, M., and Walker, M. R. 1990. Monoclonal antibodies to proteins from cocksfoot grass (*Dactylis glomerata*) pollen: Isolation and N-terminal sequence of a major allergen. *Int. Arch. Allergy Appl. Immunol.* 91: 419–425.

Walters, C. T., Niedzielski, M., Hill, L. M., Wheeler, L. J., and Puchalski, J. 2008. Temperature and moisture control of seed aging in rye. *9th International Society for Seed Science Conference on Seed Biology*. Olsztyn, Poland, pp. 281–282.

Walther, F. 1959. Fertilitätsstörungen beim Roggen. *Z. Pflanzenzüchtg.* 41: 1–32.

Wang, M. L., Atkinson, M. D., Chinoy, C. N., Devos, K. M., Harcourt, R. L., Liu, C. J., and Rogers, W. J. 1991. RFLP-based genetic map of rye (*Secale cereale* L.). Chromosome 1R. *Theor. Appl. Genet.* 82: 174–178.

Warzecha, R., and Salak-Warzecha, K. 2003. A new source of male sterility in rye (*Secale cereale* L.). *Plant Breed. Seed Sci.* 48: 61–65.

Wehling, P., Linz, A., Hackauf, B., Roux, S. R., Ruge, B., and Klocke, B. 2003. Leaf-rust resistance in rye (*Secale cereale* L.). 1. Genetic analysis and mapping of resistance genes *Pr1* and *Pr2*. *Theor. Appl. Genet.* 107: 432–438.

Weinmann, J. W. 1742. *Phytanthoza—Iconographia Oder Eigentliche Vorstellung Etlicher Tausend, So wohl Einheimisch- als Ausländischer, aus allen vier Welt-Theilen, in Verlauf vieler Jahre gesammelter Pflantzen, Bäume, Stauden, Kräuter, Blumen, Früchten und Schwämme*, Bd. 3. Hieronymus Lentz, Regensburg, Germany.

Went, F. W. 1957. The experimental control of plant growth. *Chron. Bot.* 17: 343–348.

Wenzel, G., Hoffmann, F., and Thomas, E. 1977. Increased induction and chromosome doubling of androgenetic haploid rye. *Theor. Appl. Genet.* 51: 81–86.

Wenzel, G., and Thomas, E. 1974. Observation of the growth in culture of anthers of *Secale cereale* L. *Z. Pflanzenzüchtg.* 72: 89–94.

Wilson, A. S. 1876. On wheat and rye hybrids. *Trans. Proc. Bot. Soc.* 12: 286–288.

Wodehouse, R. P. 1935. *Pollen Grains: Their Structure, Identification, and Significance in Science and Medicine*. McGraw-Hill, New York, pp. 1–318.

Wölbing, L. 1980. Die Erzeugung und Verwendung künstlich induzierter Mutanten in der Winterroggenzüchtung. PhD thesis, Akad. Landw. DDR, Berlin, Germany, pp. 1–133.

Wolski, T. 1975. The use of pair crosses in rye breeding. *Hod. Rosl. Aklim. i Nasienn.* 19: 509–520.

Wricke, G. 1978. Degree of self-fertilization under free pollination in rye populations containing a self-fertility gene. *Z. Pflanzenzücht.* 82: 281–285.

Wricke, G., and Wehling, P. 1985. Linkage between an incompatibility locus and a peroxidase isozyme locus (*Prx7*) in rye. *Theor. Appl. Genet.* 71: 289–291.

Xiao-Ming, Y. Griffith, M., and Wiseman, S. B. 2001. Ethylene induces antifreeze activity in winter rye leaves. *Plant Physiol.* 126: 1232–1240.

Xu, J., Frick, M., Laroche, A., Ni, Z.-F., Li, B.-Y., and Lu, Z.-X. 2009. Isolation and characterization of the rye waxy gene. *Genome* 52: 658–664.

Yaish, M. W. F., Doxey, A. C., McConkey, B. J., Moffatt, B. A., and Griffith, M. 2006. Cold-active winter rye glucanases with ice-binding capacity. *Plant Physiol.* 141: 1459–1472.

Yamagami, T., and Funatsu, G. 1993. Purification and some properties of three chitinases from the seeds of rye (*Secale cereale*). *Biosci. Biotechnol. Biochem.* 57: 1854–1861.

Yamamoto, M., and Mukai, Y. 2005. High-resolution physical mapping of the secalin-1 locus of rye on extended DNA fibers. *Cytogenet. Genome Res.* 109: 79–82.

Yang, B., Thorogood, D., Armstead, I. P., and Barth, S. 2008a. How far are we from unravelling self-incompatibility in grasses? *New Phytol.* 178: 740–753.

Yang, Z.-J., Li, G.-R., Jia, J.-Q., Zeng, X., Lei, M.-P., Zeng, Z.-X., Zhang, T., and Ren, Z.-L. 2008b. Molecular cytogenetic characterization of wheat–*Secale africanum* amphiploids and derived introgression lines with stripe rust resistance. *Euphytica* 167: 197–202.

Yang, Z.-J., Li, G.-R., Jiang, H.-R., and Ren, Z.-L. 2001. Expression of nucleolus, endosperm storage proteins and disease resistance in an amphiploid between *Aegilops tauschii* and *Secale silvestre*. *Euphytica* 119: 317–321.

Yang, Z.-J., Li, G.-R., and Ren, Z.-L. 2000. Identification of amphiploid between *Tritium durum* cv. Ailanmai native to Sichuan, China and *Secale africanum*. *Wheat Inf. Serv.* 91: 20–24.

Yokosho, K., Yamaji, N., and Ma, J. F. 2010. Isolation and characterization of two MATE genes in rye. *Funct. Plant Biol.* 37: 296–303.

Yu, S., Long, H., Yang, H., Zhang, J., Deng, G., Pan, Z., Zhang, E., and Yu, M. 2012. Molecular detection of rye (*Secale cereale* L.) chromatin in wheat line "Jian126" (*Triticum aestivum* L.) and its association to wheat powdery mildew resistance. *Euphytica* 186: 247–255.

Zadoks, J. C., Chang, T. T., and Konzak, C. F. 1974. A decimal code for the growth stages of cereals. *Weed Res.* 14: 415–421.

Zamani, A., Motallebi, M., Jonoubi, P., Ghafarian-Nia, N. S., and Zamani, M. R. 2012. Heterologous expression of the *Secale cereale* thaumatin-like protein in transgenic canola plants enhances resistance to stem rot disease. *Iran. J. Biotech.* 10: 87–95.

Zangenberg, M., Boskov-Hansen, H., Jorgensen, J. R., and Hellgren, L. I. 2004. Cultivar and year-to-year variation of phytosterol content in rye (*Secale cereale* L.). *J. Agric. Food Chem.* 52: 2593–2597.

Zeller, F. J. 1973. 1B/1R wheat-rye chromosome substitutions and translocations. *Proceedings of the 4th International Wheat Genetics Symposium.* Columbia, MO, August 6–11, pp. 209–221.

Zeller, F. J., and Cermeno, M.-C. 1991. Chromosome manipulations in *Secale* (Rye). In *Chromosome Engineering in Plants: Genetics, Breeding, Evolution*, vol. 2. P. K. Gupta, and T. Tsuchiya (eds.). Developments in Plant Genetics and Breeding. Elsevier Science Publication, Amsterdam, the Netherlands, pp. 313–333.

Zeller, F. J., Kimber, G., and Gill, B. S. 1977. The identification of rye trisomics by translocations and Giemsa staining. *Chromosoma* 62: 279–282.

Zhang, Q., Li, Q., Wang, X., Wang, H., Lang, S., Wang, Y., Wang, S., Chen, P., and Liu, D. 2005. Development and characterization of a *Triticum aestivum-Haynaldia villosa* translocation line T4VS·4DL conferring resistance to wheat spindle streak mosaic virus. *Euphytica* 145: 317–320.

Zhou, J. P., Yang, Z. J., Li, G. R., Liu, C., and Ren, Z. L. 2008. Discrimination of repetitive sequences polymorphism in *Secale cereale* by genomic *in situ* hybridization-banding. *J. Integrative Plant Biol.* 50: 452–456.

Zhou, J. P., Yang, Z. J., Li, G. R., Liu, C., Tang, Z. X., Zhang, Y., and Ren, Z. L. 2010. Diversified chromosomal distribution of tandemly repeated sequences revealed evolutionary trends in *Secale* (Poaceae). *Plant Syst. Evol.* 287: 49–56.

Zhou, Y., Hu, Z., Dang, B., Wang, H., Deng, X., Wang, L., and Chen, Z. 1999. Microdissection and microcloning of rye (*Secale cereale* L.) chromosome 1R. *Chromosoma* 108: 250–255.

Zohary, D. 1971. Origin of southwest Asiatic cereals: Wheats, barley, oats and rye. In *Plant Life of South-West Asia*. P. H. Davis et al. (eds.). Aberdeen University Press, Aberdeen, pp. 235–260.

Zoller, J. F., Herrmann, R. G., and Wanner, G. 2004a. Chromosome condensation in mitosis and meiosis of rye (*Secale cereale* L.). *Cytogenet. Genome Res.* 105: 134–144.

Zoller, J. F., Herrmann, R. G., and Wanner, G. 2004b. Ultrastructural analysis of chromatin in meiosis I + II of rye (*Secale cereale* L.). *Cytogenet. Genome Res.* 105: 145–156.

Zolotova, L. N., and Zolotov, A. L. 1977. Apomictic grain formation in the hybridization of tetraploid rye and winter durum wheat (Russian). *Genet. Select. Rast.* 10: 200–201.

Index

Page numbers followed by *f* and *t* indicate figures and tables, respectively

Milton Keynes UK
Ingram Content Group UK Ltd.
UKHW031140141024
449569UK00024B/1178